Post-modern Electromagnetics

Post-modern Electromagnetics

Using Intelligent MaXwell Solvers

Christian Hafner

Swiss Federal Institute of Technology (ETHZ), Zürich, Switzerland

John Wiley & Sons

Chichester • New York • Weinheim • Brisbane • Singapore • Toronto

Baffins Lane, Chichester,
West Sussex, PO19 1UD, England

National 01243 779777
International (+44) 1243 779777

e-mail (for orders and customer service enquiries): cs-books@wiley.co.uk

Visit our Home Page on http://www.wiley.co.uk or http://www.wiley.com

Other Wiley Editorial Offices

John Wiley & Sons, Inc., 605 Third Avenue,
New York, NY 10158-0012, USA

Weinheim • Brisbane • Singapore • Toronto

Library of Congress Cataloging-in-Publication Data

Hafner, Christian.
Post-modern electromagnetics : using intelligent Maxwell solvers / Christian Hafner.
Includes bibliographical references and index.
ISBN 0-471-98711-5 (alk. paper)
I. Title 98-53331
QC760.H27 1999 CIP
537′.01′13—dc21

British Library Cataloguing in Publication Data

A catalogue record for this book is available from the British Library

ISBN 0 471 98711 5

Produced from PostScript files supplied by the author
Printed and bound in Great Britain by Bookcraft Ltd, Midsomer Norton, Wiltshire
This book is printed on acid-free paper responsibly manufactured from sustainable forestry in which at least two trees are planted for each one used for paper production.

Contents

Preface

During the last 20 years, I had the chance of getting much experience with the development of novel methods for computational electromagnetics and with the design of highly reliable computer codes with a wide range of applications. This book outlines all tools I consider to be essential for creating new, powerful programs, and it illustrates how problems of practical interest may be solved.

Most textbooks on computational electromagnetics focus on a single, prominent method and explain the benefits of this method. This may be helpful for those who already have decided to work with a specific method. At the same time, such books are of little help for those who want to get an overview of existing techniques, for those who are searching for a method that is best suited for solving a specific problem, and for those who intend to find new ways. The main problem in writing a textbook that does not concentrate on a single method stems from the fact that most code designers are specialists in a certain method who are not very familiar with other techniques. Getting started with some method is quite easy, but experts on such methods are often confronted with very deep and demanding problems that require much work and do not leave much time for the study of concurrent methods. I spent most of my time with the development of boundary methods, especially the MMP code, but I always considered the study of concurrent techniques as a source of ideas that allowed me to improve my own codes. Moreover, my position at the ETH Zurich gave me enough freedom to spend some time on such studies. Although I am not a specialist in any domain method, I assume that I sufficiently well understood these methods for presenting the fundamentals in this book. However, to keep the size of the book reasonably small, I was forced concentrate on a relatively small number of methods. This was not easy, because there are always good reasons for presenting any method. First of all, I decided to focus on general Maxwell solvers, i.e., methods that can be applied to many different applications, although there are many methods that are more powerful for solving some specific problems. Such specific methods may be inspiring, but they usually require a profound knowledge of the specific problem to be solved, which is outside the scope of this book.

Pointing out the advantages of a given method is good for advertising, but the analysis of the drawbacks is the first step in the improvement of methods. One of the main goals of this book is to illustrate how a technique may be improved. Therefore, knowledge of the advantages and of disadvantages is considered to be of equal importance. Especially newcomers would appreciate very much a fair comparison of available software. Since all code designers have their personal preferences, fair comparisons can hardly be found in the literature and I am also not in a position to do such a comparison. However, I tried to be as fair as possible in the description of the various methods.

Each numerical method has its own history that goes often back to the time when no computers were available. Many methods for computational electromagnetics were originally developed within some other area of physics. During the evolution of these methods, new concepts were embedded, new interpretations were found. This typically led to more general formulations and to higher complexity. Going back to the roots and studying the process of the evolution is therefore illuminative and instructive for further development.

Students often consider electromagnetics as a difficult and dry matter of little practical value. When one considers, for example, the study of electromagnetic waves on two parallel wires by G. Mie in 1900, one has a long and extremely difficult mathematical treatment of a relatively simple model. Mie was an excellent physicist and he probably spent months obtaining a few approximate results that were of little practical interest because those who used transmission lines in practice worked with much more complicated structures. Computational electromagnetics is even more demanding than classical electromagnetics because the numerical aspects cause additional trouble and require a sufficient knowledge of numerical mathematics as an additional tool. I hope that this book demonstrates that computational electromagnetics is of great practical value, that it is very rich, and that it allows one to bring much abstract mathematical formalism alive. Computational electromagnetics allows one to handle complex models that are sufficiently close to structures of practical importance. Today, a student can design a code that simulates the propagation on transmission lines with far fewer constraints than Mie's tedious calculations. Despite modern development platforms, writing computer codes may be boring, but designing such codes is also fascinating and creative work. Finally, good computer codes should allow one to reduce the total amount of work to be done for solving a given problem.

In the current design process for new devices, computer simulations still play a secondary role, because such devices are too complex for an accurate and reliable computer analysis. Therefore, the simulation is often only a supplement to measurements. Because of the rapidly growing power of computers, one can predict that this will change drastically in the near future. One can expect that new codes linked with intelligent optimization tools will be able to design new devices that outperform devices developed by human experts. Finding a good solution for a field problem may also be considered as an optimization process. Therefore, also those who intend to improve a method for computational electromagnetics may profit from the study of optimization tools. Finally, optimization tools may be applied to improve the performance of existing techniques. For these reasons, intelligent optimization strategies are also discussed in this book.

I hope that this book will be a source of inspiration for all who want to develop new methods for computational electromagnetics.

Christian Hafner

Zurich, Switzerland

November, 1998

1 Geometry, differential and integral forms

General remarks

For a long period of time, geometry was the physicist's most important tool. The whole universe was considered to be a geometric construction. Even Newton, who invented and introduced the differential and integral calculus, geometrically described his new theory of gravitation and explained the behavior of light by geometric constructions. However, in the ensuing years, integral formulations and differential equations gradually began to replace geometry. It is well known that Newton's law of gravitation was an 'action at a distance' without any explanation for the transmission of the force from the source to the receiver. Newton was aware of this deficiency. Most of his successors, however, were too impressed by his theory to make any objections. Instead, they looked for similar formulations of other forces, such as the electric and magnetic ones.

It is remarkable that Leibniz invented the same calculus, probably independently, a little later than Newton, but his version was formally simpler than Newton's. As a consequence, British scientists who applied Newton's formalism increasingly lost their position. If one looks at the famous names in the theory of electricity before 1800, one finds only a few from Britain. This changed with Faraday, who was not satisfied with Newton's 'action at a distance' and introduced the concept of the electric and magnetic fields. Faraday believed in interactions among all kinds of forces, and showed by many experiments that such interactions exist, but he was unable to provide mathematical theory. This was done by Maxwell. Maxwell's equations were the first important system of partial differential equations in history. This theory brought the first unification of electricity, magnetism, and light, and removed the old 'action at a distance'. At first, Maxwell's equations were extremely complicated and hard to understand, but this time several British scientists created an appropriate formalism. As a consequence, spatial derivatives had to be intensively studied. This brought about a first revival of geometry. It is well known that Einstein was very impressed by both Maxwell's theory and geometry. Thus, Maxwell's theory very deeply influenced modern physics.

Newton's work on the structure of matter is almost unknown and considered to be useless, because it mainly is considered as a result of his alchemist activities. It is important to recognize that Newton built telescopes. From this, he obtained much experience with optics and also with the nature of mirrors and matter in general. He realized that mirrors have a rough surface and that polishing reduces the coarseness, but that this does not essentially change the properties of the surface, i.e., that the surface never gets really flat. This caused problems with his geometric concept of optics. Why did Newton assume that polishing does not generate a really flat surface? He obviously had a concept of matter that was very close to what has been called fractal geometry since Mandelbrot [Mandelbrot, 1977]. However, fractal geometry is outside the scope of this book although it is of increasing interest also for computational electromagnetics.

In the following, an outline of the most important concepts of classical geometry and the most common formalisms of spatial derivatives and integrals for engineers are given, together with the most important formulas. In addition, we present a formalism that illustrates the spatial derivatives and is very helpful in understanding the way in which Maxwell's equations are treated. For a more sophisticated description, see [Misner, 1973] or textbooks on electromagnetics. [Moon, 1961] contains all relevant coordinate systems and the corresponding formula of the spatial derivatives required in field theory.

3D space

Although the simplest form of Maxwell's equations is written in four-dimensional space, which includes time, engineers usually work with a more complicated formulation in the old Euclidean three-dimensional space. Among other reasons, the following are the most important: (1) 4D space as well as the appropriate 'exterior calculus' are abstract and quite hard to understand; (2) time usually plays a very special role in the experiments because the time dependence is given by devices that produce the electromagnetic field; (3) relativistic effects may be neglected in most cases.

However, 2D space is much more important for engineers because it allows one to simplify both the models and the visualization as most drawn devices are two-dimensional. In addition, 2D models are necessary for the definition of important physical terms such as the propagation constant of guided waves. For these reasons, only 2D and 3D space are considered here.

In every N-dimensional space, one can find $N+1$ objects of dimension $n=0,1,\ldots,N$ such as points, lines, surfaces, and so on. As spaces of dimensions higher than $N=3$ are not known to most people, higher dimensional objects do not have special names. Every object may be considered to be generated by movements of objects of smaller dimension.

Integrals

For scientists, a mathematical formulation and the measurement of the objects are very important. Engineers usually use coordinates and vectors to do this. As a vector is a directed straight line of finite length, it may be easily used to indicate the relative position of any two points in space. Also, vectors can be used to construct and measure objects. This is simple if the objects are partially flat. However, in general, it is necessary to shrink vectors to an infinitesimal length. To obtain the length of a curved line, for example, one can define vectors of infinitesimal length, which are tangential to the line in every point of the line. If the (infinitesimal) distance of every point to its neighbor is equal to that of the corresponding vector, one finds the length of the line by integration:

$$\ell = \int_{\ell} \vec{e}_{\tau} d\vec{\ell} = \int_{\ell} \left| d\vec{\ell} \right| , \tag{1.1}$$

where $\vec{e}_{\tau}$ is a tangential vector of unit length and $d\vec{\ell}$ is the infinitesimal tangential vector. The scalar product of these two vectors is necessary to obtain a scalar value for the length. As most measured quantities are scalars, this procedure is very important. We will encounter scalar products later in another context as well, because scalars are most important for computers. Note that the scalar product is often represented by a dot. Therefore, it is also called the dot product. The geometric interpretation of the scalar product of two vectors is a projection of one vector on the other one. By the way, there is a numerical method called the projection technique (PT), which is based on the scalar product of functions. The interpretation of a scalar product of functions as a projection of one function on another one is rather abstract and therefore not too helpful, but the procedures for functions and vectors are analogous.

Several other terms of geometry are connected to the scalar product. For example, the length of a vector is given by its scalar product with a unit vector, which points in the same direction, as in the equation above. Instead, one can use the square root of the scalar product of the vector with itself, the so-called norm[1] of a vector, which is important for another numerical method, called the error minimization technique (EMT) in this book.

As one can find n linearly independent vectors 'tangential' to an n-dimensional object in every point, one can measure any object with integrals similar to (1.1). In 3D space, most engineers use the following products of two and three vectors:

$$\vec{a} = \vec{v}_1 \times \vec{v}_2 \,, \tag{1.2}$$

$$p = (\vec{v}_1 \times \vec{v}_2) \cdot \vec{v}_3 \,, \tag{1.3}$$

where $\vec{a}$ is a vector orthogonal to $\vec{v}_1$ and $\vec{v}_2$, p is a scalar. It is known that $\vec{a}$ and p do not behave exactly like ordinary vectors and scalars: in a 'mirror-world', they have opposite signs. For this reason, $\vec{a}$ is sometimes called a pseudovector, p a pseudoscalar. These pseudoterms represent two- and three-dimensional objects of 3D space, rather than one- and zero-dimensional ones. Somehow, one uses the 'missing, orthogonal dimensions' of the objects. Or, mathematically speaking, one works in the dual space. This leads to the following measures for two- and three-dimensional objects in 3D space:

$$S = \int_S \vec{e}_v \mathrm{d}\vec{S} = \int_S \left|\mathrm{d}\vec{S}\right| , \tag{1.4}$$

$$V = \int_V e \,\mathrm{d}V = \int_V \left|\mathrm{d}V\right| , \tag{1.5}$$

where $\mathrm{d}S = \mathrm{d}\vec{v}_1 \times \mathrm{d}\vec{v}_2$ and $\vec{e}_v$ are pseudovectors, $\mathrm{d}V = (\mathrm{d}\vec{v}_1 \times \mathrm{d}\vec{v}_2) \cdot \mathrm{d}\vec{v}_3$ and e are pseudoscalars. The unit-pseudoscalar $e = \pm 1$ is necessary to obtain a scalar for the volume V.

Most engineers are not familiar with pseudoscalars and pseudovectors, but there are some important reasons for this distinction. Because scalars and pseudoscalars, vectors, and pseudovectors represent different objects of 3D space with different numbers of dimensions, it makes no sense, for example, to add scalars and pseudoscalars or vectors and pseudovectors. In the usual notation, it is quite clear that the sum of a scalar and a vector is meaningless. For example, nobody will try to add a charge (scalar) and the electric vector field. However, it is hard to understand why the sum of the electric vector field and the magnetic vector field is a physically meaningless object, unless one recognizes that the magnetic field is a pseudovector. The distinction of scalars and pseudoscalars, vectors, and pseudovectors prevents one from applying useless equations and considerably reduces, for example, the number of physically

[1] The geometric interpretation of the vector norm as the length of the vector is very simple. We will see that the norm can also be defined for functions, where the interpretation is not obvious.

meaningful equations that can be derived from Maxwell's equations. Moreover, this distinction is helpful for understanding the spatial derivatives div, curl, and grad as well as for finding an appropriate formalism for 2D problems.

The geometric meaning of scalars, vectors, pseudovectors, and pseudoscalars is very simple (Figure 1): (1) Scalars have no dimension, like points; (2) vectors have one dimension, like lines, and can be considered to be directed straight lines; (3) pseudovectors represent directed areas - they can be generated by two vectors; (4) pseudoscalars have three dimensions, like the volume, and can be generated by three vectors or a scalar product of a vector and a pseudovector. In electromagnetics, the electric charge and the energy densities as well as material properties, such as the permeability and permittivity, are considered to be scalars. From the analysis of Maxwell's equations, one finds that the electric field and the Poynting field are vector fields, whereas the magnetic field is a pseudovector field. Finally, magnetic charges would be pseudoscalars, but no such charges have yet been found.

Working with vector, pseudovector, and pseudoscalar fields, one can generalize the integrals above to obtain measures for the fields:

$$U = \int_{\ell} \vec{v}\,\mathrm{d}\vec{\ell}\,, \tag{1.6}$$

$$\Phi = \int_{S} \vec{a}\,\mathrm{d}\vec{S}\,, \tag{1.7}$$

$$I = \int_{V} p\,\mathrm{d}V\,. \tag{1.8}$$

As every object of dimension n equal to one or more has a closed boundary of dimension n-1, one can look for special fields, which can be measured by boundary integrals instead of the integrals above:

$$U = \int_{\ell} \vec{v}\,\mathrm{d}\vec{\ell} = \oint_{\partial\ell} \phi\,\mathrm{dP}\,, \tag{1.9}$$

$$\Phi = \int_{S} \vec{a}\,\mathrm{d}\vec{S} = \oint_{\partial S} \vec{A}\,\mathrm{d}\vec{\ell}\,, \tag{1.10}$$

$$I = \int_{V} p\,\mathrm{d}V = \oint_{\partial V} \vec{B}\,\mathrm{d}\vec{S}\,. \tag{1.11}$$

The term on the right-hand side of equation (1.9) is not really an integral, because the boundary of a line consists only of two points. This term has been introduced formally and is an abbreviation for the difference of ϕ in the two end points of the line ℓ. Instead of (1.9) one writes:

$$U = \int_{\ell} \vec{v}\,\mathrm{d}\vec{\ell} = \phi(P_B) - \phi(P_A)\,. \tag{1.12}$$

Note that P_A is the first end point and P_B the second end point of the line, i.e., the line is oriented from P_A toward P_B.

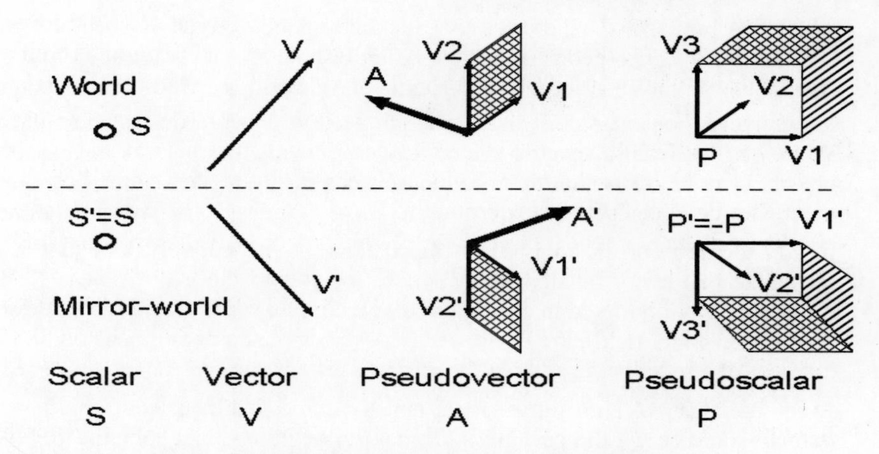

Figure 1: The elementary objects of 3D space and in a 'mirror' space.

First order spatial derivatives

The integral equations (1.9) to (1.11) allow one to define spatial differential operators, as follows:

$$U = \int_{\ell} \mathrm{grad}\, \phi \, \mathrm{d}\vec{\ell} = \oint_{\partial \ell} \phi \, \mathrm{d}P \,, \tag{1.13}$$

$$\Phi = \int_{S} \mathrm{curl}\, \vec{A} \, \mathrm{d}\vec{S} = \oint_{\partial S} \vec{A} \, \mathrm{d}\vec{\ell} \,, \tag{1.14}$$

$$I = \int_{V} \mathrm{div}\, \vec{B} \, \mathrm{d}V = \oint_{\partial V} \vec{B} \, \mathrm{d}\vec{S} \,. \tag{1.15}$$

If one integrates over an infinitesimal domain, one has the well-known, coordinate-free definitions of the 'spatial derivatives':

$$\mathrm{div}\, \vec{B} = \lim_{r \to 0} \frac{\oint_{\partial V} \vec{B} \, \mathrm{d}\vec{S}}{\int_{V} \mathrm{d}V} \,, \tag{1.16}$$

$$\vec{e}_n \cdot \mathrm{curl}\, \vec{A} = \lim_{r \to 0} \frac{\oint_{\partial S} \vec{A} \, \mathrm{d}\vec{\ell}}{\int_{S} \mathrm{d}S} \,, \tag{1.17}$$

$$\vec{e}_{\ell} \cdot \mathrm{grad}\, \phi = \lim_{r \to 0} \frac{\oint_{\partial \ell} \phi \, \mathrm{d}P}{\int_{\ell} \mathrm{d}\ell} \,, \tag{1.18}$$

where the limit $r \to 0$ indicates that the size of the volume *V*, the surface *S*, or the line ℓ shrinks to zero. The geometric shape of these objects can be quite arbitrary, but the objects should be 'simply connected' and should have a single, well-defined boundary $\partial V, \partial S, \partial \ell$ respectively. For example, *V* can be a sphere or a cube and *S* can be a circle or a rectangle. When one works in a specific system of coordinates, it is natural to select the shape of these objects in such a way that their boundaries can easily be described. For example, a cube is a natural volume for Cartesian coordinates.

In consideration of the number of dimensions of the fields in the equations above, one can recognize that every 'spatial derivative' associates to the given field a field with either one dimension more or one dimension less. For example, if the gradient operates on a scalar (i.e., a zero-dimensional field), the result will be a vector (i.e., a one-dimensional field). However, the gradient of a pseudoscalar (three-dimensional) is a pseudovector (two-dimensional).

Every boundary of an *n* dimensional object is an *n*-1 dimensional object without a boundary (which would be *n*-2 dimensional). From this, one immediately obtains the following relations:

$$\int_S \text{curl grad}\,\phi\,\mathrm{d}\vec{S} = \oint_{\partial S} \text{grad}\,\phi\,\mathrm{d}\vec{\ell} = \oint_{\partial\partial S} \phi\,\mathrm{d}P = 0\,, \tag{1.19}$$

$$\int_V \text{div curl}\,\vec{A}\,\mathrm{d}V = \oint_{\partial V} \text{curl}\,\vec{A}\,\mathrm{d}\vec{S} = \oint_{\partial\partial V} \vec{A}\,\mathrm{d}\vec{\ell} = 0\,, \tag{1.20}$$

which hold for any surface *S* or volume *V*. As a consequence, one obtains the important relations for the integrands:

$$\text{curl grad}\,\phi = 0\,, \tag{1.21}$$

$$\text{div curl}\,\vec{A} = 0\,. \tag{1.22}$$

This means that one cannot associate pseudovector fields to scalar fields or pseudoscalar fields to vector fields by 'spatial derivatives'. This considerably limits the possible spatial differential equations.

Schematic representation of spatial derivatives

It is useful to visualize (see Figure 2) the spatial derivatives with the following scheme, which has been inspired by the so-called 'exterior calculus':

$$s \xrightarrow{grad} \vec{v} \xrightarrow{curl} \vec{a} \xrightarrow{div} p \tag{1.23}$$

According to equations (1.21) and (1.22), it is better to split this scheme into two parts:

$$s \xrightarrow{grad} \vec{v} \xrightarrow{curl} 0 \tag{1.24}$$

$$\vec{v} \xrightarrow{curl} \vec{a} \xrightarrow{div} 0 \tag{1.25}$$

which makes clear that the spatial derivative of a derivative 'in the same direction' vanishes. Every spatial derivative links fields of objects of dimension *n* and *n*+1. Linking fields of objects of dimension n and $n+m$, with m larger than 1 is impossible, due to the statements at the end of the previous subsection.

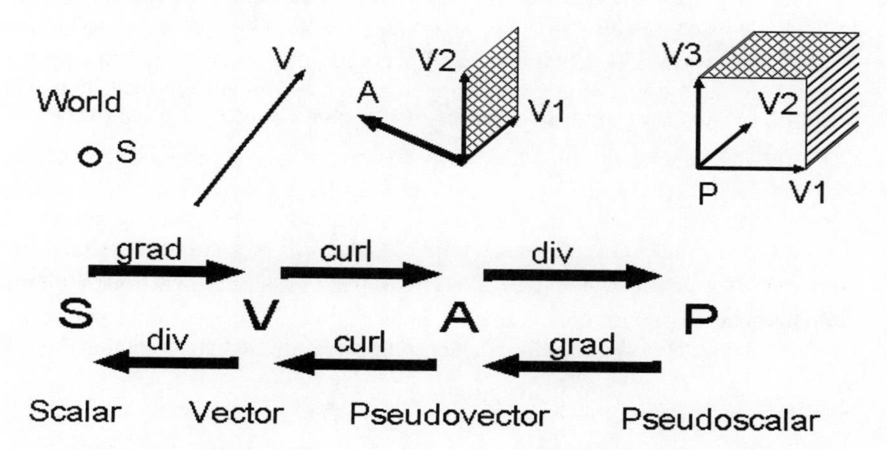

Figure 2: 3D elementary objects and the 'spatial derivatives' grad, curl, div.

Instead of the scalar integrals above, one can formally define identical pseudoscalar integrals. This generates spatial derivatives, which act in the 'opposite direction':

$$s \xleftarrow{div} \vec{v} \xleftarrow{curl} \vec{a} \xleftarrow{grad} p \tag{1.26}$$

As already mentioned, the distinction of scalars and pseudoscalars, vectors, and pseudovectors is physically important. Writing differential equations in a schematic form, which distinguishes the fields with different numbers of dimensions, prevents one from considering meaningless formulations. For example, two differential equations can be added if the corresponding fields have the same dimension and the arrows in the scheme point in the same direction. This reduces the number of meaningful equations once more.

Second-order spatial derivatives

The only way of constructing second-order spatial derivatives is to combine 'left derivatives' with 'right derivatives'. The most interesting operator of this kind is the Laplacian. Usually, one defines a scalar and a vector operator as follows:

$$\Delta\phi = \text{div grad } \phi \,, \tag{1.27}$$

$$\Delta\vec{A} = \text{grad div } \vec{A} - \text{curl curl } \vec{A} \,. \tag{1.28}$$

Precisely speaking, there are four Laplacian operators that can be applied to the four objects scalar, vector, pseudovector and pseudoscalar. It is important that these operators have the same dimension as their operand: if ϕ is a scalar, $\Delta\phi$ is a scalar as well, and so on. As we will see, the Maxwell equations are coupled, first-order differential equations. They contain 'left' as well as 'right' derivatives. Decoupling of these equations usually leads to the Laplacian operator.

It may appear strange that there is only one vector Laplacian operator, which consists of two parts. One of these parts is a 'left' derivative applied to a 'right' derivative, and the other one is a 'right' derivative applied to a 'left' derivative. The reason is the following: if one projects the vector Laplacian operator on any unit vector, one finds the scalar Laplacian operator:

$$\vec{e}\cdot(\Delta\vec{A}) = \Delta(\vec{e}\cdot\vec{A})\,. \tag{1.29}$$

If the 'left' and the 'right' derivatives of a vector $\vec{A}$ (or of a pseudovector) are given by two equations:

$$\operatorname{div}\vec{A} = P\,, \tag{1.30}$$

$$\operatorname{curl}\vec{A} = \vec{v}\,, \tag{1.31}$$

one can separate the vector field $\vec{A}$ into two parts $\vec{A}_1$ and $\vec{A}_2$, each with a homogeneous and an inhomogeneous equation:

$$\operatorname{div}\vec{A}_1 = 0\,, \tag{1.32}$$

$$\operatorname{curl}\vec{A}_1 = \vec{v}\,, \tag{1.33}$$

$$\operatorname{div}\vec{A}_2 = P\,, \tag{1.34}$$

$$\operatorname{curl}\vec{A}_2 = 0\,. \tag{1.35}$$

Because of the inhomogeneous equations, the first part of the Laplacian operator of $\vec{A}_1$ and the second part of $\vec{A}_2$ still vanish. Schematically, one can write:

$$\vec{A}_1 \xrightarrow{\;div\;} 0 \tag{1.36}$$

$$\vec{v} \xleftarrow{\;curl\;} \vec{A}_1 \tag{1.37}$$

$$\vec{A}_2 \xrightarrow{\;div\;} P \tag{1.38}$$

$$0 \xleftarrow{\;curl\;} \vec{A}_2 \tag{1.39}$$

From (1.21) and (1.22) we see that we can 'complete' the scheme (1.36)-(1.39) by introduction of scalar and vector potentials:

$$\vec{\phi} \xrightarrow{\;curl\;} \vec{A}_1 \xrightarrow{\;div\;} 0 \tag{1.40}$$

$$0 \xleftarrow{\;div\;} \vec{v} \xleftarrow{\;curl\;} \vec{A}_1 \tag{1.41}$$

$$\vec{A}_2 \xrightarrow{\;div\;} P \tag{1.42}$$

$$0 \xleftarrow{\;curl\;} \vec{A}_2 \xleftarrow{\;grad\;} -\phi \tag{1.43}$$

The negative sign of the scalar potential ϕ has mainly historical reasons. It has no influence on the general procedure. If one starts from the scalar potential ϕ, one has a simple way to the left and back to the right. This is usually written in the form:

$$\Delta\phi = -P\,, \tag{1.44}$$

which is called a Poisson equation. The homogeneous part of this equation is the Laplace equation:

$$\Delta\phi = 0. \tag{1.45}$$

The most important special solution of the Poisson equation is the Coulomb integral:

$$\phi(\vec{r}) = \int \frac{P(\vec{r}\,')}{4\pi|\vec{r} - \vec{r}\,'|} \mathrm{d}V'. \tag{1.46}$$

With equation (1.43), i.e.,

$$\vec{A}_2 = -\mathrm{grad}\,\phi, \tag{1.47}$$

one obtains a Coulomb integral for $\vec{A}_2$:

$$\vec{A}_2(\vec{r}) = -\mathrm{grad}\int \frac{P(\vec{r}\,')}{4\pi|\vec{r} - \vec{r}\,'|} \mathrm{d}V' = -\int \frac{P(\vec{r}\,')(\vec{r} - \vec{r}\,')}{4\pi|\vec{r} - \vec{r}\,'|^3} \mathrm{d}V'. \tag{1.48}$$

This equation can be found directly, without the introduction of any potential. From (1.40)-(1.43), one can see that the vector Poisson equation holds:

$$\Delta\,\vec{A}_2 = -\mathrm{grad}\,P. \tag{1.49}$$

This equation has the Coulomb integral

$$\vec{A}_2(\vec{r}) = -\int \frac{\mathrm{grad}'\,P(\vec{r}\,')}{4\pi|\vec{r} - \vec{r}\,'|} \mathrm{d}V' \tag{1.50}$$

which looks quite similar to the Coulomb integral (1.48). However, in this integral the gradient acts on the source *P*, i.e., on the coordinates x', y', z', whereas the gradient in (1.48) acts on the coordinates x, y, z of the point where $\vec{A}_2$ is measured. To verify that both integrals are equivalent, one has to consider the following. From the product rule for derivatives, one has

$$\mathrm{grad}'\frac{P(\vec{r}\,')}{|\vec{r} - \vec{r}\,'|} = \frac{\mathrm{grad}'\,P(\vec{r}\,')}{|\vec{r} - \vec{r}\,'|} + P(\vec{r}\,')\,\mathrm{grad}'\frac{1}{|\vec{r} - \vec{r}\,'|}. \tag{1.51}$$

Because of the symmetry of $|\vec{r} - \vec{r}\,'|$, one finds

$$\mathrm{grad}'\frac{1}{|\vec{r} - \vec{r}\,'|} = -\mathrm{grad}\frac{1}{|\vec{r} - \vec{r}\,'|}. \tag{1.52}$$

Finally, one can use the well known equation:

$$\int_D \mathrm{grad}'\phi(\vec{r}\,')\,\mathrm{d}V' = \oint_{\partial D} \phi(\vec{r}\,')\,\mathrm{d}\vec{S}\,'. \tag{1.53}$$

The integral on the right-hand side vanishes if $\phi = 0$ on the boundary. This is true in the case above, because all sources $\phi = P/|\vec{r} - \vec{r}\,'|$ are inside *D*. Thus, we have

$$\begin{aligned}\int_D \mathrm{grad}'\frac{P(\vec{r}\,')}{|\vec{r} - \vec{r}\,'|}\mathrm{d}V' &= \int_D \frac{\mathrm{grad}'\,P(\vec{r}\,')}{|\vec{r} - \vec{r}\,'|}\mathrm{d}V' - \int_D P(\vec{r}\,')\,\mathrm{grad}\frac{1}{|\vec{r} - \vec{r}\,'|}\mathrm{d}V' = \\ &= -\mathrm{grad}\int_D P(\vec{r}\,')\frac{1}{|\vec{r} - \vec{r}\,'|}\mathrm{d}V'\end{aligned} \tag{1.54}$$

which proves that the Coulomb integrals above are equivalent.

The advantage of the introduction of the scalar potential compared with the direct method is that one has to solve only the scalar integral (1.51) and $\vec{A}_2$ is obtained by the simple differentiation (1.52). In addition, one directly obtains an integral for $\vec{A}_2$ without any spatial derivative, because grad acts only on the Green's function $1/|\vec{r}-\vec{r}'|$ and not on the source P.

One can find a similar solution for $\vec{A}_1$ as well. In this case, the direct procedure is more simple. With (1.48), one finds

$$\Delta\vec{A}_1 = -\mathrm{curl}\,\vec{v}\,, \tag{1.55}$$

with the special solution:

$$\vec{A}_1(\vec{r}) = -\int \frac{\mathrm{curl'}\vec{v}(\vec{r}')}{4\pi|\vec{r}-\vec{r}'|}\mathrm{d}V'. \tag{1.56}$$

There is a certain difficulty involved if the vector potential $\vec{\phi}$ (defined in (1.40)) is to be used: only one derivative is given for this vector field. The derivative in the 'opposite' direction is necessary to fix $\vec{\phi}$ entirely. The simplest way of doing so is to set this derivative equal to zero. This is called the Coulomb gauge and leads to the Poisson equation:

$$\Delta\vec{\phi} = -\vec{v}\,, \tag{1.57}$$

with the special solution:

$$\vec{\phi}(\vec{r}) = -\int \frac{\vec{v}(\vec{r}')}{4\pi|\vec{r}-\vec{r}'|}\mathrm{d}V'. \tag{1.58}$$

With the first equation of (1.48):

$$\vec{A}_1 = \mathrm{curl}\,\vec{\phi}\,, \tag{1.59}$$

one obtains

$$\vec{A}_1(\vec{r}) = -\mathrm{curl}\int \frac{\vec{v}(\vec{r}')}{4\pi|\vec{r}-\vec{r}'|}\mathrm{d}V' = \int \frac{\vec{v}(\vec{r}')\times(\vec{r}-\vec{r}')}{4\pi|\vec{r}-\vec{r}'|^3}\mathrm{d}V'. \tag{1.60}$$

To show that this integral is equivalent to (1.57), one can proceed as in the case of the Coulomb integrals for $\vec{A}_2$. For this proof, equations (1.107) and (1.126) below are applied. Similar integrals exist for more complicated relations such as the Helmholtz equation for time-harmonic fields, the wave equations, and so on. The procedure for all these integrals and proofs is essentially the same as above.

Coordinates

It is well known that coordinates are a very helpful geometric tool. For analytical calculations of fields, the choice of an appropriate system of coordinates is extremely important. Unfortunately, only a few systems of coordinates allow one to find analytic solutions even of very simple differential equations. As a consequence, only geometrically simple problems can be solved analytically. For numerical computations on the other hand, it is usually sufficient to apply only:

1. Cartesian coordinates (x,y,z);
2. Circular cylindrical coordinates (r,φ,z);
3. Spherical coordinates (r,ϑ,φ).

All these coordinates are orthogonal. For the infinitesimal distance ds between two neighboring points P_1 and P_2 with the coordinates (u,v,w) and $(u+\mathrm{d}u, v+\mathrm{d}v, w+\mathrm{d}w)$ one has

$$(\mathrm{d}s)^2 = (g_u \mathrm{d}u)^2 + (g_v \mathrm{d}v)^2 + (g_w \mathrm{d}w)^2 \,, \tag{1.61}$$

where g_u, g_v, g_w are called metric coefficients. For the three coordinate systems mentioned above, the metric coefficients are:

$$g_x = 1\,, \quad g_y = 1\,, \quad g_z = 1\,, \tag{1.62}$$

$$g_r = 1\,, \quad g_\varphi = r\,, \quad g_z = 1\,, \tag{1.63}$$

$$g_r = 1\,, \quad g_\vartheta = r\,, \quad g_\varphi = r\sin\vartheta\,. \tag{1.64}$$

Obviously, the Cartesian coordinates are most simple. If the transformation between any orthogonal coordinates and the Cartesian coordinates is given, one obtains the metric coefficients as follows:

$$g_a^2 = \left(\frac{\partial x}{\partial a}\right)^2 + \left(\frac{\partial y}{\partial a}\right)^2 + \left(\frac{\partial z}{\partial a}\right)^2 ; \quad a = u, v, w. \tag{1.65}$$

The following relations exist for the first-order spatial derivatives:

$$\operatorname{div} \vec{B} = \frac{(g_v g_w B_u)_{,u} + (g_w g_u B_v)_{,v} + (g_u g_v B_w)_{,w}}{g_u g_v g_w} \,, \tag{1.66}$$

$$\operatorname{curl} \vec{B} = \frac{(g_w A_w)_{,v} - (g_v A_v)_{,w}}{g_v g_w} \vec{e}_u + \frac{(g_u A_u)_{,w} - (g_w A_w)_{,u}}{g_w g_u} \vec{e}_v + \frac{(g_v A_v)_{,u} - (g_u A_u)_{,v}}{g_u g_v} \vec{e}_w \,, \tag{1.67}$$

$$\operatorname{grad} \phi = \frac{\phi_{,u}}{g_u} \vec{e}_u + \frac{\phi_{,v}}{g_v} \vec{e}_v + \frac{\phi_{,w}}{g_w} \vec{e}_w \,. \tag{1.68}$$

Here, a short notation for derivatives with respect to coordinates has been used; for example, $\phi_{,u}$ denotes the derivative of ϕ with respect to the coordinate u. This notation is also useful for higher order derivatives. For example, $A_{w,uv}$ denotes the second order derivative $\partial^2 A_w / \partial u \partial v$.

The expressions for a specific system of coordinates are obtained when the corresponding metric coefficients are inserted. The most simple form is obtained in Cartesian coordinates. Here one sets u=x, v=y, w=z. With the metric coefficients (1.62) one obtains:

$$\operatorname{div} \vec{B} = B_{x,x} + B_{y,y} + B_{y,y} \,, \tag{1.69}$$

$$\operatorname{curl} \vec{A} = (A_{z,y} - A_{y,z})\vec{e}_x + (A_{x,z} - A_{z,x})\vec{e}_y + (A_{y,x} - A_{x,y})\vec{e}_z \,, \tag{1.70}$$

$$\operatorname{grad} \phi = \phi_{,x}\vec{e}_x + \phi_{,y}\vec{e}_y + \phi_{,z}\vec{e}_z \,. \tag{1.71}$$

In these coordinates, even the second-order derivatives (i.e., the Laplacian operators) look simple:

$$\Delta\phi = \phi_{,xx} + \phi_{,yy} + \phi_{,zz} \,, \tag{1.72}$$

$$\Delta\vec{A} = (A_{x,xx} + A_{x,yy} + A_{x,zz})\vec{e}_x + (A_{y,xx} + A_{y,yy} + A_{y,zz})\vec{e}_y + (A_{z,xx} + A_{z,yy} + A_{z,zz})\vec{e}_z \,. \tag{1.73}$$

An even shorter notation is obtained, when the coordinates of a vector in 3D space are denoted as A_i with the index i=1,2,3 indicating the three directions x, y, z. With this notation, one can write

$$\Delta\vec{A} = \sum_i \left(\vec{e}_i \sum_k A_{i,kk} \right) \tag{1.74}$$

instead of (1.73). With Einstein's convention of an automatic summation over all double indices, one can even omit the symbols for the summations and write

$$\Delta\vec{A} = \vec{e}_i A_{i,kk} \,. \tag{1.75}$$

Compact notations often hide the amount of work required for implementing the corresponding formula on a computer.

2D space

If we consider two-dimensional Euclidean space, we recognize that we have one dimension fewer. As a consequence, we have one object fewer and we get one first-order derivative fewer, but we still can identify scalars, vectors, pseudovectors, and pseudoscalars. Now, both the vector and the pseudovector may be used to describe one-dimensional objects. The pseudovector is simply orthogonal to the corresponding vector. Thus a 90 degree rotation of a vector $\vec{v}$ generates a pseudovector (see Figure 3). We write:

$$\vec{a} = \vec{v}^{\circ}. \tag{1.76}$$

To be precise, we define the rotation in the mathematically positive sense (i.e., counterclockwise). If one considers 2D space to be a subspace of a 3D space, one can define this operation by a vector product of $\vec{v}$ with a unit vector $\vec{e}$ perpendicular to 2D space:

$$\vec{v}^{\circ} = \vec{e} \times \vec{v} \,. \tag{1.77}$$

Integrals

If we proceed as in 3D space, we can define the following three integral measures for the fields in 2D space:

$$U = \int_{\ell} \vec{v} \, \mathrm{d}\vec{\ell} \,, \tag{1.78}$$

$$\Phi = -\int_{\ell} \vec{a} \, \mathrm{d}\vec{\ell}^{\circ} = \int_{\ell} \vec{a}^{\circ} \, \mathrm{d}\vec{\ell} \,, \tag{1.79}$$

$$I = \int_{S} p \, \mathrm{d}S \,. \tag{1.80}$$

Note that essentially the same line integrals are used to define the flux ϕ through a line and the voltage U along a line. As in 3D space, we can use these integrals for defining spatial derivatives. The operator o makes the difference.

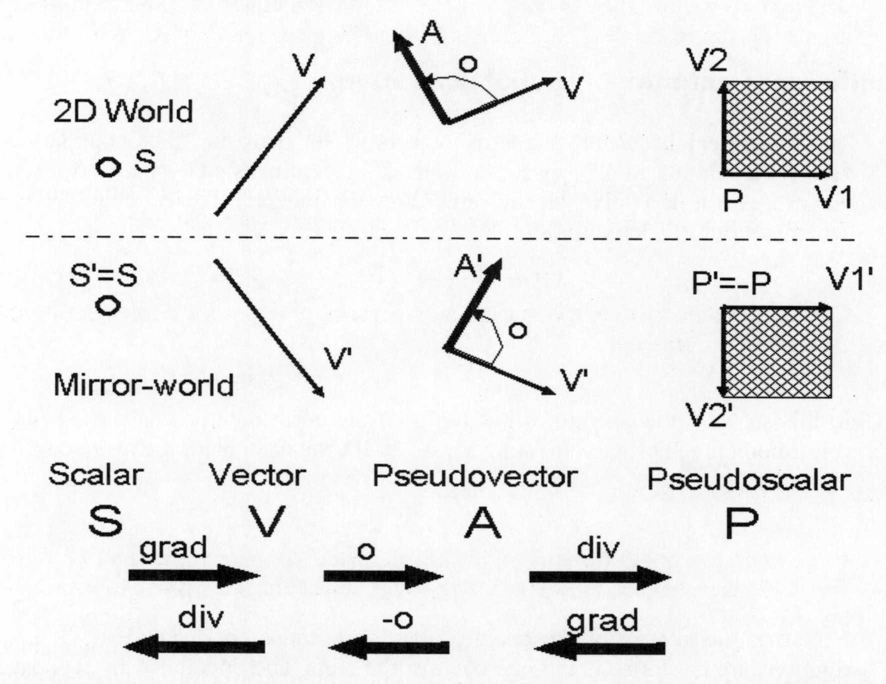

Figure 3: The elementary objects of 2D space and in the 2D mirror world. Instead of the 'spatial derivative' curl *of the 3D space, one has the orthogonality operator* o, *which turns a vector 90 degrees counter-clockwise.*

First-order spatial derivatives

We can introduce two different first-order derivatives using the integrals (1.78) and (1.80):

$$U = \int_{\ell} \mathrm{grad}_{\mathrm{T}}\, \phi \, \mathrm{d}\vec{\ell} = \oint_{\partial\ell} \phi \, \mathrm{d}P = \phi(P_E) - \phi(P_A)\,, \tag{1.81}$$

$$I = \int_{S} \mathrm{div}_{\mathrm{T}}\, \vec{B} \, \mathrm{d}S = -\oint_{\partial S} \vec{B} \, \mathrm{d}\vec{\ell}^{\circ} = \oint_{\partial S} \vec{B}^{\circ} \, \mathrm{d}\vec{\ell}\,. \tag{1.82}$$

Building a similar limit as in 3D, we obtain the definitions of the 2D spatial derivatives:

$$\mathrm{div}_{\mathrm{T}}\, \vec{B} = \lim_{r \to 0} \frac{\oint_{\partial S} \vec{B}^{\circ} \, \mathrm{d}\vec{\ell}}{\int_{S} \mathrm{d}S}\,, \tag{1.83}$$

$$\vec{e}_\ell \cdot \mathrm{grad}_\mathrm{T}\, \phi = \lim_{\mathrm{r}\to 0} \frac{\oint\limits_{\partial\ell} \phi\, \mathrm{d}P}{\int\limits_{\ell} \mathrm{d}\ell} . \tag{1.84}$$

Schematic representation of spatial derivatives

The fundamental law for geometric objects in 3D space is: "The boundary of a boundary is zero." This holds in 2D space as well as in n dimensional space. As a consequence, the derivative of a derivative 'in the same direction' is zero. To see what this means, we write a scheme of our 2D derivatives:

$$\begin{array}{ccccc} s & \xrightarrow{\mathrm{grad}_\mathrm{T}} & \vec{v} & & \\ & & \downarrow\circ & & \\ & & \vec{a} & \xrightarrow{\mathrm{div}_\mathrm{T}} & p \end{array} \tag{1.85}$$

In this scheme, the operator o has replaced the curl operator that played the most prominent role among the 3D spatial derivatives. As in 3D, we can obtain the inverse scheme

$$\begin{array}{ccccc} s & \xleftarrow{\mathrm{div}_\mathrm{T}} & \vec{v} & & \\ & & \uparrow -\circ & & \\ & & \vec{a} & \xleftarrow{\mathrm{grad}_T} & p \end{array} \tag{1.86}$$

Note that the inverse of the operator o is –o because oo is equal to –1, i.e., two 90 degree rotations are the same as the inversion of the sign. The operator o in 2D space is pretty much the same as the multiplication by $i = \sqrt{-1}$ in the complex plane.

Second-order spatial derivatives

Because

$$\int\limits_S \mathrm{div}_\mathrm{T}\, \mathrm{grad}_\mathrm{T}{}^\circ\, s\, \mathrm{d}\vec{S} = \oint\limits_{\partial S} \mathrm{grad}_\mathrm{T}{}^\circ\, s\, \mathrm{d}\vec{\ell}^\circ = \oint\limits_{\partial\partial S} s\, \mathrm{d}P = 0 , \tag{1.87}$$

we have

$$\mathrm{div}_\mathrm{T}\, \mathrm{grad}_\mathrm{T}{}^\circ\, s = 0 . \tag{1.88}$$

This means that p is zero in (1.85) if we start with a scalar s. Another consequence is that one can introduce a scalar potential s, if the 'right' derivative of any pseudovector field $\vec{a}$ is equal to zero. When we start from a pseudoscalar p, we can introduce 'left' derivatives and obtain the same equation as (1.88) for p instead of s. Therefore, we can also introduce a pseudoscalar potential p, if the 'left' derivative of a vector field $\vec{v}$ is zero.

Vector and pseudovector potentials do not exist in 2D space and we need only the scalar (or pseudoscalar) Laplacian operator:

$$\Delta_\mathrm{T} s = \mathrm{div}_\mathrm{T}\, \mathrm{grad}_\mathrm{T}\, s . \tag{1.89}$$

Note that the derivatives in (1.88) and (1.89) are very similar. Obviously, the little difference, i.e., the operator o, plays an important role. The second order derivative in (1.89) is not necessarily zero where 'almost the same' second order derivative in (1.88) is always zero.

2D subspace in 3D space

If we consider 2D space to be a subspace of a 3D space, we obtain a link between the 3D and the 2D differential operators. 2D space is usually the transverse plane of a cylindrical arrangement. In this case, z denotes the cylinder axis, which is perpendicular to the transverse plane T (i.e., 2D space). This explains the index T. When one separates all 3D vectors $\vec{v}$ into a longitudinal part $v_z\vec{e}_z$ and a transverse part $\vec{v}_\mathrm{T} = \vec{v} - v_z\vec{e}_z$, one finds

$$\operatorname{div}\vec{B} = \operatorname{div}_\mathrm{T}\vec{B}_T + B_{z,z} \,, \tag{1.90}$$

$$\operatorname{curl}\vec{A} = (\vec{A}_{T,z} - \operatorname{grad}_\mathrm{T} A_z)^\circ - (\operatorname{div}_\mathrm{T}\vec{A}_T{}^\circ)\vec{e}_z \,, \tag{1.91}$$

$$\operatorname{grad}\phi = \operatorname{grad}_\mathrm{T}\phi + \phi_{,z}\vec{e}_z \,, \tag{1.92}$$

$$\Delta\phi = \Delta_\mathrm{T}\phi + \phi_{,zz} \,. \tag{1.93}$$

Obviously, there is no need for a transverse curl operator.

Note that the metric coefficient in z direction of cylindrical coordinates u,v,z is assumed to be:

$$g_z = 1 \,. \tag{1.94}$$

It would be possible to have a metric coefficient in z direction that differs from one. This would make the equations (1.90) to (1.93) more complicated.

Coordinates

In 2D space, coordinates look much simpler. In addition, one has the method of conformal mapping as a very helpful tool. For numerical field computations, it is usually sufficient to know the two simplest orthogonal coordinate systems: (1) Cartesian coordinates and (2) polar coordinates. One obtains these coordinate systems from the 3D Cartesian and circular cylindrical coordinates simply by omitting the z coordinate. The corresponding metric coefficients are

$$g_x = g_y = 1 \,, \tag{1.95}$$

$$g_r = 1, \quad g_\varphi = r \,. \tag{1.96}$$

To compute the metric coefficients for any system of orthogonal coordinates (u,v), one has

$$g_a^2 = \left(\frac{\partial x}{\partial a}\right)^2 + \left(\frac{\partial y}{\partial a}\right)^2 ; \quad a = u,v \,. \tag{1.97}$$

The 2D spatial first-order derivatives can be found either the same way as in 3D space or directly from the corresponding 3D derivatives:

$$\operatorname{div}_\mathrm{T} \vec{B} = \frac{(g_v B_u)_{,u} + (g_u B_v)_{,v}}{g_u g_v}, \tag{1.98}$$

$$\operatorname{grad}_\mathrm{T} \phi = \frac{\phi_{,u}}{g_u}\vec{e}_u + \frac{\phi_{,v}}{g_v}\vec{e}_v . \tag{1.99}$$

As one may expect, this looks especially simple for Cartesian coordinates:

$$\operatorname{div}_\mathrm{T} \vec{B} = B_{x,x} + B_{y,y} , \tag{1.100}$$

$$\operatorname{grad}_\mathrm{T} \phi = \phi_{,x}\vec{e}_x + \phi_{,y}\vec{e}_y . \tag{1.101}$$

Finally, it is a simple matter to guess the form of the 2D Laplacian operator in Cartesian coordinates:

$$\Delta_\mathrm{T}\phi = \phi_{,xx} + \phi_{,yy} . \tag{1.102}$$

Synopsis of important equations

A list of useful equations that contain spatial derivatives is given below without proof. Most of the 3D equations are very well known and may be found in any good textbook on electrodynamics. The corresponding 2D equations are usually not listed, as most authors consider 2D space as a subspace of 3D space, and they use the 3D operators for 2D problems as well.

As all operators mentioned above are linear operators L, one has

$$L\sum a_i f_i = \sum a_i L f_i . \tag{1.103}$$

For the different products of vectors and scalars, the following equations hold:

$$\operatorname{div}(s\vec{v}) = s\operatorname{div}\vec{v} + \vec{v}(\operatorname{grad} s), \tag{1.104}$$

$$\operatorname{div}_\mathrm{T}(s\vec{v}) = s\operatorname{div}_\mathrm{T}\vec{v} + \vec{v}(\operatorname{grad}_\mathrm{T} s), \tag{1.105}$$

$$\operatorname{div}(\vec{u}\times\vec{v}) = \vec{v}(\operatorname{curl}\vec{u}) - \vec{u}(\operatorname{curl}\vec{v}), \tag{1.106}$$

$$\operatorname{curl}(s\vec{v}) = s\operatorname{curl}\vec{v} + (\operatorname{grad} s)\times\vec{v}, \tag{1.107}$$

$$\operatorname{curl}(\vec{u}\times\vec{v}) = \vec{u}\operatorname{div}\vec{v} - \vec{v}\operatorname{div}\vec{u} + (\vec{v}\operatorname{grad})\vec{u} - (\vec{u}\operatorname{grad})\vec{v}, \tag{1.108}$$

$$\operatorname{grad}(r\,s) = r\operatorname{grad} s + s\operatorname{grad} r, \tag{1.109}$$

$$\operatorname{grad}_\mathrm{T}(r\,s) = r\operatorname{grad}_\mathrm{T} s + s\operatorname{grad}_\mathrm{T} r, \tag{1.110}$$

$$\operatorname{grad}(\vec{u}\,\vec{v}) = \vec{u}\times\operatorname{curl}\vec{v} + \vec{v}\times\operatorname{curl}\vec{u} + (\vec{v}\operatorname{grad})\vec{u} + (\vec{u}\operatorname{grad})\vec{v}, \tag{1.111}$$

$$\operatorname{grad}_\mathrm{T}(\vec{u}\,\vec{v}) = \vec{u}^\circ\operatorname{div}_\mathrm{T}\vec{v}^\circ + \vec{v}^\circ\operatorname{div}_\mathrm{T}\vec{u}^\circ + (\vec{v}\operatorname{grad}_\mathrm{T})\vec{u} + (\vec{u}\operatorname{grad}_\mathrm{T})\vec{v}, \tag{1.112}$$

where $(\vec{u}\operatorname{grad})\vec{v}$ and $(\vec{u}\operatorname{grad}_\mathrm{T})\vec{v}$ are called 3D and 2D vector gradients. They are defined as follows:

$$(\vec{u}\operatorname{grad})\vec{v} = \lim_{d\to 0}\frac{\vec{v}(\vec{r} + \mathrm{d}\vec{u}) - \vec{v}(\vec{r})}{d}, \tag{1.113}$$

$$(\vec{u}\,\mathrm{grad}_\mathrm{T})\vec{v} = \lim_{d\to 0}\frac{\vec{v}(\vec{r}+\mathrm{d}\vec{u})-\vec{v}(\vec{r})}{d}\;. \tag{1.114}$$

We have already seen the following equations with two first-order spatial derivatives:

$$\mathrm{curl}\;\mathrm{grad}\;\phi = 0\,, \tag{1.115}$$

$$\mathrm{div}\;\mathrm{curl}\;\vec{A} = 0\,, \tag{1.116}$$

$$\mathrm{div}_\mathrm{T}\;\mathrm{grad}_\mathrm{T}{}^\circ\;\phi = 0\,, \tag{1.117}$$

$$\Delta\phi = \mathrm{div}\;\mathrm{grad}\;\phi\,, \tag{1.118}$$

$$\Delta\vec{A} = \mathrm{grad}\;\mathrm{div}\;\vec{A} - \mathrm{curl}\;\mathrm{curl}\;\vec{A}\,, \tag{1.119}$$

$$\Delta_\mathrm{T}\phi = \mathrm{div}_\mathrm{T}\;\mathrm{grad}_\mathrm{T}\;\phi\,, \tag{1.120}$$

and the following integrals:

$$\int_\ell \mathrm{grad}\;\phi\;\mathrm{d}\vec{\ell} = \oint_{\partial\ell}\phi\;\mathrm{d}P = \phi(P_E)-\phi(P_A)\,, \tag{1.121}$$

$$\int_S \mathrm{curl}\;\vec{A}\;\mathrm{d}\vec{S} = \oint_{\partial S}\vec{A}\;\mathrm{d}\vec{\ell}\,, \tag{1.122}$$

$$\int_V \mathrm{div}\;\vec{B}\;\mathrm{d}V = \oint_{\partial V}\vec{B}\;\mathrm{d}\vec{S}\,, \tag{1.123}$$

where (1.122) and (1.123) are the well-known theorems of Stokes and Gauss. In addition, one has the following integrals:

$$\int_V \mathrm{grad}\;\phi\;\mathrm{d}V = \oint_{\partial V}\phi\;\mathrm{d}\vec{S}\,, \tag{1.124}$$

$$\int_S \mathrm{grad}_\mathrm{T}\;\phi\;\mathrm{d}S = -\oint_{\partial S}\phi\;\mathrm{d}\vec{\ell}^\circ\,, \tag{1.125}$$

$$\int_V \mathrm{curl}\;\vec{A}\;\mathrm{d}V = -\oint_{\partial V}\vec{A}\times\mathrm{d}\vec{S}\,, \tag{1.126}$$

$$\int_V (\phi\,\Delta\psi + \mathrm{grad}\;\phi\cdot\mathrm{grad}\;\psi)\,\mathrm{d}V = \oint_{\partial V}\phi(\mathrm{grad}\,\psi)\cdot\mathrm{d}\vec{S}\,, \tag{1.127}$$

$$\int_S (\phi\,\Delta_\mathrm{T}\psi + \mathrm{grad}_\mathrm{T}\;\phi\cdot\mathrm{grad}_\mathrm{T}\;\psi)\,\mathrm{d}S = -\oint_{\partial S}\phi(\mathrm{grad}_\mathrm{T}\;\psi)\cdot\mathrm{d}\vec{\ell}^\circ\,, \tag{1.128}$$

$$\int_V (\phi\,\Delta\psi - \psi\,\Delta\phi)\,\mathrm{d}V = \oint_{\partial V}(\phi\,\mathrm{grad}\,\psi - \psi\,\mathrm{grad}\phi)\cdot\mathrm{d}\vec{S}\,, \tag{1.129}$$

$$\int_S (\phi\,\Delta_\mathrm{T}\psi - \psi\,\Delta_\mathrm{T}\phi)\,\mathrm{d}S = \oint_{\partial S}(\phi\,\mathrm{grad}_\mathrm{T}\,\psi - \psi\,\mathrm{grad}_\mathrm{T}\phi)\cdot\mathrm{d}\vec{\ell}^\circ\,, \tag{1.130}$$

where the last four equations are the theorems of Green. The first three integral equations allow one to define the operators grad, grad_T, and curl by volume integrals. Together with the theorem of Gauss, one has

$$\operatorname{div} \vec{B} = \lim_{r \to 0} \frac{\oint_{\partial V} \vec{B}\, \mathrm{d}\vec{S}}{\int_V \mathrm{d}V}, \tag{1.131}$$

$$\operatorname{curl} \vec{A} = \lim_{r \to 0} \frac{-\oint_{\partial V} \vec{A} \times \mathrm{d}\vec{S}}{\int_V \mathrm{d}V}, \tag{1.132}$$

$$\operatorname{grad} \phi = \lim_{r \to 0} \frac{\oint_{\partial V} \phi\, \mathrm{d}\vec{S}}{\int_V \mathrm{d}V}, \tag{1.133}$$

$$\operatorname{div}_{\mathrm{T}} \vec{B} = \lim_{r \to 0} \frac{\oint_{\partial S} \vec{B}^{\circ}\, \mathrm{d}\vec{\ell}}{\int_S \mathrm{d}S}, \tag{1.134}$$

$$\vec{e}_{\ell} \cdot \operatorname{grad}_{\mathrm{T}} \phi = \lim_{r \to 0} \frac{\oint_{\partial \ell} \phi\, \mathrm{d}P}{\int_{\ell} \mathrm{d}\ell}. \tag{1.135}$$

2 Function analysis

General remarks

We have already worked with fields, derivatives, and integrals in the previous chapter, i.e., with objects that are more complicated and advanced than, for example, a simple function $f(x)$ of a single variable x. In fact, it is assumed that the reader of this book is already familiar with both fields and functions. The aim of this chapter and of the previous chapter is picking out some special parts of difficult mathematical concepts that are important for computational electromagnetics. For reasons of simplicity, we will not use advanced and abstract mathematical formulations, but we will consider some important aspects that are not very well-known. It is expected that the reader is able to generalize what is presented here and will understand the consequences for the design of codes for computational electromagnetics.

For engineers, a function $f(x)$ is usually a map of a real number x on another real number f, which can easily be visualized by a curve in a plane, i.e., in 2D real space. Since we live in 3D real space, our common experience helps us to understand such functions. For mathematicians, a function is also a map, but both x and f can be elements of arbitrary sets M_x and M_f. As soon as the sets M_x and M_f are not the real number space, we have the feeling that this is 'abstract', i.e., away from everyday life. To define functions properly, one has to start with sets, groups and similar abstract objects. Then, one can proceed to the analysis of number spaces. Afterwards, one can define functions as maps. Finally, one considers sets of functions and function spaces. This procedure looks like a circle, because it plays no essential role for the analysis of a space whether the elements are numbers, functions, or something else. Therefore, a more compact overview is obtained when one anticipates the definition of functions (and integrals of functions) while one considers the properties of spaces. Since it is assumed that the reader is already familiar with the concepts of sets, spaces, functions, integrals, etc., this procedure is applied in the following. It is hoped that it will provide new insights and fill the abstract terms of function analysis with some meaning.

Collections of well-known series expansions are contained in [Abramowitz, 1970], [Gradshteyn, 1965]. For further reading, see also [Courant, 1989], [Henrici, 1974-1986], [Sommerfeld, 1950].

Number spaces, computers, and physics

Those who are familiar with computers know that real numbers must be approximated by a so-called 'real' number type because computers cannot exactly handle real numbers. The reason for this is very simple and fundamental. A computer is a machine of finite size with a finite memory. Therefore, it can only deal with objects of a finite information content. In general, the information content of real numbers is infinite. Depending on the notation we use, a finite number of real numbers can be denoted exactly. For example, in the decimal system, we can write 3.14 and we exactly know the size of this number. A simple rational number can sometimes be described exactly, for example, 1/5=0.2, but sometimes this is impossible, for example, 31/30=1.03333333 is inaccurate. Since the decimal notation of rational numbers is periodic behind a certain position, we can invent a more sophisticated notation, for example, 31/30=1.0$\underline{3}$, that allows us to denote rational numbers exactly, provided that the nominator and denominator are finite integer numbers. With this notation, we still cannot exactly denote any real number, for example, $\sqrt{2}$. We now can invent an even more sophisticated notation for square roots and other roots of rational numbers, for example, $\sqrt[4]{31/30}$=1.0$\underline{3}$r4, but we will find real numbers that cannot be exactly denoted with any sophisticated notation that uses a finite alphabet of symbols. Therefore, it is hopeless to find a general notation of real numbers for computers. However, it is important to note 1) that the accuracy of a certain real number depends on the notation, 2) that a computer 'knows' a finite number of real numbers exactly, i.e., that some of the real numbers can be represented exactly in a computer, 3) that the accuracy of a real number depends on the notation that is used in the computer, and 4) that there are different notations of real numbers. Most of the modern computers use binary representations of real numbers and there are also standard notations for real numbers, but not all computers and not even all compilers use one and the same notation. Therefore, one can obtain different results from one and the same numerical calculation when one uses different compilers or different computers.

Since computers cannot handle real numbers exactly, it is questionable whether nature works with real numbers or not. First of all, it is also impossible to measure a real number exactly. Even if we were to have absolutely accurate equipment, we could not write down the result of a measurement of a real number for the reasons mentioned above. At the beginning of the invention of quantum mechanics, it became obvious that the accuracy of observations is finite. An important difference between quantum mechanics and classical mechanics is the importance of non-real number spaces. First, it was 'observed' that some physical quantities can be described by integer numbers, then other quantities were discovered which are described by finite sets, for example, the spin. Einstein, who hated the probabilistic aspects of quantum mechanics, proposed an experiment that should demonstrate that quantum mechanics is wrong. He said that one would have two possibilities if this experiment failed: 1) accept quantum mechanics, or 2) abandon the idea of real valued space and time. Today, quantum mechanics is widely accepted. It even seems that the experiment proposed by Einstein failed. Therefore, the remaining question is whether space and time can be represented by real numbers. For classical physics, real numbers were essential. Quantum mechanics introduced new number spaces and sets with the corresponding arithmetics, but it did not remove real numbers. Moreover, the probabilistic aspect of quantum mechanics made the computation of random numbers very important. Therefore, quantum mechanics is far away from the heart of computers, i.e., from a finite, deterministic machine.

Although inaccuracies of numerical simulations caused by the approximation of real numbers are fundamental, the typical accuracy of modern computers is much higher than the accuracy of measurements. Therefore, the approximation of real numbers is usually a minor problem.

Special properties of spaces

Mathematicians think about very general and abstract sets and spaces that are far away from 'reality'. Afterwards, they think about special properties that are typical for more 'realistic' spaces. In the following, some of the most important properties of different spaces are outlined.

Metric

For most of us, it is 'natural' that one can define a metric, i.e., a distance $d(f_1,f_2)$ between two points (or elements) f_1 and f_2 of a space, but it is good to know that there are also sets without a metric. The most important properties of the metric are the following: 1) The metric (distance) is a real number associated with two points of a set or space. On computers, this causes the fundamental problems mentioned above. 2) If the metric (distance) of two points is zero, the points coincide, i.e., are identical. 3) If one has three points, the distance between two points is smaller than or equal to the sum of the distances from the two points to the third point. This is evident when we consider standard geometry, but we can also define the metric for spaces of functions, i.e., we can ask and answer the question: What is the distance of $\sin(x)$ and $\cos(x)$? Answering such questions is simpler in spaces with additional properties that are outlined below. Fortunately, engineers usually work in such spaces that will be outlined in the following sections.

Why is it important to define and compute the distance of functions? What is a metric good for? Assume that you have two different results obtained with two different codes. How can you know which of the results is the better one? If the results are points in a metric space, you can compute the distance from these points to the point that corresponds to the correct solution. Without a metric, you cannot do such comparisons, or you cannot do them in a simple and natural way. Therefore, the metric is important for all sort of numerical computations.

It is important to note that the metric is required for comparisons, therefore it is desirable that the results are points in a metric space. This does not mean that all quantities we are working with need to be points in a metric space. For example, in genetic programming one can approximate a function by another function that is not necessarily defined in a metric space of functions. To decide which of the possible solutions is the best one, one associates a real number called the fitness value to each function and searches for the function with the highest fitness. In general, the fitness numbers do not define a metric for the function space because completely different functions may have similar or even identical fitness values. Despite of this, genetic programming and other modern algorithms working on non-metric spaces may lead to useful results.

Linear space

Assume that f_i are points in a metric space. Now, build the following linear combination:

$$f = \sum_{i=1}^{I} a_i f_i \,, \tag{2.1}$$

where a_i are real numbers. Again, we have the real numbers that will cause fundamental problems for our computer, but linear combinations are very frequently used. As we will see later on, this is the first step toward linear systems of equations that can efficiently be handled by computers. Therefore, most computational methods work implicitly or explicitly with linear systems of equations. Consequently, we almost always work in linear space and we do not even ask the question whether the space is linear or not.

Some readers might assume that f denotes a function. If f is a function, the f_i are also functions, so-called basis functions. Instead of this, f and f_i can be numbers, vectors, fields, etc., i.e., the linearity is a property of many different mathematical objects. To understand the meaning of the abstract definition, one can think about a specific, well-known object, for example, a function, but one should always keep in mind that the definition is more general. When we assumed f to be a function, we anticipated the main question: Is f also a point in the given space, spanned by the basis f_i ? This question cannot be answered in general for arbitrary spaces. If it is answered in the affirmative, the corresponding space is linear. To answer the question, the definition of the metric is helpful. If the space is metric, one can compute the distance between f and the sum in (2.1). Since we know that a computer approximates real numbers by 'real' ones using a certain convention, we guess that the 'real' number space is not exactly linear. For example, we can set $a_1 = \pi$, $f_1 = 1$, $I = 1$ and obtain $f = \pi$, which is not a point in the 'real' number space of our computer when π is not in this space, whereas 1 is in it.

The main purpose of (2.1) is to guarantee that the space we are working in is complete with respect to linear combinations. Therefore, it is reasonable to restrict the coefficients a_i to the 'real' space of our computer, rather than the real space of mathematicians. Since the 'real' space is necessarily finite, we can set $a_1 = b$, $f_1 = c$, $I = 1$ and obtain $f=bc$, which is outside the 'real' space when the two numbers b and c are sufficiently large. This is usually called numerical overflow.

To avoid numerical overflow, one can invent so-called 'protected' operations. For example, one can define the protected multiplication in such a way that the result is set equal to one if an overflow occurs. Protected operations are preferred in genetic programming and allow one to obtain codes that never crash because of overflow and other numeric 'exceptions'. Is this a way to make the 'real' space linear? The answer is negative. The resulting space is not even metric because different numbers that cause an overflow are mapped on one and the same number. One can try more sophisticated definitions of protected operations for 'real' numbers, but none of them allows one to make sure that the space becomes metric. Note that protected operations are favorable in genetic programming, where non-metric, non-linear spaces are usual.

Since the numbers observed in nature are finite and of a size that is usually small compared with the size of big 'real' numbers that are available on computers, overflow problems can often be avoided without protected operations. Therefore, the incompleteness of the 'real' numbers caused because of overflows is of minor importance. Are the 'real' numbers complete when no overflows occur? Unfortunately, we have to answer even this question in the negative, because the product of two 'real' numbers b and c is not always a 'real' number when no overflow is encountered. To understand this, assume that the computer is working in the decimal system, using a notation of the following form: 1.234E+01. For example, one gets 1.234E+01*1.234E+01=152.2756. To obtain the same notation as before, the result must be truncated or rounded, i.e., the resulting 'real' number is 1.522E+03 (truncated) or 1.523E+03 (rounded). The situation does not change in principle when we use a binary notation or when we increase the number of digits of our notation. Truncation or rounding is required in the 'real' number space of our computers and this is a permanent source of inaccuracies. Since these errors accumulate when several operations are performed, they are important, even if the accuracy of the 'real' number notation is high. This is especially true when subtractions are involved. For example, *a*b-c*d* may be different from zero, but it becomes numerically zero when *a*b* is equal to *c*d* because of truncation or rounding errors. This effect is called cancellation. Cancellation can drastically reduce the numerical accuracy.

Norm

In some linear spaces it is possible to associate a real number, the so-called norm $\|f\|$, to each element f of the space. To visualize the meaning of the norm, one best considers a simple vector space, which is also a linear space. Here, the norm of a vector is nothing else than its length. In general, the norm has the following properties: 1) It is positive or zero. If it is zero, one has f=0, i.e., f is the zero element. 2) $\|f+g\| \leq \|f\| + \|g\|$, which corresponds to the third property of the metric. 3) $\|af\| = |a|\|f\|$ holds for any real number a.

The norm allows one to easily define a metric d(f,g):

$$\mathrm{d}(f,g) = \|f-g\| \,. \tag{2.2}$$

When f and g are two points in a vector space, the metric d is nothing else than the distance between these points. When f is the desired solution of a problem and g its numeric approximation, the metric d is a measure for the error of the approximation. Without the definition of the norm, it is hard to make any statements about the quality of a numeric approximation.

Note that the definition of the norm is usually not unique at all. For example, consider the N-dimensional real space. The most common definition of the norm is the so-called square norm:

$$\|f\| = \left(\sum_{n=1}^{N} x_n^2 \right)^{1/2} , \tag{2.3}$$

where x_n denotes the n-th component of f. A more general definition is the p norm:

$$\|f\| = \left(\sum_{n=1}^{N} |x_n|^p \right)^{1/p} . \tag{2.4}$$

Assume that you are searching for an approximation f^M of a point f in N-dimensional space in such a way that f^M is close to a given point f. The error of the approximation can be defined with the norm:

$$\|f - f^M\| = \left(\sum_{n=1}^{N} |x_n - y_n|^p \right)^{1/p} = \left(\sum_{n=1}^{M} |x_n - y_n|^p + \sum_{n=M+1}^{N} |x_n|^p \right)^{1/p} . \tag{2.5}$$

You now are searching for the values y_n that minimize the error. Since the error is a positive number, you must search for the values y_n that minimize

$$\left\|f - f^{M}\right\|^{p} = \sum_{n=1}^{M}\left|x_n - y_n\right|^{p} + \sum_{n=M+1}^{N}\left|x_n\right|^{p} . \quad (2.6)$$

Obviously, the minimum error is reached if $y_n=x_n$, for $n=1,2,\ldots,M$. This is almost trivial. To be less trivial, let us consider an approximation of the function $f(x)$ in the interval $x=0\ldots1$ by a series expansion of the form

$$f^{M}(x) = \sum_{n=1}^{M} a_n f_n(x) , \quad (2.7)$$

where f_n are given basis functions and a_n are unknown parameters to be computed in such a way that the error

$$\left\|f(x) - f^{M}(x)\right\| = \left\|f(x) - \sum_{n=1}^{M} a_n f_n(x)\right\| \quad (2.8)$$

is minimized. With the definition of the square norm

$$\left\|f(x)\right\| = \left(\int_0^1 f^2(x)\,\mathrm{d}x\right)^{1/2} , \quad (2.9)$$

we obtain

$$\left\|f(x) - \sum_{n=1}^{M} a_n f_n(x)\right\|^2 = \int_0^1 \left(f(x) - \sum_{n=1}^{M} a_n f_n(x)\right)^2 \mathrm{d}x = \min . \quad (2.10)$$

Setting the derivatives of (2.10) with respect to all unknowns a_n equal to zero, leads to a linear system of M equations for M unknowns. Later, we will consider this so-called error minimization technique in detail. Note that this technique is mainly based on the definition of the norm. When you replace the square norm by a p norm with p different from 2, you do not obtain a linear system of equations, which makes the computation of the unknown parameters difficult and time-consuming. Thus, finding an appropriate norm is crucial for the error minimization technique.

Projection, inner product or scalar product, orthogonality

Engineers are used to linear spaces that have additional features, especially unitary spaces that allow the definition of inner products or scalar products. We have already seen that a scalar product in a vector space may be interpreted as a projection. In general, the inner product (f,g) of two elements f and g of a linear space is an associated, complex number with the following properties: 1) $(f,g)=(g,f)^*$, where * denotes the conjugate complex. 2) $(f,ag)=a(f,g)$, where a is a complex number. 3) $(f,g+h)=(f,g)+(f,h)$. Together with the second property, this means that the inner product is linear. 4) (f,f) is bigger than or equal to zero. Note that (f,f) is real because of the first property. Obviously, (f,f) is a norm of f. Once the inner product is defined, it is natural to define the norm as (f,f).

There were different ways to define a norm and there are also different ways to define inner products or scalar products. It is important to note that the usual formalism of the Method of Moments (MoM) [Harrington, 1968] defines a product $<f,g>$ with $<f,g>=<g,f>$ instead of the first property mentioned above. This little modification becomes important when (f,g) is complex. For mathematicians, it is hard to understand why the usual inner product is replaced by a modified product that does not directly define a norm.

The product (f,g) defines a projection of f on g. This is easily understood in 3D space, when f and g are 3D vectors, for example, $\vec{u}$, $\vec{v}$. In this case, one has

$$\vec{u} \cdot \vec{v} = |\vec{u}||\vec{v}|\cos\alpha, \tag{2.11}$$

where α denotes the angle between the two vectors. This geometric definition of the scalar product can be generalized for vector spaces with arbitrary dimension. Linear function spaces have usually infinite dimension, but the projection is still a good picture for what the scalar or inner product does.

The norm is the mathematical vehicle of the error minimization technique. As we will see, the inner product is the mathematical vehicle of the projection technique and the modified product mentioned above is the mathematical vehicle of the method of moments. To get a first impression, consider the approximation of the function $f(x)$ in the interval x=0...1 by a series expansion of the form (2.7) again. When you project $f(x)$ on M so-called testing functions $g_m(x)$, you immediately obtain M linear equations:

$$\left(\sum_{n=1}^{M} a_n f_n(x), g_m(x)\right) = \sum_{n=1}^{M} a_n (f_n(x), g_m(x)) = (f(x), g_m(x)) \cdot \tag{2.12}$$

On one hand it is much easier to obtain a system of equations with the projection technique than with the error minimization technique. On the other hand, the selection of the testing function has a big impact on the results and it can be difficult to find an optimal set of testing functions that guarantees accurate results. Since the scalar product allows one to define a norm, one has a direct link to the error minimization technique. With the special choice $g_m(x)$=$f_m(x)$, i.e., the testing functions are the same as the basis functions, one obtains the same equations as with the error minimization technique that uses the square norm:

$$\sum_{n=1}^{M} a_n (f_n(x), f_m(x)) = (f(x), f_m(x)) \cdot \tag{2.13}$$

This choice of testing functions is the Galerkin choice, the corresponding method being called the Galerkin method.

The scalar product allows one to define orthogonality as a special property of a pair of functions. Two functions f and g are orthogonal with respect to each other when the scalar product (f,g) is zero. Although this property does not seem to be very exciting, it can be very helpful, as one knows from vector analysis. If one has a basis f_m of a linear space, one can orthogonalize it in such a way that all pairs of basis elements are orthogonal to each other, i.e., (f_n,f_m)=0 if n and m are different. An orthogonal basis has considerable consequences, for example, for the Galerkin method mentioned above. Obviously, most of the scalar products in equation (2.13) vanish and one easily obtains

$$a_m = (f(x), f_m(x)) / (f_m(x), f_m(x)) = (f(x), f_m(x)) / \|f_m(x)\|^2 , \tag{2.14}$$

i.e., one can explicitly compute the parameters a_m without solving the system of equations (2.13). Such simplifications are typical for an orthogonal basis. A further simplification of (2.14) is obtained when the norm of each basis function is equal to one. One then obtains the unknown parameters as simple projections of the given function f on the corresponding basis function:

$$a_m = (f(x), f_m(x)) \ . \tag{2.15}$$

Note that essentially the same procedure is useful not only in function spaces, but also in vector spaces, when f is a field, and so on. A basis that is both orthogonal and normed is called orthonormal. Although such a basis has nice properties and can be constructed in unitary spaces, the construction of an orthonormal basis may be more time-consuming than the solution of a system of equations of the form (2.13).

Convergence and completeness

In a metric space, one can consider sequences $\{f_n\}$ of enumerably many elements f_n. Such a sequence is convergent toward an element f, when the distance $d(f,f_n)$ tends to zero for an increasing index n. Usually, one writes:

$$\lim_{n\to\infty} d(f, f_n) = 0 \ . \tag{2.16}$$

A more sophisticated notation is:

$$d(f, f_n) < \varepsilon, \quad \text{if } n \geq N \equiv N(\varepsilon) \ . \tag{2.17}$$

This means that the distance between f and f_n is smaller than a given (arbitrarily small) quantity ε, as soon as the index n is bigger than or equal to than a certain number N which depends on the size of ε. The element f is called the limit of the sequence.

For the fundamental Cauchy sequence, one has a similar condition for the distances between the elements of the sequence with sufficiently large index:

$$d(f_m, f_n) < \varepsilon, \quad \text{if } m,n \geq N(\varepsilon) \ . \tag{2.18}$$

The question now is: Is the limit of a Cauchy sequence also contained in the space where the sequence is defined? If this question is answered in the affirmative for all Cauchy sequences, the space is called complete. Note that both convergence and completeness depend on the metric. Since the definition of the metric in a space is not unique, it may happen that a space is complete when a certain definition of the metric is used, but it is incomplete when another definition of the metric is used.

Convergence is considered to an important property of a numerical algorithm. How can one understand the convergence of an algorithm? Assume that you have an iterative procedure to compute the value of a function f. The output of the algorithm after n iterations is equal to f_n. Therefore, you get a sequence of output data. In electromagnetics, f may denote the electromagnetic field or some derived quantity. Since the information content of the field is infinite, numeric algorithms can only compute derived quantities, for example, components of the field in some points of space and time, the impedance of an antenna, and so on. The output of the algorithm after n iterations is therefore typically an array or a set of 'real' numbers. Incidentally, the reduction of the information of an entire field to the information of a finite number of 'real' numbers is a tremendous data compression.

Convergence of sequences and completeness are nice properties from the mathematical point of view. For numeric algorithms, these terms may be too strong. Because of the limited accuracy of computers, ε may be finite, and because of the limited speed of computers, one can not compute an infinite sequence. On the other hand, convergence is not sufficient for useful algorithms. To obtain a sufficiently accurate approximation of f with an efficient algorithm, the convergence of the series must be quick enough. Therefore, one is interested in sequences with

$$\mathrm{d}(f, f_n) = d_n < d_{n-1}, \quad \text{for} \quad 1 < n \leq m, \quad \text{with} \quad d_m < d \ . \tag{2.19}$$

Here m denotes the maximum number of elements of the series to be computed, d is the maximum error that is allowed. Such a sequence should be convergent up to the element f_m, but for $n>m$ it may diverge. Very often one can find such semi-convergent sequences are more powerful than convergent ones. It is important to keep this in mind when one implements algorithms. Most of the sequences discussed in literature are convergent. Watching out for other sequences may be the first step toward more powerful algorithms. Note that even (2.19) is not necessary for sequences obtained from useful numeric algorithms. For example, the sequence may diverge for small n, converge for medium n and diverge again for big n. The only goal to be achieved is finding a sufficiently accurate approximation.

One of the main numerical problems is finding an appropriate stopping criterion. This is obvious when the series does not converge, but it is also a problem for convergent series, because the limit f is usually unknown. Therefore, one cannot compute d_n after having computed f_n. Finding an estimate of d_n may be tricky.

Although iterative algorithms are frequently implemented, one should note that many algorithms are more complex and cannot be described by a single number n that increases with the computation time. For such algorithms, it may be hard or even impossible to define their convergence.

We have already discussed some problems arising from the implementation of 'real' numbers in our computers. From this, it is quite evident that completeness in the strict mathematical sense cannot be obtained on computers. If we consider the output of an algorithm, completeness in a weaker sense means that the algorithm is able to compute all possible solutions in the metric space of the results with the desired accuracy. Because of overflow and similar problems, each numeric algorithm has a limited range, where it is valid. Therefore, even the weaker formulation of completeness is too strict. Moreover, the space of the results is usually not metric because the 'real' number space is not metric. This makes the definition and verification of the completeness of an algorithm more difficult. From the practical point of view, the completeness of an algorithm is not required at all. All the user wants to obtain is a sufficiently accurate result for a problem of interest.

To get rid of the problems caused by the limitations of 'real' computers, one can imagine an 'ideal' computer with an infinite memory, speed, and accuracy. On such computers, the convergence and completeness of a numerical method would be most important. Many people who are talking about completeness have either an 'ideal' machine or a very 'weak' definition of completeness in mind.

Tools

Having outlined important properties of different spaces, we now outline some important tools for working in such spaces.

Series

When we introduced linear spaces, we started with a linear combination, i.e., a sum of the form (2.1). With a slight modification of (2.1), we obtain a sequence $\{F_n\}$ in the linear space:

$$F_n = \sum_{i=1}^{n} a_i f_i \ . \tag{2.20}$$

Such sequences lead to so-called series expansions, a powerful tool for the numeric computation of functions, fields, etc. A series expansion is obtained in the limit of n towards infinity:

$$f = \sum_{i=1}^{\infty} a_i f_i \ . \tag{2.21}$$

Usually, one says that the basis f_i is complete when all elements of the space can be expanded by (2.21). This definition of completeness is often too strict. To illustrate this, let us consider a function $f(x)$ defined in a finite interval, for example, x=-1...+1. If the function is infinitely many times differentiable, it may be expanded with many different series expansion of the form (2.21), for example, power series, Fourier series, and so on. The basis functions of all these series are infinitely many times differentiable. What happens when we expand, for example, the function abs(x)? This function is continuous everywhere. It is equal to x for x>0, equal to $-x$ for x<0 and equal to 0 for x=0. Consequently, the first derivative of abs(x) is +1 for x>0, -1 for x<0, and undefined for x=0. When we expand abs(x) with a series of infinitely many times differentiable functions, we obtain an approximation $f(x)$ that equals abs(x) everywhere, but the first and all higher order derivatives of $f(x)$ are continuous and well defined everywhere, where the first and all higher order derivatives are discontinuous and not defined at x=0. For a mathematician, this is an essential difference between $f(x)$ and abs(x), but for an engineer and especially those working on computers, this difference is probably negligible. When a series expansion allows one to obtain such an approximation, the corresponding basis is an approximation basis. It is not complete in the strict mathematical sense. Engineers often work with functions or fields that are not infinitely many times differentiable everywhere. In such cases, it is almost impossible to find a complete basis, but it may be quite easy to find an approximation basis.

The space of infinitely many times differentiable functions is unitary. This gives one the opportunity to work with an orthogonal basis, which allows one to simplify the computation of the coefficients a_i. Note that the orthogonality depends on the definition of the scalar product and this usually depends on the interval where the functions are defined. When one has found an orthonormal basis, for example, on the interval x=-1...+1, this basis is usually not orthogonal on a part of that interval, for example, on the interval x=0...+1. This situation becomes more delicate for functions with several variables and for fields. Therefore, one often applies non-orthogonal series expansions.

It is evident that the limited capacity of computers does not allow one to compute infinite sums, i.e., series expansions. Therefore, the series expansions must be truncated. This is usually done by evaluating (2.20) instead of (2.21). Although truncation is simple, it is not necessarily efficient. We will consider the most important series expansions later in this chapter and we will discuss alternatives there.

Integrals

It is well known that integrals of usual functions can be approximated by sums. From the numerical point of view, there are very different algorithms based on different sums. These algorithms have different advantages and drawbacks. Their efficiency and accuracy usually depend on the function to be integrated. [Press, 1992] gives a good overview.

Mathematicians often prefer the most simple approximation of an integral by a Riemann sum of the form:

$$\int_a^b f(x)\,\mathrm{d}x = \frac{b-a}{n}\sum_{i=1}^{n} f(x_i) + d_n = I_n + d_n \,. \tag{2.22}$$

Here, it is assumed that the points x_i, where the function is evaluated, are uniformly distributed over the interval $a...b$ of the integration. Obviously, the sums I_n are a sequence and the distances d_n from I_n to the correct value of the integral denote the errors of the n-th approximation. We let mathematicians prove that or when this sequence is converging.

On one hand, it is quite obvious that a uniform distribution of the sampling points x_i is far from being numerically optimal for most cases. On the other hand, it is impossible to find an optimal distribution for all kind of functions. Therefore, one often uses adaptive algorithms that start with an initial discretization of the interval $a...b$ and refine this discretization in those areas where the function has a complicated shape. As soon as one tries to design an adaptive algorithm, one becomes aware of several difficulties that do not allow one to write an algorithm that is simple, efficient, and robust. Therefore, it is impossible to find an algorithm that is optimal for everything. A relatively general form for adaptive and non-adaptive integration is the following:

$$I_n = \sum_{i=1}^{n} w_i \cdot f(x_i) \,, \tag{2.23}$$

where the w_i are 'weights' that depend on the method and on the location of the sampling points x_i. In simple techniques, w_i only depends on the locations of the point x_i (with the same index) and its neighbors, x_{i-1} and x_{i+1}. For more advanced methods, the computation of the weights can be difficult and time-consuming. This may be reasonable when the number of function calls can be reduced with such a technique and when the evaluation of the function f in the sampling points is time-consuming.

When one can find a series expansion of the integrand $f(x)$ with a basis that may be integrated analytically, one obtains a numerical integration that is different from (2.23):

$$I_n = \int_a^b F_n(x)\,\mathrm{d}x = \sum_{i=1}^{n} a_i \int_a^b f_i(x)\,\mathrm{d}x \,. \tag{2.24}$$

In fact, also I_n in (2.22) can be obtained from (2.24). First, one selects a set of basis functions f_i that are equal to one in a finite interval $x_i...x_{i+1}$ and zero outside. The integral over such a basis function is equal to x_{i+1}-x_i . When one properly selects the sampling points and the parameters a_i, one obtains the I_n in (2.22).

Basis functions that are different from zero only in a part of the interval where the original function is defined, are called subdomain basis functions. Such functions may be advantageous because they can be simple and allow an analytic integration. The drawback of subdomain basis functions is that at least higher order derivatives of them must be discontinuous. The subdomain basis used above for obtaining the Riemann sum was discontinuous in two points. When it is replaced by more sophisticated basis functions, one can obtain better algorithms. For example, a 'rooftop' basis is continuous everywhere and piecewise linear. Inside a finite interval, it has the shape of a triangle, outside it is zero. With an appropriate rooftop basis, one obtains the well-known trapezoidal integration from (2.24). This algorithm can also be written in the form (2.23). Actually, one can often use both notations (2.23) and (2.24). The former is more general, but it does not tell one how to evaluate the weights, where the correct evaluation of the parameters and integrals in (2.24) is usually known.

Later on, we will encounter very similar techniques for computing fields. A subdomain basis is typical for finite element methods and it is also frequently used in the method of moments. The formulation (2.23) is typical for the generalized point matching technique, where the field is also sampled in a discrete set of sampling points.

So far, we have considered definite integrals, i.e., integrals defined on a well defined interval *a...b*. For analytic integration, indefinite integrals are important. The indefinite integral of a function is a function of the variable of the function that is only defined up to a constant C:

$$I(x) = \int f(x)\,\mathrm{d}x = F(x) + C \; . \tag{2.25}$$

When the indefinite integral is known, the definite integral can be obtained for an arbitrary interval:

$$I = \int_a^b f(x)\,\mathrm{d}x = I(b) - I(a) = F(b) - F(a) \; . \tag{2.26}$$

It is evident that it is useful to know the indefinite integrals of subdomain basis functions, because these functions are usually defined in several different subdomains, i.e., on several different intervals.

The integration of functions with more than one variable causes additional problems, especially when the geometry of the domains of the integration is complicated. This makes the implementation of algorithms for multi-dimensional integration difficult, but it does not provide much deeper insights that are relevant for computational electromagnetics.

Derivatives

It is well-known that indefinite integrals are the inverse of derivatives of a function. If (2.25) holds, one automatically has

$$I'(x) = \frac{\mathrm{d}}{\mathrm{d}x} I(x) = \frac{\mathrm{d}}{\mathrm{d}x} \int f(x)\,\mathrm{d}x = \frac{\mathrm{d}}{\mathrm{d}x} F(x) = f(x) \; . \tag{2.27}$$

Usually, the derivative of a function is defined as the limit:

$$f'(x) = \frac{\mathrm{d}}{\mathrm{d}x} f(x) = \lim_{\mathrm{d}x \to 0} \frac{f(x + \mathrm{d}x) - f(x)}{\mathrm{d}x} \tag{2.28}$$

and the indefinite integral is then defined as the inverse of the derivative. Numerically, one can easily approximate a derivative by a finite difference, i.e., by setting dx in (2.28) sufficiently small and by omitting the limit. This is the origin of the finite difference method. This is almost trivial and it is quite obvious from the definition (2.28) that such a method should converge, but there are several numerical problems that can cause inaccuracies. First of all, for sufficiently small values of dx, one will often encounter severe cancellations. For example, when one computes the derivative of cos(x) near x=0, cos(x+dx)-cos(x) becomes numerically zero when dx is sufficiently small, even when the correct solution cos'(x)=–sin(x) is numerically different from zero. To avoid such problems, one can apply similar techniques as for the numerical integration, i.e., one can expand the given function with a series of basis functions with analytically known derivatives. From this, one obtains a link of numerical techniques based on differential notations with numerical techniques based on integral formulations.

The definition and handling of higher order derivatives and of partial derivatives of functions with several variables provides no essential difficulties. Therefore, it is assumed that the reader can generalize the formalisms that have been outlined above.

Operators

Both indefinite integrals and derivatives associate a new function to a given function. An operator is a more general object that associates an element of a space to another element of a space. The two spaces may coincide, but they may also be different. For example, the derivative of a function of a space with n times differentiable functions is only n-1 times differentiable, i.e., the two spaces are not identical. Obviously, the spaces of a definite integral are different because the integral operates on a function, whereas the result is a number.

To apply a certain operator, one should know about the spaces the operator is acting on. For engineers, this is often too sophisticated, but it can become important when one is designing numerical methods. For example, when one decides to use rooftop basis functions with a discontinuous first order derivative, one should know that problems will occur as soon as one wants to work with second order derivatives.

In computational electromagnetics, one is mainly interested in linear operators L. For such operators, one has 1) L(af)=aL(f) and 2) L(f+g)=L(f)+L(g), where a is a real or complex number. When a linear operator is applied to a function that is expanded by a series, one can therefore write:

$$\mathrm{L}f = \mathrm{L}\sum_{i=1}^{\infty} a_i f_i = \sum_{i=1}^{\infty} a_i \mathrm{L} f_i \,. \tag{2.29}$$

This considerably simplifies working with linear operators.

It is easy to show that derivatives and integrals are linear operators. Linear operators are most common in physics. Up to now, Einstein's theory of general relativity is the only non-linear theory. In computational electromagnetics, one usually works with the classical Maxwell equations, i.e., with a linear set of equations, but these equations do not fully describe electromagnetics. The material properties are additional equations that may be non-linear. In fact linearity is not a property of nature, it is a simplification that is often sufficiently accurate. Since it is almost impossible to solve non-linear systems of differential equations, such problems have been ignored for a long time, but with the increasing power of computers, non-linear problems are of increasing interest. Despite this trend, the focus of this book is on linear problems.

Classes of functions

Since Newton, function analysis has been one of the main pillars of physics. Mathematicians have done a huge amount of work and have studied many different classes of functions. Therefore, it is impossible to give a quick overview. In the following, only some classes of functions that are important for computational electromagnetics are outlined.

First, we need a more explicit definition of what a function is. Mathematicians use several, sophisticated notations, for example,

$$f: \mathrm{D} \rightarrow \mathrm{V}, \quad x \rightarrow y = f(x), \quad x \in D, y \in V \ , \tag{2.30}$$

where D is the domain of definition, V the domain of values. The argument x of the function is an element of the space D, the result y is an element of the space V. (2.30) indicates that the function f is a map of D on V that associates an element y of V to each element x of D.

Often, D and V are subspaces of one and the same space, for example, the real number space or the complex number space. For example, the well-known function sin can be a map of all real numbers on the real interval –1…+1. In this case, V is a subspace of D. We can define sin also as a map of all complex numbers. To be precise, this is a function different from sin defined on the real number space, but usually, the same symbol is used. What is the space V of sin, when D is the complex number space? This question cannot be answered easily, but for engineers, it seems to be more or less unimportant. They usually simply write y=sin(x) and note that x and y are complex. This is also done when a function is implemented on a computer. Let f : D → V denote the mathematical function and f' : D' → V' its implementation. Typically, the domain of definition D' is different from D because of the discrete and limited 'real' and 'complex' number spaces the computer is working with. Then, the user wants to know the domain of definition D' and the type of the result rather than V'. The knowledge of V' becomes important as soon as the inverse function f'^{-1} is implemented and used. Note that the domain of definition depends on the numerical implementation. Therefore, the domain of definition D'^{-1} of f'^{-1} must not coincide with V'. Moreover, one can have different implementations of the 'same' function with different domains of definition. This is often the case when the 'same' function is implemented on different machines or with different compilers. One has to keep that in mind when one is replacing the compiler or the computer by another one.

Continuous functions

When we consider physical theories, we have the impression that differential and other equations describe all important physical laws. In fact, one of the oldest laws is the observation that 'nature makes no jumps', i.e., that there is some fundamental continuity in nature. This law is not contained in an explicit equation. It usually is contained in the implicit assumptions that 1) space, time, and other physical quantities are continuous, i.e., elements of a continuous number space such as the real number space, and 2) the functions or fields that solve the physical equations are continuous.

The definition of the continuity of a function uses the concept of the limit. Therefore, a metric must be defined. A function is continuous when

$$\lim_{d \to 0} \mathrm{d}(f(x+d), f(x)) = 0 \,, \tag{2.31}$$

where d denotes the metric or distance. Remember that the completeness of a space depends on the definition of the metric. Engineers often work with a metric defined by a square norm. The space of continuous functions turns out to be incomplete when this norm is used, but it is complete when the so-called maximum norm is used. The maximum norm is a special case of the p norm with an infinite value p. Maybe that is too sophisticated, but one should keep that in mind when one is talking about completeness. When one expands continuous functions by series, one cannot expect to find a complete basis when the square norm is used for computing the parameters of the series expansion. For the numerical analysis, this is not dramatic, because an approximation basis would be sufficient. Can one find an approximation basis that allows one to approximate all continuous functions with an arbitrarily high precision? This question cannot be easily answered. Fortunately, most of the functions we are working with have additional properties.

Differentiable functions

Since many of the fundamental laws of physics are described by differential equations, it is necessary that the functions are differentiable. Remember that a limit is involved in the definition of the derivative. From that, it is obvious that differentiable functions must be continuous. When we need to evaluate the second order derivative, the given function and its first order derivative must be continuous. The continuity of a function does not guarantee the continuity of its derivative. Therefore, the space of functions with a continuous first order derivative is a subspace of the continuous functions. Consequently, the space of n+1 times differentiable functions is a subspace of the space of n times differentiable functions. Mathematically, the most attractive space is the space of infinitely many times differentiable functions. These functions have very nice properties and are therefore frequently used.

If the function $f(x)$ and all derivatives $f^{(n)}(x)$ (of a function that is infinitely many times differentiable) are known in a single point $x=a$, the function is entirely known and can be evaluated everywhere. This means that a precise, but completely local knowledge is sufficient to entirely describe such a function. This allows one to extrapolate the value of x arbitrarily far away from a. In practical measurements, it is impossible to determine the function and all derivatives in a single point precisely. From the measurements of the function in n+1 points, one can estimate the function and its first n derivatives only, i.e., the information one can obtain in practice is finite and also of finite accuracy. Thus, it is practically impossible to find out from measurements how many times a function can be differentiated. Assuming that nature works with infinitely many times differentiable functions is therefore an extended application of the observation that ‘nature makes no jumps’. That this assumption is often successful may be surprising, because most of the differential equations in physics use second order derivatives. This assumption is also fundamental for our expectation that we can extrapolate our local observations toward infinity or at least to galaxies that are extremely far away.

Most of the basis functions of well-known series expansions are infinitely many times differentiable. Prominent examples are power series and Fourier series. Therefore, the result of such series expansions is also infinitely many times differentiable. Does that mean that only infinitely many times differentiable functions can be handled with such series expansions? Once more, the answer to this question is not simple. First of all, it is obvious that only functions of the same function space can be exactly expanded by a series, but for the numerical analysis, an approximation basis is sufficient. Therefore, the question is: For which kind of functions can one use an approximation basis consisting of infinitely many times differentiable functions?

Piecewise continuous functions

We have already considered the continuous function abs(x). This function is infinitely many times differentiable everywhere, except at the point x=0. At this point, already the first order derivative is discontinuous and the second and higher order derivatives are undefined. Remember that the function abs is involved in the definition of the p norm (2.4). If p is an even number, the abs may be omitted because an even power of a real number is always positive. Therefore, the p norm with odd p can cause problems, for example when the derivative of the norm with respect to a parameter is evaluated in the error minimization technique.

In engineering, it is often assumed that some functions are discontinuous at certain points. For example, functions describing material properties are assumed to be discontinuous on the surface of a body. Maybe such assumptions are wrong because of the principle that 'nature makes no jumps', but we know from experience that we may still obtain accurate results. As soon as we make such assumptions, we also have to think about the mathematical consequences.

How can one handle a discontinuity? When a function has a discontinuity at $x=a$, one can split the domain of definition D of the function into two parts D_1 and D_2, one with $x<a$ and one with $x>a$. This means that one also splits the function into two parts. What happens at the point $x=a$? Either one can include that point in one of the two domains of definition or one can explicitly define the value of the function at that point. For example, the first derivative of abs(x) is –1 for x<0, +1 for x>0, and one can set abs'(x) equal to –1, i.e., in D_1, equal to +1, i.e., in D_2, or anything else. What happens when a series approximation of abs(x) is applied to compute abs'(x)? If the basis functions are infinitely many times differentiable, one can approximate abs(x) with any desired accuracy using (2.21). From this, one obtains an arbitrarily accurate approximation (in the sense of the square norm) of the derivative abs'(x) when one replaces the basis functions in (2.21) by their derivatives. Finally, one sets x=0 in the series for abs'(x) and obtains a well defined value. This value depends on the choice of the basis functions and on the method for computing the parameters of the series expansion. Thus, one can handle a discontinuity with an appropriate series expansion, provided that one is ready to accept the result obtained from the expansion at the discontinuity. For engineers, this is probably a minor problem.

Numerically, a discontinuity has another consequence that is much more important. One typically observes that the approximation of a truncated series near the discontinuity is worst near the discontinuity. This also holds when the function itself is continuous, but the first or a higher order derivative is discontinuous. Near a point with a discontinuity in an n-th order derivative, one has a reduced convergence and therefore one has higher errors near that point. The convergence is the better the higher n is.

Obviously, one can adapt the procedure of splitting the function at the discontinuity also to functions with more than one discontinuity and to functions with discontinuities in n-th order derivatives. Practically, the number of discontinuities must be finite or at least countable.

Dirac delta functions

Newton invented the concept of a point mass and Coulomb used the same concept for charges. Since all objects in Maxwell's theory are fields or continuous functions, special problems occur with the description of point charges. A point charge is a charge distribution that is zero everywhere in the space, except at a single point. At this point the charge density is infinite. Such an object is not a true function in the usual sense. To obtain a mathematically correct formulation, one has to consider limits of sequences of functions. The limit of such a sequence itself is no longer a function; it is a distribution or a functional that is linear and continuous.

The delta function $\delta(x)$ invented by Dirac is the most *prominent* distribution. $\delta(x)$ is zero everywhere except at x=0, where it is infinite in such a way that any definite integral is equal to one when x=0 is within the interval of the integration:

$$\int_a^b \delta(x)\,\mathrm{d}x = 1, \quad \text{if} \quad a < 0 < b \,. \tag{2.32}$$

Note that the Riemann definition of integrals cannot be used here. It must be replaced by the Lebesgue integral. The mathematical theory of distributions is quite difficult and cannot be outlined here.

The most important property of the Dirac delta function is the following:

$$\int_a^b f(x)\delta(x-x_0)\,\mathrm{d}x = f(x_0), \quad \text{if} \quad a < x_0 < b \,. \tag{2.33}$$

This property is used in Maxwell theory as well as in many computational methods. If you interpret the integral in (2.33) as a scalar product, you see that the projection of a function on a delta function is equal to the value of the function at the position where the delta function is non-zero, i.e., the function $f(x)$ is sampled at the point $x=x_0$. Sampling is a well-known technique in engineering. Measuring a function is usually the same as sampling it in several points. When delta functions are used as testing functions in the projection technique, one obtains the collocation or the point matching technique, i.e., the delta function is not only a bridge from continuous charge distributions to discrete sets of point charges, it is also a bridge between different numerical techniques.

It is quite easy to extend the definition of the delta function to functions with several variables, for example, functions defined in 3D real space. One simply sets the value of the delta function equal to zero, except at a single point, where its argument is equal to the zero element in the domain of definition. Moreover, one defines an integral similar to (2.32) that equals one, provided that the zero element is within the range of the integration.

Pathologic functions

Since the Dirac delta function is not a true function, it is somehow pathologic. However, with a proper theory, one can handle such distributions. With some experience, working with delta functions turns out to be even quite easy.

We have seen that problems arise from discontinuities. When the number of discontinuities is countable, these problems can be overcome by splitting such functions. What happens, when there are infinitely many discontinuities that cannot be counted? A typical example is a random function that is discontinuous everywhere. Obviously, it is impossible to exactly define such a function everywhere. Instead, one can only define some statistical properties. Despite of this, random functions are important in modern physics and also for computational methods.

It has already been mentioned that the numerical implementation of random functions is very difficult because computers are deterministic machines. Therefore, one often uses pseudo-random generators that produce almost random numbers, i.e., functions that have almost the same statistical properties as the desired true random function. It is interesting to see that the code of pseudo-random generators can be very short. Therefore, it is no surprise that other pathologic functions can often be implemented with short algorithms. Such algorithms are typically recursive. Since the mathematical analysis of such functions is extremely demanding, it is no surprise that they rarely had been considered by mathematicians before sufficiently fast computers became available. Meanwhile, the design of so-called fractals has become one of the favorite sports of hackers because visualizations of fractals can be very nice and because the corresponding algorithms can be very simple. Such algorithms allow one to easily generate objects that look like clouds, trees, mountains, stars in the sky, and so on, i.e., like objects that can hardly be described with classical geometry. On one hand, fractals can only be approximated by recursive algorithms on computers because of the limited capacity and speed that limits the recursion depth. On the other hand, recursive algorithms are richer than fractals, because many nice pathologic functions can be obtained that do not correspond to fractals.

Series expansions

We have already worked with series expansions in the previous section and we have seen that linearity is the most important requirement. Since there are many excellent books on series [Gradshteyn, 1965], we quickly review the most important series and focus on possible extensions of this concept.

Fourier series

The basis functions of Fourier series are harmonic functions of the type $\cos(\omega t)$ and $\sin(\omega t)$. Since Fourier series are often applied to time-dependent functions, we have replaced the variable x by t. Harmonic functions are solutions of the harmonic differential equation

$$\frac{\partial^2 f}{\partial t^2} + \omega^2 f = 0 \,, \tag{2.34}$$

where the angular frequency ω is a constant. Since this is a linear, second order differential equation, one obtains two types of solutions, cos and sin. The former is symmetric, the latter anti-symmetric with respect to the origin t=0. Both are infinitely many times differentiable and the derivatives of cos and sin are mainly sin and cos again. For symmetric functions, we can easily build Fourier series of the form (2.21):

$$f^{symm}(t) = \sum_{i=1}^{\infty} a_i \cos(\omega_i t) \ . \tag{2.35}$$

Similarly, we obtain for anti-symmetric functions

$$f^{anti}(t) = \sum_{i=1}^{\infty} b_i \sin(\omega_i t) \ . \tag{2.36}$$

An arbitrary function can be split into a symmetric and an anti-symmetric part and we obtain

$$f(t) = f^{symm}(t) + f^{anti}(t) = \sum_{i=1}^{\infty} a_i \cos(\omega_i t) + \sum_{i=1}^{\infty} b_i \sin(\omega_i t) \ . \tag{2.37}$$

We could reorganize (2.37) in such a way that it has the form (2.21) with a single sum, but this is not necessary and makes the formalism more sophisticated. Instead, we now want to find an orthonormal basis. This requires the definition of a scalar product. Usually, an integral over the domain of definition of the function is helpful. So far, we have not specified the domain of definition of $f(t)$. Obviously, our basis can be defined for arbitrary real or even complex arguments. Therefore, we can assume that $f(t)$ is defined in an arbitrary sub-domain, for example, on the real interval $t_a \ldots t_b$ of the finite length $T=t_b-t_a$. Then we can define the scalar product, for example,

$$(f,g) = \frac{2}{T}\int_{t_a}^{t_b} f(t)\cdot g(t)\,\mathrm{d}t \ . \tag{2.38}$$

What is the factor in front of the integral good for? We will see that as soon as we have found the desired orthonormal basis. How can we find such a basis? We could start with an arbitrary basis, i.e., an arbitrary selection of harmonic functions and apply a standard orthogonalization procedure, but this would probably not lead to the standard Fourier series. Standard Fourier series are obtained by starting with the simplest harmonic function that is available, i.e., $\cos(0t)=1$. The square norm (f,f) of this function is equal to 2 when we use (2.38). Therefore, the simplest harmonic function with unit norm is ½. Now, we take those harmonic functions that are orthogonal to ½, i.e., for which

$$(f,1/2) = \frac{1}{T}\int_{t_a}^{t_b} f(t)\,\mathrm{d}t = 0 \ . \tag{2.39}$$

Equation (2.39) holds for all harmonic functions with a frequency that is a multiple of the fundamental frequency $\omega_0=2\pi/T$. When we start with the lowest frequency ω_0, we obtain two additional basis functions, $\cos(\omega_0 t)$ and $\sin(\omega_0 t)$ and we can verify that these are already orthogonal to each other by inserting them into the scalar product (2.38). Moreover, one can easily verify that the norm is equal to one. This explains the factor in (2.38). Now, we take the next higher frequency $2\omega_0$ and obtain two additional basis functions that are already orthogonal to each other and also to the previous basis functions. By repeating the procedure, we obtain the standard Fourier series

$$f(t) = a_0 / 2 + \sum_{k=1}^{\infty} a_k \cos(k\omega_0 t) + b_k \sin(k\omega_0 t) \; . \tag{2.40}$$

The set of basis functions in (2.40) is well ordered. The higher the order k, the more rapidly the corresponding basis functions oscillate. This is important for proofs of convergence and completeness.

For computing the parameters, one can project (2.40) on each of the basis functions and obtain:

$$a_k = (f(t), \cos(k\omega_0 t)) = \frac{1}{T}\int_{t_a}^{t_b} f(t)\cos(k\omega_0 t)\,\mathrm{d}t \; , \tag{2.41}$$

$$b_k = (f(t), \sin(k\omega_0 t)) = \frac{1}{T}\int_{t_a}^{t_b} f(t)\sin(k\omega_0 t)\,\mathrm{d}t \; . \tag{2.42}$$

In complex space, one has $\exp(\pm ix) = \cos(ix) \pm i\sin(ix)$, where $i = \sqrt{-1}$. This allows one to replace (2.40) by the more compact notations

$$f(t) = \sum_{k=-\infty}^{\infty} c_k \exp(ik\omega_0 t) = \sum_{k=-\infty}^{\infty} c_k e^{ik\omega_0 t} = \sum_{k=-\infty}^{\infty} d_k e^{-ik\omega_0 t} \; . \tag{2.43}$$

From this, we obtain the Fourier integral in the limit of an infinite definition interval:

$$f(t) = \frac{1}{\sqrt{2\pi}}\int_{-\infty}^{\infty} F(\omega)\, e^{-i\omega t}\,\mathrm{d}t \; , \tag{2.44}$$

and the inverse

$$F(\omega) = \frac{1}{\sqrt{2\pi}}\int_{-\infty}^{\infty} f(t)\, e^{+i\omega t}\,\mathrm{d}t \; . \tag{2.45}$$

As in equation (2.38), the factors in front of the integrals are not uniquely defined. Similarly, the sign in the exponential function is more or less a matter of taste.

It is interesting to note that (2.40) can be evaluated anywhere outside the definition interval of $f(t)$. Therefore, one can use (2.40) to extend the definition of $f(t)$. What are the properties of the extended function? One can easily recognize that it is periodic. The periodicity is described by the fundamental angular frequency $\omega_0 = 2\pi/T$. The length of the period is equal to T, i.e., the length of the original definition interval. By a different selection of the fundamental frequency one can easily obtain another Fourier series, for example, with a period equal to $2T$. One simply replaces $\omega_0 = 2\pi / T$ by $\omega_0 = \pi/T$. This means that one inserts additional frequencies. What does that mean for the completeness? Is the Fourier basis with $\omega_0 = 2\pi/T$ incomplete? A mathematically correct answer to this question is very sophisticated. It requires the careful definition of the function space. For engineers, it is usually sufficient to know that one has an approximation basis for all piecewise continuous and infinitely many times differentiable functions with a finite number of discontinuities. What happens when we insert additional frequencies by setting $\omega_0 = \pi/T$? First of all, the resulting basis is no longer orthogonal. This makes the computation of the parameters in (2.40) more difficult. When the original basis with

$\omega_0=2\pi/T$ is complete for the given function, this means that one can set the parameters of the additional frequencies equal to zero. Numerically, one must truncate the Fourier series because one can only compute a finite number of terms. In this case, the parameters of the additional frequencies must not be zero at all. It then depends on the given function whether a better approximation is achieved with the original Fourier series with $\omega_0=2\pi/T$ or with the non-orthogonal Fourier series with $\omega_0=\pi/T$.

Power series and Taylor expansions

When the definition interval of a function is finite, we can assume that this is a sub-domain of a larger interval and we can set up a Fourier series for the larger interval to expand the given function. We have already done this at the end of the previous subsection when we doubled the original interval. We saw that this makes the basis non-orthogonal, but not necessarily useless. What happens when we make the larger interval infinite? In this limit, the fundamental Fourier frequency ω_0 tends to zero. When we do not replace the sum in (2.40) by a Fourier integral (2.44), we observe that all cosine terms become equal to one. Therefore, we can drop all cos terms. What happens with the sin terms? We can write

$$\lim_{\omega_0\to 0}\sin(k\omega_0 t) = \lim_{\omega_0\to 0} k\omega_0 t = 0\ . \tag{2.46}$$

Therefore, we have only one basis function, the constant one, left. In fact, ω_0 is a constant and we can combine it with the corresponding parameter and obtain for the first term

$$\lim_{\omega_0\to 0} b_1\sin(\omega_0 t) = \lim_{\omega_0\to 0} b'_1\, t \neq 0\ . \tag{2.47}$$

This means that we have obtained a basis function $f_1(t)=t$ that is independent from the constant, zero order basis function $f_0(t)=1$. Using the well known formula for harmonic functions, we can write for the second sin term:

$$\lim_{\omega_0\to 0} b_2\sin(2\omega_0 t) = \lim_{\omega_0\to 0} b_2 2(\sin(\omega_0 t)\cos(\omega_0 t)) = \lim_{\omega_0\to 0} b'_2\, t \neq 0\ . \tag{2.48}$$

We therefore get the basis function f_1 again. Now, we can continue with the second cos term:

$$\lim_{\omega_0\to 0} a_2\cos(2\omega_0 t) = \lim_{\omega_0\to 0} a_2(1-2\sin^2(\omega_0 t)) = a_2\ . \tag{2.49}$$

At first sight, this is nothing new, but when we make a little linear combination with the zero order basis function, we find

$$\lim_{\omega_0\to 0}(a_2 1 - a_2\cos(2\omega_0 t)) = \lim_{\omega_0\to 0} a_2(2\sin^2(\omega_0 t)) = a'_2\, t^2\ , \tag{2.50}$$

i.e., we have one more basis function $f_1(t)=t^2$. When we continue the procedure, we obtain the well known power series expansion:

$$f(t) = \sum_{i=0}^{\infty} a_i t^i\ . \tag{2.51}$$

This is a strange link from Fourier series to power series that helps us to understand how Fourier series can be generalized and why non-orthogonal Fourier series can be more powerful than standard ones.

Obviously, the basis of power series is not orthogonal in general. Remember that orthogonality depends on the definition of a scalar product and this usually depends on the interval where the function is defined. With the usual definition of the scalar product, one has

$$(f_i, f_k) = \int_{t_a}^{t_b} f_i(t) f_k(t)\,\mathrm{d}t = \int_{t_a}^{t_b} t^{i+k}\,\mathrm{d}t = \frac{t_b^{i+k+1} - t_a^{i+k+1}}{i+k+1} \,. \tag{2.52}$$

In general, this is not zero for a non-zero interval, but it may happen that (2.52) is zero, for example, when t_b=-t_a and when i+k is an odd number. Therefore, the parameters a_i in (2.51) have to be computed by solving a linear system of equations obtained, for example, by the projection technique. This can be done numerically, when the series is truncated. Although the basis of a power series is not orthogonal, there is a way to compute each parameter explicitly. Note that some sort of orthogonality is achieved when the definition interval shrinks to zero length with t_b=-t_a. On such an interval, i.e., at the point t=0, one still can define derivatives. If the given function is infinitely many times differentiable, one can compute the parameters in (2.51). For the n-th derivative of (2.51) one obtains

$$f^{(n)}(t) = \sum_{i=n}^{\infty} a_i \frac{i!\,t^{i-n}}{(i-n-1)!} \,. \tag{2.53}$$

From the limit of t toward zero, one obtains a non-zero term for i=n:

$$a_n = \frac{f^{(n)}(0)}{n!} \,. \tag{2.54}$$

When one inserts (2.54) in (2.51), one obtains the well-known Taylor expansion around the point t=0. By a simple substitution, one can obtain the Taylor expansion around an arbitrary point t_0:

$$f(t) = \sum_{n=0}^{\infty} \frac{f^{(n)}(t_0)}{n!} (t-t_0)^n \,. \tag{2.55}$$

Note that one can rearrange the terms in (2.55) in such a way that one obtains the form (2.51) of a simple power series, i.e., a Taylor series specifies the parameter set of the power series depending on the location of t_0. Obviously, the convergence depends on the choice of this point. For example, one can easily obtain the Taylor expansion of cos(t) near t=0:

$$\cos(t) = \sum_{k=0}^{\infty} \frac{(-1)^k}{(2k)!} t^{2k} \,. \tag{2.56}$$

Note that all odd orders are zero because of the symmetry of cos with respect to t=0.

To evaluate a Taylor expansion numerically, it must be truncated. Therefore, it is important to know something about the accuracy or the error. First, one can define the error function:

$$e(t) = f(t) - \sum_{n=0}^{N} \frac{f^{(n)}(t_0)}{n!} (t-t_0)^n \,. \tag{2.57}$$

Figure 4 shows the truncated Taylor approximation $\cos(\pi x)=1+0x+(\pi x)^2/2+0x^3+e(x)$ in the interval –1…+3.

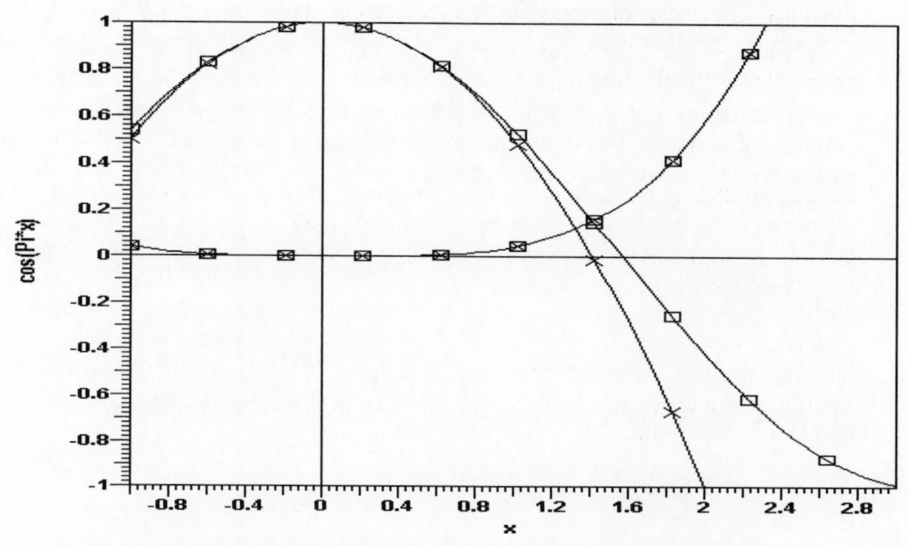

Figure 4: Truncated Taylor approximation of a given function. Square markers: Original function $\cos(\pi x)$, X *markers: truncated Taylor approximation* $1+0x+(\pi x)^2/2+0x^3$, *square plus* X *markers: error of the approximation.*

For $t=t_0$ one obtains $e(t_0)=0$. Near t_0, the error function is almost linear, i.e., the absolute error is increasing with the distance from t_0. The error is based on the definition of a norm and the norm depends on the definition interval of the function that is expanded. Using the p norm, one finds the error of a truncated Taylor expansion:

$$\|e\|^p = \int_{t_a}^{t_b} \left| f(t) - \sum_{n=0}^{N} \frac{f^{(n)}(t_0)}{n!}(t-t_0)^n \right|^p \mathrm{d}t \cdot \tag{2.58}$$

Obviously, it is very hard to select the point t_0 in such a way that the p norm error of a truncated Taylor expansion for a given function in a given definition interval is minimized. Usually, one simply sets t_0 in the center of the definition interval.

To illustrate the procedure, let us consider a simple example. Expand $\cos(t)$ in the interval $-\pi/2\ldots+\pi/2$ by a second order truncated Taylor expansion, i.e.,

$$\cos(t) = 1 - \frac{t^2}{2} + e(t) \cdot \tag{2.59}$$

For example, for the square error we find

$$\|e\|^2 = \int_{-\pi/2}^{+\pi/2} \left(\cos(t) - 1 + \frac{t^2}{2} \right)^2 \mathrm{d}t \cdot \tag{2.60}$$

The numeric result of this integral is approximately 0.02. What can be said about the maximum error? It is reached at the end points of the interval and is approximately ten times bigger. This means that the error distribution along the definition interval is not very well balanced. When you shift the origin of the Taylor expansion away from $t=0$, the error distribution becomes even less balanced. Does this mean that this is the optimal second order power series expansion for cos(t) in the interval -π/2…+π/2 or can you find the parameters a_0, a_1, and a_2 of the expansion (2.51) in such a way that the square or the maximum errors are reduced? Note that one can set a_1=0 because of the symmetry with respect to t=0. One way to determine the remaining parameters a_0 and a_2 is the projection technique. This requires the definition and evaluation of scalar products and of testing functions. With Galerkin's choice of testing functions, one obtains the two equations

$$(a_0 + a_2 t^2, 1) = (\cos(t), 1) \quad \text{or} \quad a_0 \int_{-\pi/2}^{+\pi/2} 1\, \mathrm{d}t + a_2 \int_{-\pi/2}^{+\pi/2} t^2\, \mathrm{d}t = \int_{-\pi/2}^{+\pi/2} \cos(t)\, \mathrm{d}t \tag{2.61}$$

and

$$(a_0 + a_2 t^2, t^2) = (\cos(t), t^2) \quad \text{or} \quad a_0 \int_{-\pi/2}^{+\pi/2} t^2\, \mathrm{d}t + a_2 \int_{-\pi/2}^{+\pi/2} t^4\, \mathrm{d}t = \int_{-\pi/2}^{+\pi/2} \cos(t)\, t^2 \mathrm{d}t\,. \tag{2.62}$$

The corresponding integrals may be evaluated and one obtains the two linear equations

$$a_0 \pi + a_2 \frac{\pi^3}{12} = 2 \tag{2.63}$$

$$a_0 \frac{\pi^3}{12} + a_2 \frac{\pi^5}{80} = \frac{\pi^2}{2} - 4\,. \tag{2.64}$$

Obviously, the parameters a_0 and a_2 obtained from solving this system of equations are different from the parameters obtained from the Taylor series. When other testing functions are used, one can obtain other systems of equations and other parameters of the power series. A careful analysis of the results shows that power series can be obtained that are more accurate over a given interval than the Taylor expansion. When the interval of the approximation is not symmetric with respect to t=0, the odd parameters a_1, a_3, etc. must not be zero. Figure 5 shows the truncated series approximation

$$\cos(\pi x) = 0.99463 + 0.059789x - 0.64854x^2 + 0.1325x^3 + e(x) \tag{2.65}$$

in the interval –1…+3. The parameters of this approximation have been computed by minimizing the integral over the square error in the interval 0…2. As one can see, the error is well balanced within the interval 0…2, but it is quickly increasing outside the interval.

Non-orthogonal Fourier series

We have seen that the orthogonality of basis functions allows one to obtain the parameters of a series expansion without solving a system of equations. Although this is agreeable, it does not lead to optimal series expansions for the numerical approximation of a given function. Working with non-orthogonal series is more difficult, but it gives one the chance to find better expansions. An early example of this kind is Sommerfeld's anharmonic analysis [Sommerfeld, 1950].

Let us consider a simple example of the Fourier approximation of the function $f(t)=t^2$, for example, in the interval -1…+1. This is more or less the opposite of the example in the previous subsection. For a most simple, third order Fourier series approximation

$$t^2 = a_0/2 + a_1 \cos(\omega_0 t) + a_2 \cos(2\omega_0 t) + a_3 \cos(3\omega_0 t) + e(t)\ , \tag{2.66}$$

one cannot expect very accurate results. Remember that the fundamental frequency ω_0 is well defined from the size of the definition interval when one works with an orthogonal basis. Figure 6 shows the approximation obtained from a minimization of the square error over the interval –1…+1. Obviously, the Fourier approximation is periodic with respect to this interval, whereas the original function is not periodic. Therefore, the error outside the interval –1…1 increases rapidly.

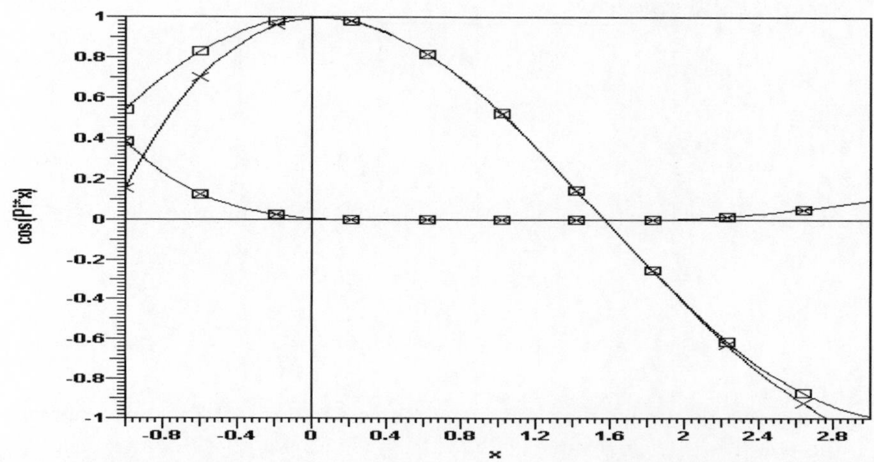

Figure 5: Truncated power series approximation of a given function. Square markers: Original function $\cos(\pi x)$, X *markers: truncated series approximation* $0.99463+0.059789x-0.64854x^2+0.1325x^3+e(x)$, *square plus* X *markers: error of the approximation.*

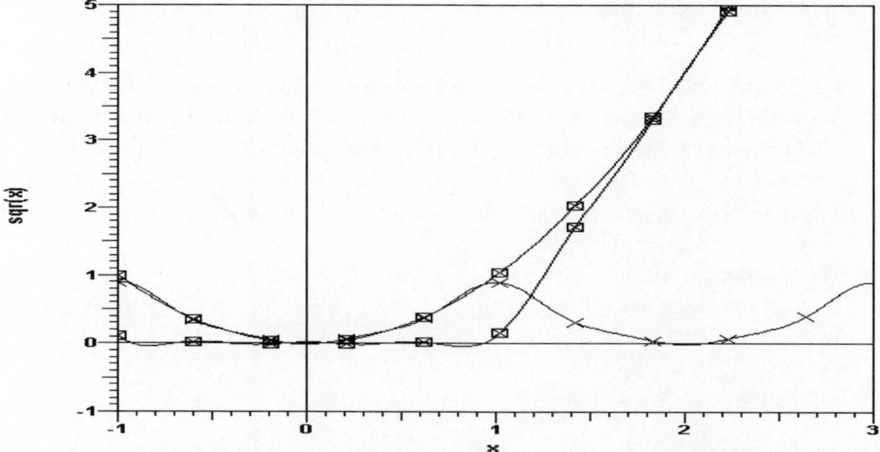

Figure 6: Truncated orthogonal Fourier series approximation of the given function x^2 *in the interval –1…1. Square markers: Original function* x^2, X *markers: truncated orthogonal Fourier series approximation* $0.33448 - 0.40758\cos(\pi x) + 0.10362\cos(2\pi x) - 0.047329\cos(3\pi x) + e(x)$, *square plus* X *markers: error of the approximation. The coefficients of the series have been computed in such a way that the square error over the interval –1…1 is minimized.*

When one reduces ω_0, the basis becomes non-orthogonal, but this does not mean that the resulting expansion becomes less accurate. In contrary, with an appropriate computation of the parameters, the error function $e(t)$ in (2.66) tends to zero! Figure 7 shows that one obtains an excellent approximation when ω_0 is divided by 10. To understand this, remember the limit at the beginning of the previous subsection.

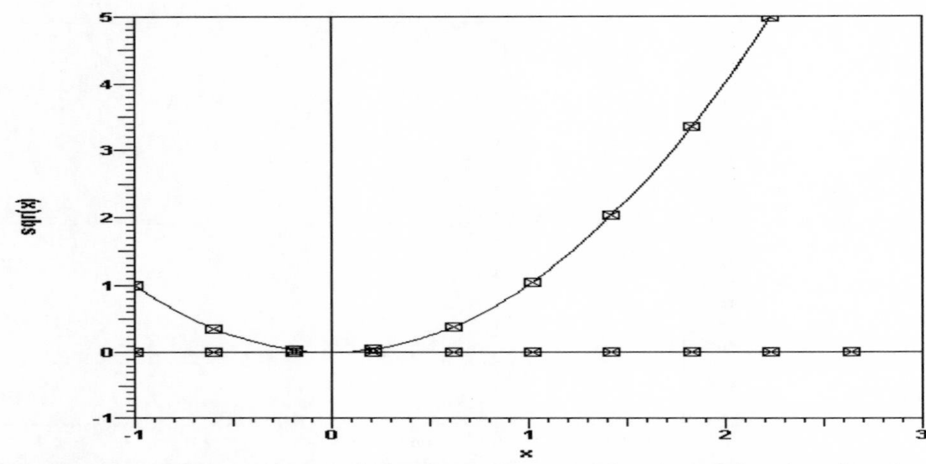

Figure 7: Truncated non-orthogonal Fourier series approximation of the given function x^2 in the interval –1…1. Square markers: Original function x^2, x *markers: truncated non-orthogonal Fourier series approximation* 27.65 - 30.498 cos(πx/10) + 3.0806 cos(2πx/10) - 0.23206 cos(3πx/10) + $e(x)$, *square plus* x *markers: error of the approximation. The coefficients of the series have been computed in such a way that the square error over the interval –1…1 is minimized.*

Of course, the limit of a zero fundamental frequency does not allow one to improve the accuracy of a Fourier approximation. A typical example for this is the approximation of a harmonic function, for example, $\cos(\omega t)$, in an interval of a length T_0 that is different from the period $T = 2\pi / \omega$ of the given function. One then can obtain exact results with non-orthogonal Fourier series, when the fundamental frequency ω_0 is set either equal to ω or equal to ω/N, where N is an integer number.

What happens when the given function is more complicated, for example, a superposition of several harmonic functions? In this case, the given function has the form of a non-orthogonal Fourier series with arbitrary frequencies (2.37). When the Fourier frequencies ω_i are not known in advance, one can consider them as parameters that have to be evaluated. These parameters are much more difficult than the parameters a_i and b_i. The latter can be computed by solving a linear system of equations; for the former, a non-linear optimization procedure is required. Even on modern computers, non-linear optimization is a tricky and time-consuming task. This might explain why such generalized, non-orthogonal Fourier series are rarely applied. Note that the frequencies in Sommerfeld's anharmonic analysis are not optimized. Instead of this, Sommerfeld computes them as the roots of a transcendental equation and shows that the harmonic expansions with these frequencies are orthogonal when the scalar product is appropriately defined. Consequently, Sommerfeld avoids the problems of non-orthogonal Fourier series, but his procedure is less general and very sophisticated.

A good example of a simple non-periodic function is sin(x)/x. Figure 8 shows the approximation obtained with a truncated orthogonal Fourier series in the interval -π...π.

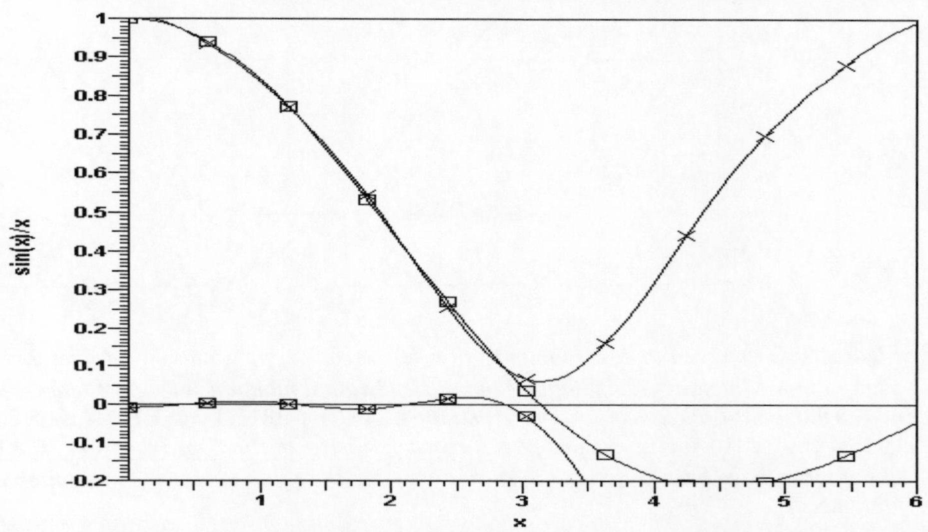

Figure 8: Truncated orthogonal Fourier series approximation of the given function sin(x)/x *in the interval* $-\pi...\pi$. *Square markers: Original function* sin(x)/x, X *markers: truncated orthogonal Fourier series approximation* 0.58981 – 0.45049 cos(x) –0.055756 cos($2x$) - 0.02264 cos($3x$) + $e(x)$, *square plus* X *markers: error of the approximation. The coefficients of the series have been computed in such a way that the square error over the interval* $-\pi...\pi$ *is minimized.*

The approximation behaves very much like the approximation of x^2. The error rapidly grows outside the approximation interval. What happens when the fundamental Fourier frequency is divided by 10? As one can see from Figure 9, the approximation is considerably better, but not as much as in the previous example. In both examples, the coefficients of the non-orthogonal Fourier approximations are considerably bigger than those of the orthogonal Fourier approximations. When the fundamental Fourier frequency is divided by a very big factor, these coefficients become huge. This causes not only numeric overflow problems, but also numerical inaccuracies due to cancellations in the computation of the coefficients and of the series.

Although a considerable improvement of the approximation has been obtained by the division of the fundamental Fourier frequency by a factor of 10 or more, one can try to obtain an even better approximation by an optimization of the Fourier frequencies. Since this is a non-linear optimization problem, this is numerically very time consuming. Moreover, problems occur because good approximations can be obtained when very similar frequencies are used. Figure 10 shows a result that has been obtained with a 'quick and dirty' parameter optimization. Note that severe cancellations are caused by the numerical similarity of the first and fourth Fourier term of this approximation.

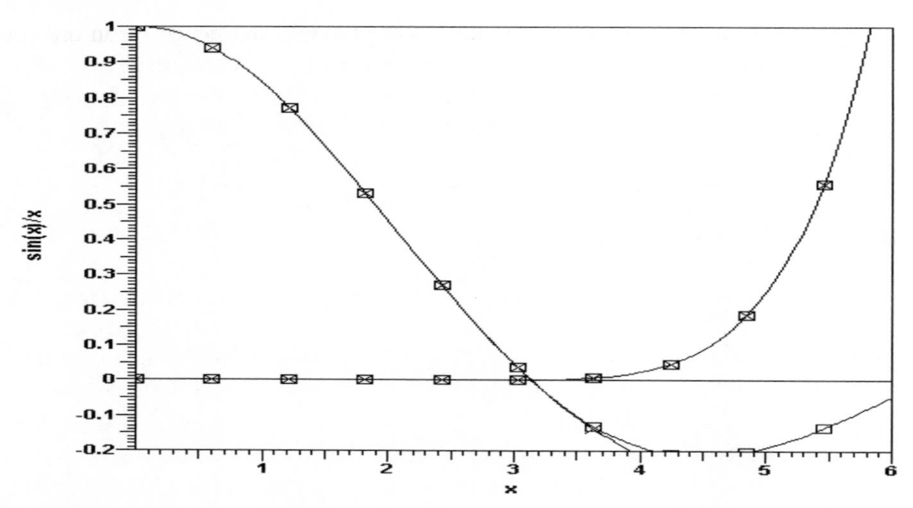

Figure 9: Truncated non-orthogonal Fourier series approximation of the given function $\sin(x)/x$ *in the interval* $-\pi\ldots\pi$. *Square markers: Original function* $\sin(x)/x$, X *markers: truncated non-orthogonal Fourier series approximation* $-2499.1 + 3813.4 \cos(x/10) - 1608 \cos(2x/10) + 294.64 \cos(3x/10) + e(x)$, *square plus* X *markers: error of the approximation. The coefficients of the series have been computed in such a way that the square error over the interval* $-\pi\ldots\pi$ *is minimized.*

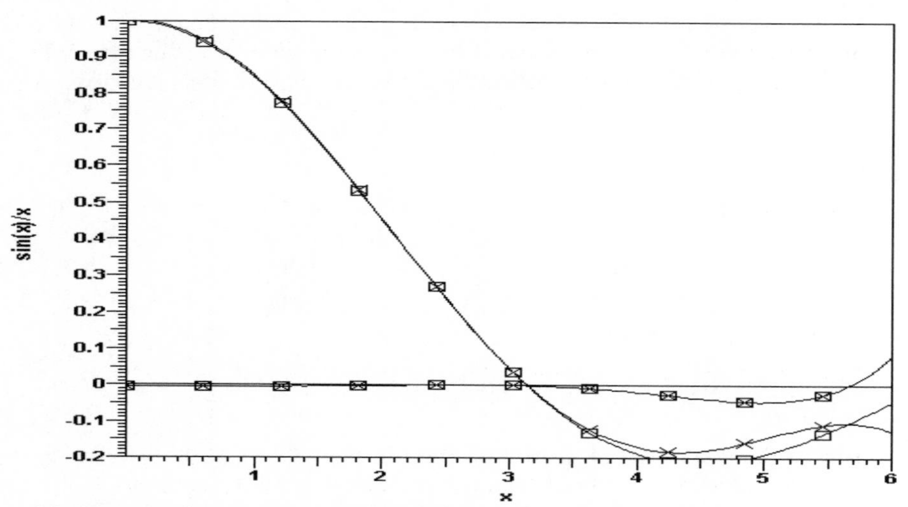

Figure 10: Truncated non-orthogonal Fourier series approximation of the given function $\sin(x)/x$ *in the interval* $-\pi\ldots\pi$. *Square markers: Original function* $\sin(x)/x$, X *markers: truncated non-orthogonal Fourier series approximation* $-3.1049\text{E}{+}012 + 4.4412 \cos(0.55384\, x) - 20.418 \cos(0.32504\, x) + 3.1049\text{E}{+}012 \cos(6.0276\text{E}{-}007\, x) + e(x)$, *square plus* X *markers: error of the approximation. The coefficients of the series have been computed in such a way that the square error over the interval* $-\pi\ldots\pi$ *is minimized. The frequencies of the* cos *terms have been optimized by a non-linear optimization algorithm.*

Non-orthogonal Fourier series are much more flexible and general than orthogonal ones. This allows one to obtain more accurate approximations of functions within a given interval with the same number of terms. Despite the problems caused by the non-linear optimization of the Fourier frequencies, one can invent even more general series expansions that are both more flexible and more difficult to apply.

Generalized series expansions

Generalized, non-orthogonal Fourier series can be used not only to expand functions that are known to be superpositions of harmonic functions, but also to approximate arbitrary functions. Although the procedure to numerically obtain the frequencies is time-consuming, these series allow one to obtain much better approximations than standard Fourier series.

A further generalization of the non-orthogonal Fourier series is straightforward. One obtains much more general series expansions by admitting not only harmonic basis functions, but also basis functions that belong to other classes of functions. This makes the procedure even more difficult, although the notation is simple:

$$f(t) = \sum_{i=0}^{\infty} a_i f_i(p_{i1}, p_{i2}, \ldots, p_{in}, \ldots, p_{iN}, t) \ . \tag{2.67}$$

To use such functions, one has 1) to evaluate the linear parameters a_i by solving a linear system of equations, 2) to compute the non-linear parameters p_{in} with a non-linear optimization procedure, and 3) to select the appropriate type of the basis functions f_i. The latter is most difficult because the type of the basis functions is usually not an element of a metric space. Since natural evolution is assumed to work with some selection principle, one can invent evolutionary strategies for selecting the appropriate type of basis functions. A possible strategy will be presented below. Before this is done, we must consider some typical applications of series expansions.

Sub-domain basis functions

So far, we have considered series expansions with so-called entire domain basis functions, i.e., with basis functions that are non-zero in the entire domain of definition, except in a finite number of points. Moreover, these basis functions were infinitely many times differentiable. We have seen that it is quite easy to obtain good approximations of sufficiently smooth functions within a sufficiently short interval. In the extreme case of Taylor series, the information in a short interval of zero length is used to compute the coefficients of the series. The resulting approximation is also useful in the vicinity of the interval, but it becomes less accurate further away. Therefore, it is natural to split a function defined on a long interval into several parts and to approximate it by different Taylor series in each sub-domain.

The most simple subdomain basis function is one that is zero everywhere, except in the subdomain, where it is equal to one. Such a basis function has the shape of a pulse. If the subdomains where the different basis functions are non-zero are not overlapping, the scalar product of two different basis functions is always zero, i.e., the basis is orthogonal. Consequently, one can compute the amplitude of each basis function directly without the explicit solution of a linear system of equations. For example, one can set the amplitude of each basis function equal to the average of the given function within the subdomain of the corresponding basis function. An even more simple alternative is to set the amplitude equal to the value of the given function in the center of the subdomain. When the domain of definition of the given function is subdivided into finitely many subdomains and when one pulsive basis function per subdomain is defined, the approximation has the shape of an irregular stair. It is a piecewise constant function with a finite number of discontinuities. Although the average error of such an approximation may tend to zero when the number of subdomains tends toward infinity, the number of discontinuities tends also toward infinity. Discontinuities cause severe problems as soon as one needs to compute the derivative. Therefore, one often wants to have a more sophisticated approximation with subdomain basis functions that guarantee that the approximation is continuous everywhere.

A prominent example of this type is the approximation of a given function by a polygonal, i.e., by a continuous, piecewise linear function. In other words, one uses a subdomain basis with first order Taylor expansions in the subdomains, i.e., a piecewise linear approximation. The first derivative of such a function is discontinuous at some points. At these discontinuity points one usually sets the value of the approximation equal to the value of the function. Figure 11 shows such an approximation of sin(x)/x. Does such an approximation minimize, for example, the square error or the maximum error? In fact, such an approximation minimizes neither the square error nor the maximum error and also no other error norm. The situation is a bit tricky, because of the discontinuities in the derivative of the approximation. First, let us consider an interval between two discontinuity points, for example, t_i and t_{i+1}. If the given function is $f(t)$, the linear approximation in the interval $t_i \ldots t_{i+1}$ has the form:

$$a_i + b_i t = f(t_i) + \frac{t - t_i}{t_{i+1} - t_i}\left(f(t_{i+1}) - f(t_i)\right) = \left(f(t_i) - \frac{f(t_{i+1}) - f(t_i)}{t_{i+1} - t_i} t_i\right) + \left(\frac{f(t_{i+1}) - f(t_i)}{t_{i+1} - t_i}\right) t. \quad (2.68)$$

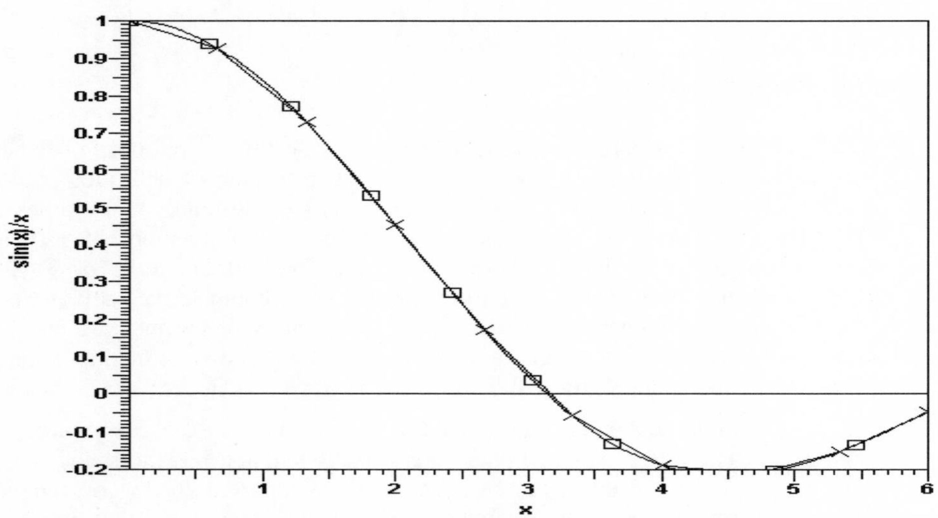

Figure 11: Polygonal or piecewise linear approximation of the given function sin(x)/x *in the interval –1…1. Square markers: Original function* sin(x)/x, x *markers: approximation.*

Obviously, the error is zero at the end points of the interval, i.e., at t_i and t_{i+1}. When the function $f(t)$ is sufficiently smooth and when the interval is sufficiently short, the error is either strictly positive or negative in most of the intervals. When the error is strictly positive in an interval, one can always reduce both the square error and the maximum error by increasing the parameter a_i in (2.68). Similarly, one can reduce both the square error and the maximum error by decreasing the parameter a_i in (2.68) when the error is strictly negative. Note that the resulting approximation is no longer a continuous function, but it is more accurate than the original one. By properly setting both parameters a_i and b_i in (2.68) one can also construct a continuous, piecewise linear function that is more accurate than the one with the parameters evaluated according to (2.68).

In order to be able to use the usual framework of series expansions, i.e., the series notation (2.21), one can introduce so-called rooftop functions. Rooftop functions are piecewise linear functions $f_i(t)$ that are zero everywhere, except in two adjacent intervals $t_{i-1} \ldots t_i$ and $t_i \ldots t_{i+1}$ (see Figure 12). Moreover, these functions are continuous everywhere. Therefore, one has $f_i(t_{i-1})=0$ and $f_i(t_{i+1})=0$. Finally, the value $f_i(t_i)$ in the center point t_i is set equal to one to entirely define the function, which looks like a roof. When one subdivides a given interval into several sub-domains, i.e., intervals $t_i \ldots t_{i+1}$, one can define a rooftop basis function for each pair of neighbor sub-domains. The superposition of the form (2.21) then leads to a polygonal approximation of the given function. Rooftop functions are frequently applied because they are very simple to handle. When the given function is known analytically, one can often analytically evaluate the scalar product of the given function with the rooftop functions. This may simplify the numerical procedure. However, it is important to note that a complete set of rooftop functions is a non-orthogonal basis. The scalar product of two neighbor rooftop functions is non-zero because these rooftop functions share an interval where they are non-zero. Therefore, a general procedure to compute the amplitudes of rooftop functions leads to a matrix equation to be solved. Since each rooftop function shares an interval with two neighbor functions, the matrix is sparse of the tridiagonal type and can easily and quickly be solved with iterative procedures. By selecting the collocation method with collocation or matching points in the points t_i, one can even compute the amplitudes of the rooftop functions without explicitly solving a matrix equation. Such a simplified procedure leads to an approximation that is less accurate than the approximation obtained by a more advanced and numerically more time and memory consuming technique, that minimizes, for example, the average of the square error.

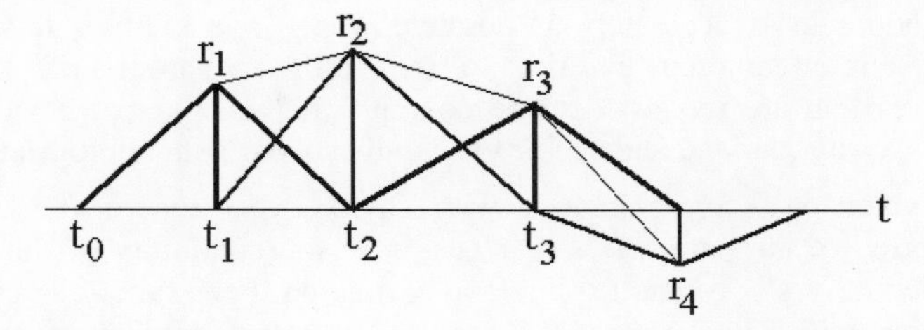

Figure 12: Four rooftop functions $f_i(t)$ with different amplitudes r_i at t_i. Note that the interval size is variable. Neighbor rooftop functions share a common interval. The thin line indicates the polygonal function that is obtained from the superposition of the rooftop functions.

The generalization of rooftop functions is straightforward. One simply can replace the linear sub-domain basis functions by other types of basis functions within each sub-domain.

Obviously, this may destroy the possibility of an analytic evaluation of the scalar product when one is too generous. Moreover, it is obvious that the resulting matrix becomes less sparse when the basis functions are non-zero in several subdomains.

Spline approximations

In the previous subsection, we started with a piecewise constant approximation of a given function. This approximation had the drawback of discontinuities. With a piecewise linear approximation, we could remove this drawback, but obviously the first order derivative of this approximation is still discontinuous. When we want to have an approximation with a continuous n-th order derivative, we can generalize the procedure. We subdivide the domain of definition of the given function into several subdomains. In each subdomain, we approximate the function with an $n+1$ order Taylor expansion, i.e., a power series of order $n+1$.

$$f(t) = \sum_{k=0}^{n+1} a_{ik} t^k + e_i(t), \quad for \quad t_i \le t < t_{i+1}. \tag{2.69}$$

Now, we compute the parameters a_{ik} in such a way that $f(t)$ and its derivatives up to the n-th order are continuous at the points t_i, i.e., on the borders of the subdomains. From the continuity of $f(t)$ at t_i one obtains the equation

$$\sum_{k=0}^{n+1} a_{ik} t_i^{\,k} = \sum_{k=0}^{n+1} a_{i-1,k} t_i^{\,k}. \tag{2.70}$$

A similar equation is obtained for the continuity of the first derivative by deriving (2.70) once:

$$\sum_{k=1}^{n+1} a_{ik}\, k\, t_i^{\,k-1} = \sum_{k=1}^{n+1} a_{i-1,k}\, k\, t_i^{\,k-1}. \tag{2.71}$$

Note that the summation in (2.71) starts at k=1 rather than at k=0 because the first derivative of the zero order term vanishes. From the derivative of (2.71) one obtains the condition for the continuity of the second order derivative, and so on. Writing down the continuity conditions for $f(t)$ and its first n derivatives at t_i leads to $n+1$ linear equations for each point t_i. Since the approximation (2.69) has $n+2$ unknown parameters for each point t_i, one has one condition per subdomain left. The missing condition is that the given function should be approximated in each subdomain as accurately as possible. When one imposes this condition, one obtains a linear system of equations that is characterized by a square matrix. The higher the order of the derivatives that are requested to be continuous, the less sparse the matrix is. However, the matrix is always banded and can be efficiently solved with appropriate iterative methods.

A subdomain basis approximation with Taylor expansions in the subdomains with a special computation of the parameters that guarantee the continuity of the approximation and of its first n derivatives is called a spline approximation. For example, a cubic spline approximation is working with third order Taylor expansions that guarantee the second derivative to be continuous.

Approximation, interpolation, extrapolation

We have seen that there are different methods for obtaining the parameters of a series expansion to be used for a numeric approximation of a given function. When the error minimization technique is applied, one most frequently works with the square norm. Similarly, one implicitly uses the square norm when one applies the projection technique with Galerkin's choice of testing functions. The user of a numeric approximation of a given function usually wants to be sure that the maximum absolute or relative error is smaller than a given quantity. To obtain the maximum error, the maximum norm is required and we have seen that the maximum error can be much bigger than the square error. Therefore, one is interested in series expansions with a balanced error distribution over the entire range. The ratio of the maximum error and the square error is a good measure of the balance of a series expansion. We have considered simple as well as generalized series expansions. The error distributions of these expansions may look completely different, even if the square error is the same.

Generalized series expansions allow one to reduce the number of terms that guarantee the desired accuracy. Especially when the desired numerical accuracy is high and when the approximation is frequently used, it is important to find a series expansion that can be quickly evaluated, i.e., that 1) has as few terms as possible and 2) uses basis functions that can quickly be computed. In this case, it is of minor importance that finding an appropriate, generalized series expansion may be time-consuming.

A very old and quick way to numerically compute values of a certain function uses simple tables. Assume that a table contains the values $f_i(x_i)$ of a given function $f(x)$ for a discrete set of arguments x_i. When you want to find an approximation of $f(x)$ for a value x that is not contained in the list of arguments x_i, you can find the closest argument x_c, i.e., the argument with abs(x-x_c)<abs(x-x_i) for all i that are different from c. Once you have found x_c, $f_c(x_c)$ is an approximation of the desired $f(x)$.

When the desired accuracy is high and when the function is complicated, the table becomes long and requires a big amount of memory. In this case, it is convenient 1) to store only a rough table and 2) to provide an appropriate interpolation technique that allows one to find the desired result more accurately. The most famous interpolation techniques are based on power series, i.e., on an approximation of the form

$$f(x) = \sum_{n=0}^{N} a_n (x - x_c)^n + e_N(x) \ , \tag{2.72}$$

where x_c is a point that should be close to x and where $e_N(x)$ is the error distribution of the N-th order approximation. Since there are N+1 parameters in (2.72), at least N+1 equations are required for computing these parameters. These equations may be obtained from the values in the table near x. For example, for a first order interpolation, two points are required and you should select the two points in the table with $x_c<x<x_{c+1}$. When you insert $f_c(x_c)$ and $f_{c+1}(x_{c+1})$ in (2.72), you obtain the two equations

$$f_c(x_c) = \sum_{n=0}^{N} a_n (x_c - x_c)^n + e_N(x_c) = a_0 + e_1(x_c) \ , \tag{2.73}$$

$$f_{c+1}(x_{c+1}) = \sum_{n=0}^{N} a_n (x_{c+1} - x_c)^n + e_N(x_{c+1}) = a_0 + a_1(x_{c+1} - x_c) + e_1(x_{c+1}) \,. \qquad (2.74)$$

One can easily solve these two equations by setting the two error terms $e_1(x_c)$ and $e_1(x_{c+1})$ equal to zero. This is the usual and well-known linear interpolation, i.e., an interpolation based on a first order power series. Obviously, there are alternatives. Instead of $x_c<x<x_{c+1}$, one can write $x_{c-1}<x<x_c$ which leads to the same linear interpolation. Are there better, i.e., more accurate, alternatives? First, one probably expects that the interpolation is the more accurate the higher its order is. Unfortunately, this is not true in general. The main problem is that higher order interpolations require data from points in the table that are further away from x. When one proceeds as for the linear interpolation, one picks up N+1 points of the table and writes down the N+1 linear equations of the form:

$$f_{o+j}(x_{o+j}) = \sum_{n=0}^{N} a_n (x_{o+j} - x_c)^n + e_N(x_{o+j}) = \sum_{n=0}^{N} a_n (x_{o+j} - x_c)^n, \quad j = 0 \ldots N \,. \qquad (2.75)$$

Again, one sets the error terms equal to zero. Moreover, one selects a 'closest' point x_c. Finally, one will select the values x_{o+j} in the table as close as possible. When N is sufficiently big, x_o or x_{o+N} is too far away from x. When a higher order interpolation fails, one must either use a finer table, or find a better interpolation technique. Generalized series expansions provide a straightforward way to such techniques.

It is natural to select the point x_o at such a position that x is somewhere near the center of the interval $x_o \ldots x_{o+N}$. When x is entirely outside the interval $x_o \ldots x_{o+N}$, you still can apply the same formalism, but you will find much less accurate results. Since x is outside the interval, this is called extrapolation. Numerical extrapolation is an extremely demanding task. Linear extrapolation is usually too inaccurate and higher order extrapolation fails almost always. Also other traditional series expansions, such as Fourier series, are useless. Here, generalized series expansions turn out to be most interesting and helpful.

The design of an extrapolation technique is similar to the design of a physical theory. A theory with sufficiently many free parameters allows one to tune it in such a way that all known observations are 'explained'. According to Einstein, a theory should be simple, which means that it should have as few free parameters as possible. It turns out that theories with many parameters can 'explain' what is already known, but they are unable to correctly predict future observations. This means that such theories provide good interpolations but bad extrapolations. 'Simple' theories that accurately explain the known observations often predict future observations much more precisely. Predictions are the most important tests for physical theories. Predictions, i.e., extrapolations, are both difficult and desirable. These statements also hold for series expansions. A strategy to find generalized series expansions that provide good extrapolations is presented in the next section.

Symbolic regression

In the previous sections we have worked with series expansions to approximate given functions. Mathematicians provide more sophisticated methods to approximate functions. These methods use constructions that are more complicated than series, but all of these techniques are characterized by an infinite set of parameters. Typical examples are rational functions of the form

$$f(x) = \frac{p(x)}{q(x)} = \frac{\sum_{k=0}^{\infty} a_k x^k}{\sum_{k=0}^{\infty} b_k x^k} \tag{2.76}$$

or continued fractions. Continued fractions can have different forms, for example

$$f(x) = a_1 / \left(1 + a_2 x / \left(1 + a_3 x / \left(1 + \ldots\right)\right)\right). \tag{2.77}$$

We now can generalize these formalisms in the same way as we have generalized series expansions. Obviously, we can generalize (2.76) by replacing the powers of x by arbitrary basis functions. Continued fractions can be generalized in several ways. For example, we can take the structure (2.77) and replace the terms $a_k x$ by arbitrary functions $f_k(x)$. Another generalization of (2.77) would be obtained when we replace the operator / by arbitrary functions with two arguments $f_k(.,.)$. Note that the main problem in the generalized formalisms consists in finding sequences of 'basis' functions f_k. Moreover, it must be proven that the solution converges toward the desired function.

Obviously, the numeric evaluation of all these infinite constructions requires some truncation to obtain finite expressions. The disadvantage of truncation is that some error is introduced, but truncation offers also an advantage that should not be underestimated: finite expressions allow for a much wider generalization that may lead to much more accurate approximations when the number of 'tunable' parameters is fixed. First of all, no sequence of 'basis' functions is required. Instead, one can select each basis function out of a pool of known basis functions. Since we can replace all arithmetic operators, +,-,*,/, etc. by functions, for example $add(a,b){:=}a+b$, we can write arbitrary formulae of finite size as nested functions. For example, when we truncate (2.77) after the third term and define $mul(a,b){:=}a*b$, $div(a,b){:=}a/b$, $add(a,b){:=}a+b$, we can write:

$$f(x) \approx div(a_1, add(1, mul(a_2, div(x, add(1, mul(a_3, x)))))). \tag{2.78}$$

A better visualization of the structure of this formula is obtained by the following tree representation:

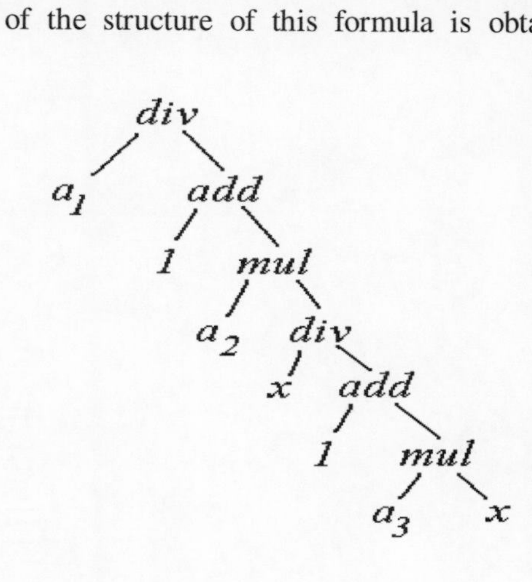

(2.79)

Symbolic regression designs a formula of finite size that approximates a given function within a given interval. The formula consists of 'elementary' functions or operators and of so-called terminals. In (2.79) one has three elementary operators that can be considered as functions with two arguments and three different types of terminals: 1) the parameters a, 2) the constants 1, and 3) the variable x. To approximate a given function, one can design the geometric shape of the tree, fill in elementary functions at each node of the tree and fill in constants, parameters, or variables at each terminal. Obviously, the space of possible formulae is huge, even when the corresponding trees are small and when the number of available elementary functions is small.

As soon as the error of an approximation has been defined, finding an accurate approximation of a given function is an optimization process. The optimization of formula trees is much more demanding than the linear optimization required for finding linear parameters in standard series expansions and also more demanding than the non-linear parameter optimization required for generalized series expansions. Some appropriate optimization procedures will be outlined in the following chapter.

One of the important benefits of symbolic regression is its capability to find solutions that provide excellent extrapolations of a given function.

3 Optimization

General remarks

We are confronted with optimization not only in mathematics and engineering, but also in almost every aspect of life. Maybe optimization is the most important process of life. When we learn walking, talking, or anything else, our brain is doing some optimization. When we observe natural evolution, we have the impression that some optimization is done. When we look at the development of theories, we can consider this as an optimization process. These observations are fascinating and also inspiring for the development of numerical optimization procedures, especially when difficult, highly complex problems have to be tackled.

In the design of codes for electromagnetics, one can distinguish three different types of optimizations. First of all, such a code should compute the field or some derived quantity in such a way that 'good' results are obtained. When we have defined some error of the result, we look for a code that minimizes the error, which is an optimization procedure. Secondly, we are interested in 'good' codes that are sufficiently accurate, reliable, efficient, and so on. Therefore, code designers try to optimize codes. Instead of a design optimization, one can even try to implement codes that optimize themselves. Such codes exhibit some learning capability. Finally, codes are applied to speed up the design process of electrical engineers who develop new devices, which is again an optimization procedure. The device designer can either do this optimization using his experience and his brain or he can run an optimization procedure on a computer. The latter requires the coupling of a sufficiently reliable and efficient electromagnetics code with an efficient optimization code.

The numerical optimization of code or device design is very demanding, but the quickly growing speed and memory size of modern computers allows us to solve increasing number of increasingly difficult optimization problems within a sufficiently short time. One can foresee that this will cause a revolution in the design process in the near future, because numerical optimization can explore a much wider area of the search space than a human designer. Therefore, numerical optimization is able to 'invent' unconventional and more efficient solutions that clearly outperform conventional solutions of human designers.

The aim of optimization is finding the 'best' or at least 'good' solutions. For computer codes, the goal of the optimization must be precisely defined. In general, a 'cost' or 'fitness' function must be specified. The goal is to minimize the 'cost' or to maximize the 'fitness'. Cost and fitness functions are real valued and may have many variables. The variables represent parameters that characterize the system, code, or device to be optimized. In complex situations it may be hard to define cost or fitness functions because several concurrent properties of the system might be desired. For example, one might want to develop a sensor that is cheap, accurate, and robust. The desired properties must have an influence on the fitness function, but the definition of the function is not unique and has a big influence on what is considered to be optimal. Moreover, each of the properties depends on some of the design parameters. These dependencies are often quite unclear and complicated. Usual optimization procedures assume that the fitness function is a known and well-defined function of the design parameters.

Design parameters may be complex, real, integer, binary, or other numbers as well as more complicated objects. With an appropriate coding one can map even complicated objects on simple number sets. Therefore, we may consider fitness or cost functions as real functions of one or several parameters that are elements of a number space.

For more information on optimization, see [Polak, 1971], [Press, 1992]. Evolutionary and genetic strategies are discussed in [Bäck, 1996], [Fogel, 1995], [Fröhlich, 1997], [Goldberg, 1989], [Holland, 1975], [Koza, 1992].

Linear optimization

In the previous chapter we have considered series expansions that approximate a given function. These series expansions contain a set of linear parameters that can be computed in such a way that the least squares error of the approximation is minimized. This means that a linear optimization problem is solved. The simplest solution consists of the following steps: 1) Define the goal of the optimization, i.e., the cost function as a real function of the linear parameters. 2) Set the derivatives of these parameters with respect to each of the parameters equal to zero. This leads to a linear system of equations that is characterized by a symmetric matrix. 3) Solve the matrix equation numerically. Note that this method may lead to non-linear systems of equations when the cost function is not defined appropriately. When the cost function is set equal to the square norm of the error, a linear system is obtained. This is the method of least squares.

Note that we have encountered other methods to compute the linear parameters of series expansions. These methods may lead to sub-optimal solutions. Sub-optimal solutions are often acceptable when the computational effort can be reduced. Note that one can have different definitions of cost functions that lead to linear systems of equations. Therefore, a sub-optimal solution can turn out to be optimal when the error definition is modified, but most of the sub-optimal solutions remain sub-optimal for any reasonable error definition. Since the numerical matrix evaluation is always inaccurate, also the optimal solutions are not really optimal in a very strict sense. Despite this, it makes sense to distinguish between optimal and sub-optimal solutions.

We will see in the following chapters that many of the methods for computational electromagnetics can be written as matrix methods. These methods usually compute the linear parameters in either an optimal or sub-optimal way. However, such methods solve a linear optimization problem.

Linear optimization problems can always be reduced to matrix equations. There are many numerical algorithms for solving such equations. Note that the design of an appropriate algorithm for solving a given class of matrix equations can also be considered as a highly non-linear optimization process. This process is very demanding. Therefore, one can usually not expect to find an optimal algorithm. Finding at least a sub-optimal matrix solver is important for the design of linear optimization codes because the matrix solver is the most time-consuming part of such codes.

For more information on matrix equations, see the next chapter and textbooks [Arfken, 1985], [Golub, 1983].

Non-linear optimization

Non-linear optimization is much more difficult and usually also much more time-consuming than linear optimization. There is no unique way to solve non-linear optimization problems. Usually, iterative techniques are applied that converge toward the global minimum or at least toward a local minimum of the cost function. Note that finding the minima of cost functions is essentially the same as finding the maxima of fitness functions. For reasons of simplicity, we focus on the cost functions in this section. Finding the global minimum is much more difficult then finding a local minimum – except in situations with a single minimum. As soon as the number of minima is infinite, finding the global minimum is impossible.

Fortunately, finding the global minimum is rarely required in engineering. Usually, it is even sufficient to find an acceptable solution that meets some specifications. This means that one searches for an area where the cost function is below a certain barrier. Thus, it is not even necessary to detect local minima. Despite this, many of the traditional algorithms converge toward a local minimum. When this minimum is above the barrier, one must restart the search. Since the detected minimum depends on the start point of the search, one must modify the start point and hope that the search will find a deeper minimum.

Real parameter optimization

To illustrate the procedure, we can look at the most simple case of a one-dimensional cost function $f(p)$ with a real parameter p defined in a finite interval $p_a...p_b$. To start with, let us assume that the function is very nice in the sense that it can be differentiated infinitely many times and that its derivatives are continuous everywhere. A typical example is shown in Figure 13. Assume that we already have an algorithm that finds a local minimum when it is started anywhere. When the function has more than one minimum, we do not know which of them will be found, and when the algorithm has found a minimum, we don't know whether there is another, deeper minimum somewhere else. When we run the same algorithm with another start point, we do not know whether it will find another minimum or the same one as before.

Rough search

Therefore, it may be reasonable to start with a rough search that scans the entire interval in a certain number of points. An alternative would be a search that starts at one side of the interval. As soon as this search has found the first minimum, it continues and now searches for a maximum. After detecting a maximum, it searches for a minimum again, until it reaches the other end of the interval. An initial rough search allows the user to obtain some graphical insight that may be helpful for selecting an appropriate minimum search routine. The main problem of the rough search is that one should test the function at a sufficiently high number of points for capturing all minima. This is far from being trivial, because some functions have very sharp minima that require an extremely high resolution. Therefore, the rough search never guarantees that all minima have been detected when one does not have some additional knowledge of the behavior of the function.

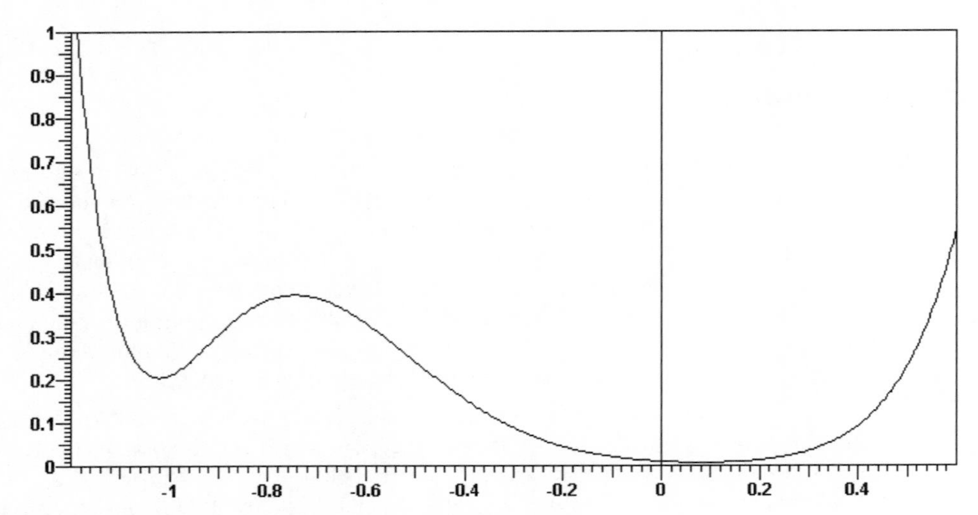

Figure 13: Simple cost function with two local minima.

Beside very sharp minima, a cost function can also have an infinite number of minima. Figure 14 shows a cost function with an infinite number of minima. Despite this, the global minimum may be detected with an initial rough search, provided that the initial scan is sufficiently fine.

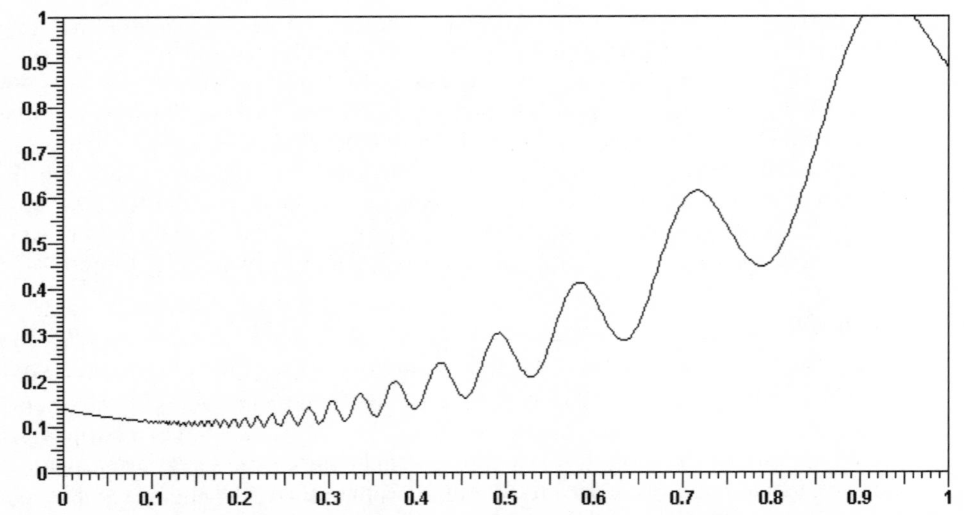

Figure 14: Cost function with infinitely many local minima. Despite this, the search for the global minimum may be successful.

Figure 15 shows another cost function with infinitely many minima that are so sharp that it is even impossible to show a correct picture. Since the minima are located near the origin, you may 'zoom in'. From this, you may guess that the global minimum is $f(0)=0$. Despite this, all classical numeric search routines will run into severe problems when you let them analyze this function. Note that even the rough search alone will allow you to find non-minimal solutions below a given barrier.

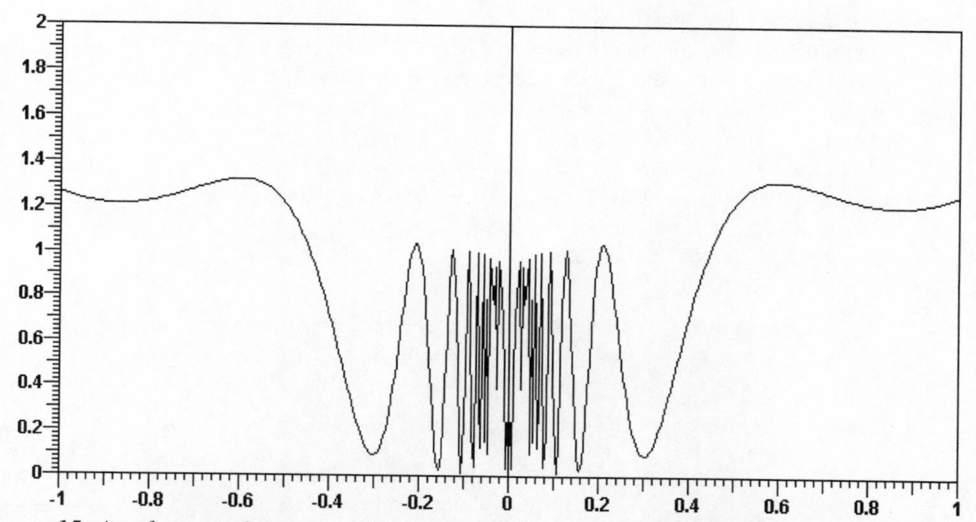

Figure 15: Another cost function with infinitely many local minima. The numeric search for the global minimum will fail.

Obviously, the rough search becomes much more demanding when the number of parameters of the cost functions becomes bigger. When you require N test points for the rough search of each parameter, you require N^m test points for the rough search of a cost function with m parameters. Thus, it is hopeless to perform a rough search when the number of parameters is too high or when the cost function has too sharp minima.

Fine search

After the initial rough search, one usually starts a more or less sophisticated fine search. Even when the rough search cannot be performed (for the reasons mentioned above) one may start the fine search at an arbitrary point in the parameter space and observe what happens. When one searches for values where the cost function is below a given barrier, one will stop the fine search when either such a value has been found or a local minimum has been detected. When the local minimum is above the barrier, one can restart the fine search from another start point.

There are two main classes of fine search algorithms: 1) algorithms that need to know the values of the functions and of their n-th order derivatives and 2) algorithms that work with the function values only. The latter provide additional numerical problems, but they can be applied to more general problems. The most simple algorithms of the first class use the first order derivatives only, i.e., the gradient information of the cost function. When we know that the second order derivatives are continuous, we also know that the gradient of the cost function is zero at each minimum. Therefore, gradient search allows one to replace the minimum search algorithm by a zero search algorithm that is numerically much easier.

To illustrate that even zero search of simple functions is not trivial, let us consider the search for the zeros of the function z^n-1 in the complex plane. You may consider this as the search of the zeros of a function with two real parameters, but working in the complex plane is simpler. The function z^n-1 has n zeros at the positions $e^{i2k\pi/n}$, where k=1,2,...,n. When we start a classical Newton [Press, 1992] algorithm for finding the zeros at a point z_0 of the complex plane, it will find one of the zeros, z_k. Which of the zeros will it find? Probably, you guess that it will find the zero that is closest to the start point z_0. If this would be correct, you could find the complex z_0 plane of the possible start points in n sectors. When starting in the sector number k, you would find the solution number k. It turns out that the complex plane is subdivided into n domains that are fractals [Mandelbrot, 1977] rather than simple sectors. Figure 16 illustrates this for n=5. The structure obtained is characteristic of the search algorithm that is applied, but exploring this structure is extremely demanding both from the numeric and analytic points of view.

Figure 16: Newton search of the zeros of z^5-1 in the complex plane. The different gray scales indicate the areas of the plane that lead to one and the same zero when the Newton algorithm is started from there. The resulting geometric structure is a fractal.

The Newton search algorithm is very simple. The zero that is found depends only on the start point. More sophisticated search algorithms contain many user-definable parameters that have a big influence on the search behavior. Analyzing such algorithms is much more difficult and time consuming. Therefore, one often applies search algorithms without precise information on the search pattern. This makes the selection of appropriate algorithms and the development of such algorithms very demanding.

A description of real parameter search algorithms is outside the scope of this book. For more information, consult a textbook, for example, [Press, 1992].

So far, we have only considered relatively 'nice' cost functions. It is obvious that discontinuities and other 'ugly' properties of the cost function cause additional problems that can considerably disturb classical search algorithms. For example, the accuracy of the numerical evaluation of the cost function (and its derivatives) is limited. This causes some disturbing noise. For example, near a zero of the first derivative of a cost function, the noise can cause additional zeros. This can disturb a gradient search algorithm. There are several ways to overcome such problems, but this makes the algorithms more complicated.

Finite parameter sets

One might believe that the minimum search of a cost function with non-real parameters is simpler when the parameter set is finite. This is correct when the parameter set is so small that one can evaluate the cost function for all possible parameters. Unfortunately, the power of computers is limited and the size of parameter sets of typical engineering applications is so big that one can explore only a small part of the entire search space. Even in relatively simple situations the search space can have in the order of 10^{100} points. Moreover, the numerical evaluation of the cost function may be time-consuming. For example, one may have to solve a field computation problem that takes several seconds even on a fast machine. Therefore, one rarely can compute the cost function at more than 10^6 points.

Another difficulty arises from the structure of the search space. An n-dimensional search space with real parameters usually has nice properties. First of all, one can define a metric. This is essential for search algorithms that should know which points of the space are close to a given point and which points are far away. When one talks of 'neighborhood search', this means that the metric of the search space is known. Moreover, one can usually define scalar products. We have seen the benefits of scalar products in the previous chapters. As soon as one works with cost functions with finite parameter sets, one is confronted with non-metric spaces. How can one explore a non-metric space?

The most simple search algorithm that works in any situation is random search. In complicated engineering applications, we do not expect that 'blind' random search is able to find some solution that outperforms the solution of an experienced human designer. Such a solution might be obtained by chance, but a more goal-oriented search should be possible and more promising. We are tempted to say that random search will be far from efficient and that there should be much more efficient, 'intelligent' search procedures. Unfortunately, experience demonstrates that such statements are not correct at all. The efficiency of search procedures depends very much on the problem to be solved. It seems that we are often confronted with problems that can be solved more efficiently than by random search, but there is no search procedure that is efficient for everything. Classical gradient searches are most efficient for an extreme class of very nice cost functions, whereas random search is most efficient for another extreme class of very ugly cost functions. In the following, we try to find more intelligent algorithms that can outperform random search in highly complex but not completely random problems.

Random search

Classical non-linear optimization procedures work well when the cost function is 'nice', i.e., sufficiently smooth, differentiable, when it has a small number of minima, etc. Discontinuities at unknown positions in the cost function cause severe problems. From this point of view, a random cost function is most demanding because it is discontinuous everywhere. Finding the absolute minimum of a random function is completely impossible. Although there is a small chance of hitting it, one never can know whether the minimum has been found or not. Despite this, one can search for values of the random function that are sufficiently small, i.e., below a certain barrier defined by the user. Finding such values may have no practical interest, but complex optimization problems may have cost functions with many discontinuities at unknown positions. For such problems, non-deterministic algorithms may be more efficient than classical, deterministic search. Random search is the most simple non-deterministic algorithm: a random generator decides at which point of the search space the cost function shall be evaluated. As soon as a value below the given barrier is found, the algorithm can stop.

It can be proven that random search is the best method for finding sufficiently small values of a random function rnd(p). Random search means that one randomly selects the parameter p. What does it mean when we say that random search is the best method? The properties of rnd(p) do not allow one to say that any method is able to find a solution with rnd(p) $< e$ within a finite number of steps, but it is obvious that each method has a certain chance of finding such a solution within a certain number of steps. Therefore, a proper mathematical definition of the quality of a search procedure is quite sophisticated.

Unfortunately, random search is extremely inefficient when we search for minima of more 'brave' functions. For example, to find the minimum of a parabolic function $f(p) = p^2$ in the interval p=0…1, random search would randomly set n points in the given interval. One has a good chance that the procedure would set one point in the interval 0…$1/n$. To obtain a solution with $f(p) < e$, one can expect that approximately $1/\sqrt{e}$ evaluations of $f(p)$ are required. When e is small, this is a huge number of function calls. Classical search procedures such as gradient search algorithms, algorithms based on parabolic interpolation, etc. could find the minimum within machine precision with a finite number of function calls. In the best case, two function calls and two gradient evaluations would be sufficient – beside algorithms that find the minimum 'by chance' within fewer steps.

Since classical optimization procedures have severe problems with more complex cost functions, it makes sense to combine random search methods with classical ones. For example, one can let the random procedure set a point within the search interval and explore the neighborhood afterwards by a classical algorithm – provided that some metric is defined. As soon as this algorithm has detected either a local minimum or numeric problems, one can let the random procedure set another point, and so on. Such algorithms are promising when the cost function is piecewise continuous. For more complex situations, other algorithms must be designed, especially for non-metric search spaces.

Evolutionary Algorithms (EA)

When one observes nature, one obtains the impression that one can see different optimization procedures working on different levels of a highly complex system. The optimization problems 'solved' in nature seem to be much more complex and demanding than difficult engineering problems. Therefore, we may get hints from nature.

The most prominent example is the natural evolution observed in biology. It seems that even simple cells could not have been generated by a random combination of molecules that were available in the atmosphere. In fact, such a statement is not precise enough and even when the chance of a random generation of cells is extremely small, this does not prove that no random generation took place. It is well known that Darwin proposed a theory of evolution that is based on some simple concepts, such as 'natural selection', 'survival of the fittest', etc. There is still an extended discussion on how evolution works and we cannot claim that we really know the mechanisms of evolution. Despite of this, we can implement optimization procedures that are based on simplified models of natural evolution and of other optimizations that seem to be observed in biology, economy, etc.

In order to demonstrate the procedure, the implementation of genetic optimization is outlined in the following. This type of optimization has successfully been applied to symbolic regression and other complicated optimization problems [Fröhlich, 1997], [Quagliarella, 1998], [Winter, 1995].

Genetic Algorithms (GA)

The first Genetic Algorithm (GA) was developed by Holland in 1975 [Holland, 1975]. Since then, GA's have been applied to many problems in various areas of science and engineering [Fröhlich, 1997], [Quagliarella, 1998], [Winter, 1995]. In the following, the GA concept is outlined and some hints are given how more efficient GA codes might be developed. Standard GA's roughly imitate the process of natural evolution. Improved GA's can be created by better imitations of the process observed in nature, by generalizations of these procedures, or by adding new concepts that are not observed in natural evolution.

Inside a computer, everything is represented by a simple bitstring. Therefore, also the solution of any problem must be represented by a bitstring. A GA does nothing else than finding such bitstrings. It plays no role whether we search for a number, a set of numbers, a function, or even an algorithm; we can code this into a bitstring and we can let a GA search for an appropriate bitstring. The GA is not much more than a machine that produces bitstrings of a desired length. The GA user defines the coding and tells the GA the desired length of the bitstrings. When a GA outputs a bitstring, the user has to evaluate the quality of the bitstring and to return a fitness number to the GA. When the user wants to let the GA minimize a cost function, he can define the fitness as the inverse of the cost. The GA will use the fitness evaluation to construct more promising bitstrings.

Although GA's can be applied to any optimization problem, we focus on the problem of symbolic regression to demonstrate the procedure. Note that symbolic regression may include non-linear optimization of real parameters, as well as finite parameter sets. The goal of symbolic regression is finding a formula that is a good approximation of a given formula. To apply a GA, we first have to define the coding.

Coding

Assume that we already have some coding that defines how a formula is mapped on a bitstring. Then, we can let our computer generate bitstrings that describe formulae. This situation is similar to the genetic information that describes an animal. The former is called genotype, the latter is the phenotype. In a first approach, it plays no role for us whether the genotype defines the phenotype exactly and uniquely. We can ignore the fact that there may be differences between different individuals with identical genotype caused by the influence of the environment. When we want to solve a symbolic regression problem with a GA, a bitstring corresponds to the genotype and the corresponding formula corresponds to the phenotype.

Obviously, there are many ways to code a formula into a bitstring and it is also quite obvious that the coding can have an important influence on the GA performance. Assume that we want to code a set of most simple formulae that consist of a single function with two arguments, i.e., we have a tree with one node and two branches. Assume that we have four elementary functions that correspond to the operators +,-,*,/. Moreover, the terminals shall be either the variable x or a constant. First, we decide how to code the constant. Depending on the accuracy and range of this constant, we must reserve a certain number of bits in the string. For example, when the constant is an integer value in the range 0…9, we need at least four bits for coding it, but with four bits we can code more than 10 different integers. Four bits correspond to an integer number in the range 0…15. When the result is bigger than 9, we decide to insert the variable x, otherwise we insert the constant. Since we have two terminals, we need 8 bits. To decide which of the four possible operators should be applied, we need another two bits. Thus, the first two bits could characterize the operator, the bits 3 up to 6 the first terminal, and the bits 7 up to 10 the second terminal. For example, `1011110011` might be the genotype of the formula x*3. Since x*3 is equal to 3*x, `1000111111` is another genotype with the same phenotype. Also `1000111101` has the same genotype because both `1111`=15 and `1101`=13 correspond to numbers >9 that cause the variable terminal x.

When we generate bitstrings by a random procedure, the coding has an influence on how frequently a certain genotype is generated. In the previous example, one could use one or several extra bits for each terminal. This would considerably increase the probability of having a variable terminal instead of a constant one. Have you an idea how you could code the formula with a reduced probability of having a variable terminal? This is not very difficult. If you wish, you can also code the formula in such a way that the probability of having the operation * is higher than the probability of having another operation. A high probability of a certain phenotype is obtained when many different genotypes have the same phenotype. Note that such a coding with different probabilities leads to longer bitstrings.

Note that the genotype of an animal does not entirely define the phenotype in nature; nutrition, education, etc. can have a strong influence. Such facts are not implemented in standard GA's.

Although coding has a big influence on the GA performance, it is somehow outside the GA business. GA codes generate the bitstrings and not the phenotypes. The coding is the most demanding task for the GA user.

Mutation

How does nature 'design' and 'optimize' animals? First, we can observe that animals create 'children'. The children of primitive animals (without sexual reproduction) have almost the same genotype as their parents. Typically, each child has one parent and mutation is the mechanism that modifies the genotype. The implementation of bitstring mutation is obvious and simple: one flips one or several bits of the string. For example, for a parent with a bitstring of length 6 and a mutation of the third bit, you might have the following:

`Parent:010011` → `Child:011011`

Mutation of the third bit

Crossover

After mutation, nature has invented a more complicated, 'sexual' process that modifies the genotype: crossover. Crossover requires at least two parents to create children. Crossover means that the genetic information of the parents is split into (at least) two parts. The child then obtains a new genotype that is a mixture of the genotypes of its parents. It seems that crossover speeds up the progress of evolution. The implementation of crossover mechanisms for bitstrings is also simple. Unlike in nature, one can work with more than two parents. The simplest crossover is obtained when one has two parents with bitstrings of the same length and when one cuts these strings at one and the same position. For example, for two parents with bitstrings of length 6 and crossover after the second bit, you might have the following:

```
Parent1:010011 → Part11:01, Part12:0011 →
     Child1:=Part11+Part22:011010

Parent2:111010 → Part21:11, Part22:1010 →
     Child2:=Part11+Part22:110011
```

Crossover after the second bit

You may easily generalize the crossover procedure for more than two parents and more than one crossover points. Note that this does not allow you to create children that cannot be generated (at least indirectly) by one-point crossover with two parents. For example, you may have the following:

```
Parent1:010011 → Part11:01, Part12:00, Part 13:11

Parent2:111010 → Part21:11, Part22:10, Part 23:10

Parent3:001100 → Part31:00, Part32:11, Part 33:00
```

Crossover after the second and fourth bits

One of the children is, for example, `Part11+Part32+Part23:011011`. There is no way to select two of these parents in such a way that the child is directly generated by a one-point crossover, but there are several ways to generate the child within two steps. For example, with a crossover after the first bit you can generate a first child with the bitstring `011010` by mating parent1 and parent2. Now, you can combine this child again with parent1, crossover after bit 5 for obtaining `011011`. Obviously, you might have created this also with a two-point crossover of the first two parents.

Randomness

Multi-parent crossover and multi-point crossover allow one to skip intermediate steps required by one-point crossover with two parents to reach a certain bitstring from a given population of parents. Thus, one might reach optimal bitstrings more quickly. At the same time, the chance of generating much worse individuals is also increased. Therefore, it is not obvious how many parents and how many crossover points should be admitted. The more parents and the more crossover points one has, the bigger the jumps in the search space of the bitstrings caused by crossover. Is this desired? Note that one could also reach every possible bitstring from any bitstring by multi-point mutation. Such 'far' jumps therefore are similar to a random search. There is no reason for a Genetic Algorithm (GA) that does not outperform random search. Therefore, we should be careful when we introduce generalization such as multi-parent, multi-point crossover. Note that a proper definition of terms like 'near' and 'far' would require the definition of a metric, which is not always possible. Therefore, precise formulations can be tedious.

Remember that it is impossible to outperform random search in extreme situations, such as finding minima of a random function. Depending on the problem to be solved, it may be reasonable to push the GA search toward random search, but in most of the engineering applications the randomness inherent in simple GA search is already sufficiently high or even too high.

The most simple way to control the randomness of a GA is a combination of crossover with mutation. After crossover one can apply mutation to the children with a certain rate. In most of the practical applications, it turns out that the mutation rate should be low, but not equal to zero. The optimal mutation rate depends on the problem to be solved and on the GA implementation. For so-called steady-state GA a mutation rate of approximately 0.1 is appropriate, whereas lower mutation rates are optimal for other implementations.

Population

The individuals of a GA are members of a population. Working with big populations means using much memory. The limited population size in nature as well as in GA implementations implies that individuals must die. When an individual dies, the corresponding information and experience is lost. Although this is a drawback at first sight, it also avoids accumulating information contained in the population, which would cause problems in the selection process (see below). In nature, one can observe that smaller populations may be more flexible, whereas large populations can become more stable or even conservative. To perform crossover, at least two parents must be available, i.e., the minimum population size is two. It is quite obvious that a GA with such a minimal population is not optimal. Finding an optimal population size is not easy at all. This requires much experience and depends on the complexity of the problem to be solved and on the specific GA implementation. In engineering problems, the fitness evaluation required for each individual may be time-consuming. In this case, one is forced to work with relatively small populations of about 100 individuals, even if larger populations would improve the GA performance.

Instead of working with populations of fixed size, one can also work with populations of varying size, which is also observed in nature. This is especially promising when the fitness evaluation is time-consuming. In this case, it seems to be natural to start with a small population and to let the population grow until a certain limit is reached. However, classical GA's work with fixed size populations.

Initialization

When a GA is started, one usually assumes that no information is available. Therefore, all well-known GA implementations start with a pure random initialization of the population. The initial population is called generation number 0. With a random initialization, one obtains high randomness of the GA search when one works with a big initial population, whereas GA's with smaller populations are further away from random search.

Note that non-random initializations might be attractive in practice. For example, one might start with a user-defined initialization. This is not easy because the user certainly would hate to define initial bitstrings. Instead of this, he might be able to 'propose' some initial phenotypes that are promising. Therefore, user-defined initialization requires a coding of given phenotypes into bitstrings.

In general, one can consider a GA as an iterative process. Any iterative process can be supported by an appropriate initialization that takes prior knowledge into account. This knowledge may be obtained from analytical considerations, from previous computations, from known solutions of problems similar to the one to be solved, from practical experience, and so on.

Generations

Starting with the initial generation 0, a GA creates a new generation using the operators copy, crossover, and mutation applied to the parents. This procedure is repeated iteratively.

Most GA implementations work with generations with a fixed population size. This means that the size of generation n is the same as the size of generation n-1. Because of the copy operator, some individuals of generation n-1 are also contained in generation n, but others 'die'.

Note that one has neither a constant population size nor a strict separation of generations in nature. Moreover, the lifetime of different individuals can be very different. Therefore, standard GA implementations mimic natural procedures very roughly. Since we cannot say that a natural optimization procedure is optimal, this does not mean that a standard GA must perform badly.

Elitism

One of the most important drawbacks of the usual fixed-size generation implementation is that even the best individuals of a generation can die when the next generation is created. As a consequence, the best individual of the new generation may even be worse than the best that one had found before. To avoid this, one can always keep the best (or the m best) individuals of a generation by copying the corresponding bitstrings into the new generation. This is called elitism.

Completeness

When a new generation is created, it is hoped that the individuals of this generation concentrate more on those areas of the search space where the individuals of the previous generation were successful. As in classical minimum search algorithms, there is a certain chance that a GA is 'trapped' by some local minima. Because of the randomness contained in the GA mutations, escaping the trap is always possible, but in this case, the GA becomes even slower than random search. In this case, special procedures might help the GA to escape the trap. How can one detect that a GA is trapped? Mutations allow the GA to escape a trap in a relatively inefficient way. Moreover, the mutation rate is usually much smaller than the crossover rate. Therefore, we should focus on crossover.

First of all, one can create all possible bitstrings of a finite length from two parents only, provided that the corresponding two bitstrings are complementary, i.e., when the logic XOR operation of the two bitstrings results in a bitstring of the form `111111`. When one has a population of $N > 1$ different parents, it is always possible to generate an arbitrary bitstring by crossover when one has at least one `0` and one `1` bit in at least one of the parents for each of the positions of the bitstrings. Therefore, it is relatively simple to make sure that crossover can generate all possible bitstrings, i.e., that it can theoretically explore the entire search space. Unfortunately, this does not guarantee a quick escape from traps.

Diversity

To get more information on a generation, a statistical analysis of the bitstrings within a generation may be helpful. This analysis should clarify how diverse the individuals are. Obviously, there are many different ways to define diversity. Without a precise definition of diversity, we can guess that it might be good when one has a similar number of `0`'s and `1`'s in each bit position of all individuals. For more precise definitions of diversity see [Fröhlich, 1997].

To achieve a sufficiently diverse generation, one might first use crossover and mutation for creating a part of the generation only. Afterwards, one can fill the rest of the generation with individuals that increase the diversity.

Fitness

From observations we know that mutations and crossover provide a certain chance of generating a 'better' child that is fitter than its parent, but also the danger that the child is not able to survive. The terms 'good', 'bad', 'better', 'best', etc. are vague and need a proper definition that permits a GA implementation.

In biology, it is not clear at all what these terms mean. The principle of the 'survival of the fittest' replaces this term by another one that seems to be more scientific and precise, but the principle does not define the term 'fitness' at all. When we observe animals, we are usually unable to say which of them is the fittest. Thus, we are unable to apply the principle of the 'survival of the fittest' to foresee which of the animals will not survive, which of them will have many children, and so on. When we observe that an animal dies in a specific situation, we might be tempted to say that it was less fit than another one that survived. But in another situation, it might have survived, whereas the other one would have died. However, a simple fitness definition could be obtained from the number of children produced by an individual. Would you agree with the statement that mice are fitter than humans because they have more children on the average? Maybe you want to restrict such a fitness definition to one species only, but would you agree with the statement that people without children have a zero fitness? What about the fitness of an individual with many children when all of the children are sterile? When we look at bees, we see that even sterile individuals can have great value for a society. When looking at birds, we find birds that can fly and swim very well and we find other birds that can neither fly nor swim, but they still survive. What is the influence of some specific skills on the fitness of an individual? It seems that the environment has a big influence on the fitness of animals. The ability to fly may be useful within a certain environment but useless in another one. Thus, the fitness depends not only on the properties of an individual. Finally, it is even questionable whether attaching a fitness value to each individual is reasonable.

The dependence of fitness on the environment is much stronger than one might expect. For example, when we decide to keep bird as a pet, attributes that were important for the survival of the bird in nature may become unimportant and vice versa. Breeding of animals may modify their skills within a few generations in such a way that the animal might even be unable to survive without human support. Breeding means that we redefine the fitness of an animal in a very specific way and that we impose a selection mechanism that is much more rigorous than what we observe in nature. Obviously, breeding is extremely successful in some sense. It speeds up the development of some animal in a direction we have specified by deciding what we consider to be nice attributes of the animal. Of course, we cannot expect that breeding can create a flying cat that could catch birds more efficiently even if we think that such an animal might be possible. However, breeding really works within some limitations.

In computer optimization, one typically has a mathematical formalism that allows one to define the quality of an individual much more precisely, but the fitness definition is never unique and it has a considerable impact on the GA performance. For example, when a GA is applied to symbolic regression of a given function $f(x)$, each bitstring represents a function $f_n(x)$ that is an approximation of $f(x)$. In the previous sections we usually applied the square norm of the difference $f(x)$-$f_n(x)$ to define an error number e_n. The square norm was important because it usually led to a linear system of equations. Now, we can use any norm because we do not even intend to obtain a linear system of equations. Since small errors are desired, the fitness definition $fit_n=1/e_n$ seems to be reasonable, but this is also not unique. If you wish, you can apply some fitness filter of the form $fit_n=Filter(e_n)$. In this case, *Filter* should be a continuous function that has a maximum when the argument is zero.

Since the fitness is obtained from an analysis of the phenotype, the fitness definition and the fitness evaluation are also outside the GA itself. The GA user must provide not only the coding of the genotype into the phenotype, but also the computation of the fitness for a given bitstring. A GA does little more than produce bitstrings. It needs the fitness of all individuals of a generation to decide how to create the next generation.

Selection

Assume that the GA has created a generation. It then passes the bitstrings of all individuals to an external routine provided by the GA user. This routine computes and returns the fitness values of all individuals to the GA. Now, the GA creates the next generation using crossover and mutation. For deciding which of the individuals of a generation should become parents, the GA uses the principle of the 'survival of the fittest', i.e., it will more frequently select individuals with higher fitness than individuals with lower fitness. There are several ways to perform this selection. Fitness proportional selection and tournament selection are most frequently applied, but other selection mechanisms could also be designed. Note that a strict selection should be avoided; giving 'poor' individuals at least some chance is important for obtaining good performance. Obviously, fitness filters have a big impact on the selection.

Once the GA has selected parents, it applies crossover to create children. Crossover requires the specification of the crossover point. This point can be randomly selected. Of course, one could use some statistics on the success of the crossover points. This would allow one to attach another type of fitness values to the crossover points and to select the fittest crossover points more frequently than other ones. There are certainly much more sophisticated routines for selecting crossover points, but such routines are not considered here. A prominent example is so-called uniform crossover.

Remember that the GA will apply mutations with a certain rate to all children of the new generation. If an elitist strategy is implemented, it can copy the best individuals of the previous generation and, if desired, it can create special individuals that increase the diversity of the new generation.

Termination

A GA produces one population after another. This can be done in an infinite loop. When the user specifies a fitness value to be reached, the procedure can be stopped as soon as at least one individual has a higher fitness than the desired one.

Often, the user does not exactly know how big the fitness value of an acceptable solution should be. Therefore, one would like to stop a GA when one can no longer expect to obtain much better solutions. From the experience with traditional optimization algorithms one might be tempted do observe convergence and to stop the GA when the maximum fitness remains more or less constant over some populations. This should not be done because the GA search exhibits a pronounced staircase behavior. The maximum fitness may be almost constant over many generations and then it can be suddenly increased drastically within a few additional generations. Incidentally, it seems that the same staircase behavior can be observed in natural evolution. Note that the GA staircase is a random staircase rather than a regular one. It is impossible to predict the next step of this staircase. Therefore, an analysis of the GA convergence is useless for obtaining an appropriate stopping criterion.

In most GA applications, the user has some ideas about the desired performance of the object to be optimized by the GA, i.e., the user knows the desired properties of the genotypes that correspond to the bitstrings produced by the GA. In this case, the GA can output the best bitstrings of a generation and the user may study the corresponding phenotypes. As soon as he is convinced that one of the phenotypes is sufficiently good, the user can stop the GA.

Conventional GA structure

The following is a pseudo code of a simple conventional GA. Many of the GA codes that are available free are mainly based on this structure.

```
I=0
Initialize Population P(I)
Do {
  Evaluate the fitness values of population P(I)
  Stop if terminal condition is met
  While (next population P(I+1) is not full) {
    Select two individuals Ind1, Ind2 out of P(I)
    Copy(Ind1,Ind2) => Offspring1,Offspring2
    If(random<c-rate)
       Crossover(Ind1,Ind2) => Offspring1,Offspring2
    If(random<m-rate) Mutation(Offspring1) => Offspring1
    If(random<m-rate) Mutation(Offspring2) => Offspring2
    Copy Offsprings into the next population P(I+1)
  }
}
```

Figure 17: Pseudo code of a simple, conventional GA. Note that the crossover and the mutation operations are performed with some probabilities depending on the crossover rate and mutation rate specified by the user. When the crossover is not done and when no mutations are applied, the offsprings are simple copies of the parents.

Note that some randomness affects most of the steps of this GA and that the implementation of these steps is not defined in detail. Therefore, the same pseudo code can be used for different implementations with different behavior. For example, the routine Crossover can randomly select the crossover points. When crossover points before the first bit or after the last bit of the strings are permitted, Crossover also can generate simple copies of the parents. Similarly, the routine Mutation may select the locations of the bits to be flipped. If this location is allowed to be outside the strings, pure copies of the parents can be obtained. This automatically leads to a weak form of elitism because the fittest individuals are selected most frequently. Therefore, the fittest individual may have a good chance of being reproduced. To implement a strict elitism, the pseudo code must be modified. In addition to elitism, we have outlined several ideas that can be implemented in the GA to improve its performance. When you analyze these ideas and the GA structure, you will see that some ideas may be embedded without any modification of the main GA structures.

Hybrid GA

When a real parameter is optimized by a GA, it is coded into a bitstring. Now, when mutation is applied to one of the bits, this can have either a small or a big influence on the size of the corresponding parameter, depending on the position of the bit. For example, you may code a real number as follows: 1) The first bit specifies the sign. 2) The bits 2 up to 5 specify the mantissa. 3) The bit 6 specifies the sign of the exponent. 4) The bits 7 and 8 specify the absolute value of the exponent. Assume that a given bitstring contains the number –12E+3, i.e., the bitstring is 11100011. When the mutation flips bit number 5, one has 11101011, which corresponds to –13E+3, i.e. a number similar to the original one, whereas flipping bit 6 would generate the number –12E-3, which is very different. This means that the same genetic operation will sometimes correspond to little modifications of the phenotype and sometimes to huge modifications. The same happens with crossover. The main reason for this behavior is the fact that the GA itself has no information on the meaning of the different bits of the string.

To achieve a better neighborhood search of real parameters to be optimized by a GA, one can invent several extensions of the conventional GA. Instead of coding a real parameter into the bitstring, one can code a placeholder and apply a traditional parameter optimization to it. For example, assume that a GA shall solve a symbolic regression problem with real parameters to be optimized. For reasons of simplicity, try a simple tree with one node, two terminals and four operators +,-,*,/. The terminals may be either the variable or a parameter to be optimized. Instead of coding the numeric value of the terminal, you can use a single bit for each terminal. When the bit is zero, the terminal is the variable x; when it is one, the terminal is a parameter p_i to be optimized externally where i is the number of the terminal. For example, the bitstring `1110` might represent the formula $p_1 * x$. The routine that evaluates the fitness of this formula would have to apply some non-linear parameter optimization to evaluate the parameter and the maximum fitness to be reached with this formula. If you wish, you can interpret the individual $p_1 * x$ as an individual that is able to 'learn'. The external parameter optimization then corresponds to some learning process.

Non-binary GA

Often, the coding of finite parameter sets into binary strings is not trivial. For example, assume that you have one parameter that can have three states, i.e., three different values. This might happen in the symbolic regression problem when you admit three operators +,-,* only. Direct binary coding of this parameter requires at least two bits. When you reserve two bits in the GA bitstring for this parameter, you obtain four states. What can you do with the fourth extra state?

The simplest way to overcome coding problems consists in the design of non-binary GA. This is a very simple generalization of binary GA. One now has strings of a finite number of elements. Each element can have a finite number of states. The number of states can be one and the same for all elements or it can differ from element to element. The former is simpler, but less general and less flexible.

The combination of non-binary GA with traditional optimization procedures is essentially the same as the combination of binary GA with traditional optimization procedures. For complex optimization problems, such hybrid non-binary GA implementations are most promising.

Genetic Programming (GP)

Although we know that computers work with bitstrings, no computer user likes input and output in the form of bitstrings. Several computer languages have been designed to make communication with computers easier. These languages use symbols. Moreover, computer codes are subdivided into subroutines that do a meaningful part of the job. Such a subroutine might be used for the evaluation of a specific mathematical function. For example, assume that the function $\sin(x)$ is needed, but not contained in the pool of intrinsic functions of the computer language. Now, you can write a subroutine that evaluates $\sin(x)$ within a certain range of the argument x with a sufficient precision. To do this, you start with some mathematical literature and you find some series expansion, for example, power series.

When you truncate such a series, you obtain an approximate formula that contains the four arithmetic operators +,-,*,/, the variable x, and several constants, for example, $\exp(x) \approx 1+x+x*x/2+x*x*x/6$. This formula might have been obtained as a solution of the symbolic regression problem obtained by an appropriate GA. However, humans who search for such a formula will avoid bitstring coding and use algebra instead. Maybe, they will even apply some symbolic mathematics code such as Maple [Char, 1991] or Mathematica [Wolfram, 1991]. Symbolic math codes demonstrate that one can design codes that handle formulae more or less directly. Therefore, one also can write computer codes for symbolic optimization that design formulae directly instead of coding them into bitstrings. When we apply the genetic concept, this means that the genotype of such codes is a formula instead of a bitstring. Such formulae can easily be represented by trees as at the end of the previous chapter.

Conventional GP structure

Genetic Programming (GP) [Koza, 1992] is an optimization procedure working on symbolic formulae using essentially the same procedure and genetic manipulations as GA. Consequently, the structure of conventional GP is the same as the structure of conventional GA (see Figure 17). All that has to be done to obtain such a GP implementation is to define the individuals and the genetic operations (crossover and mutation) that are to act on them. The fitness evaluation, selection process, etc. are the same as for conventional GA. Therefore, you easily get hints for generalizations when you study the sections on GA.

Tree representation

To illustrate the tree representation of two individuals, let us consider two simple formulae, for example, $x-x*x*x/2$ and $1+x*x$ (see Figure 18).

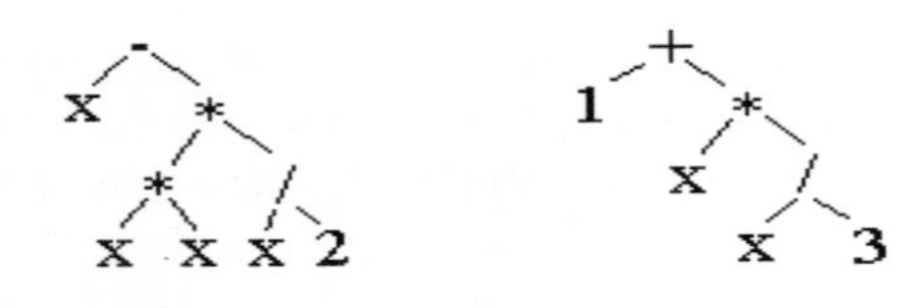

*Figure 18: Tree representations of two formulae, Individual 1: x-x*x*x/2 and Ind. 2: 1+x*x/3.*

Note that these two individuals have a different topology, i.e., a different number of nodes and a different number of terminals. The depth of both trees is the same, i.e., each of the trees has three different levels of operators, i.e., nodes. In general, the depths of different trees can be different. Because of limited storage capabilities, the maximum tree depth must be limited. This is an important difference between standard GP and GA. The size of the GA bitstrings is typically one and the same for all individuals. Note that also the size of chromosomes is constant for those animals that can be crossed. However, in computers, crossover of individuals with completely different genotypes can easily be defined.

Terminals

Standard GP implementations work with two types of terminals: variables and random constants. With real random constants one is very unlikely to obtain integer values as in the two formulae of the example above. Thus, one is also unlikely to obtain accurate approximations of truncated series expansions such as $\exp(x) \approx 1+x+x*x/2+x*x*x/6$. In general, it is expected that appropriate parameter values may be constructed by a branch of the tree with random constant terminals. A typical construction of an approximation of the constant 2 in the formula above with a branch with three random terminals might be (0.783+0.154)/0.912. With sufficiently big trees and sufficiently many random number terminals, one has some chance to approximate a certain constant. When the GP procedure is smart enough, one can hope that it is able to find a sufficiently accurate solution. However, this kind of parameter optimization is extremely inefficient.

The Generalized Genetic Programming concept that is outlined in the following section, avoids the difficulties of appropriate constructions of constants or parameters by adding other types of terminals.

Nodes and elementary functions

To each node of a GP formula tree, some elementary operator or function is attached. Modern programming languages know many of them and mathematicians know even more. When one implements a GP code, one has to provide a certain set of elementary functions that are available. The bigger this set is, the more difficult the implementation and the bigger the search space becomes.

Somehow, the symbolic regression problem is the same as the problem of a scientist who analyzes some measured data and tries to find a simple and useful formula to explain the measurements. Mathematics is often considered as a pure, abstract formalism without any intrinsic correlation to physics or even engineering. When one considers the history of mathematics, physics, and engineering, one observes that some co-evolution took place. Therefore, the best-known mathematical tools are those best suited to analyzing most engineering problems. For example, well-known arithmetic operations and mathematical functions (sin, cos, exp, log) are often very successful in describing engineering problems. The frequent use of a specific function in a special class of problems makes it likely that this function is also involved in a successful solution of a new problem of the same class. The frequent use of certain functions in the solution of engineering problems makes it reasonable to teach such functions to engineers. Consequently, engineers will first try using such functions for solving everyday problems. This is some sort of selection mechanism on mathematical functions that took place in the past. Consequently, an experienced engineer who wants to apply a GP code for solving one of his problems can associate some probability to each of the elementary functions provided by the code. When these probabilities are used for the initialization and mutation of the GP individuals, the efficiency of the GP may be increased.

Mutation

Mutation affects one branch of the genotype. This may be either a node or a terminal. Consider the terminal mutation first. One randomly selects one of the terminals. Then one randomly specifies its new type. If its type is a random constant, one can replace it by another random number (see Figure 19).

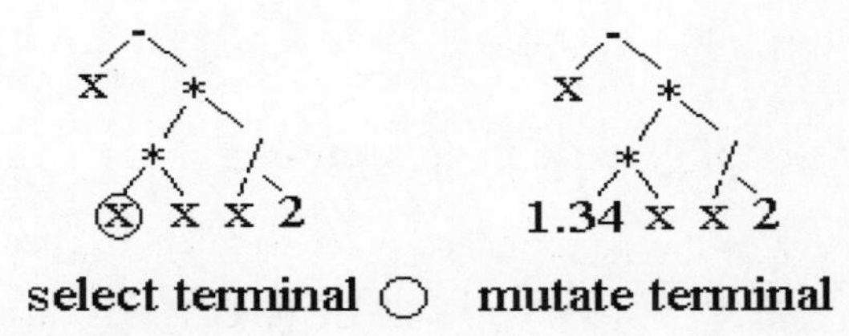

Figure 19: Terminal mutation.

For node mutation, one can randomly select a node and replace the corresponding branch by another one that is randomly constructed (see Figure 20). This mutation is very drastic, especially when the selected node is far away from the terminal. In this case, there is not much difference between a node mutation and a pure random generation of the individual. Therefore, mutations of standard GP implementations are not very helpful. For this reason, the mutation rate in standard GP is often set to a very low value.

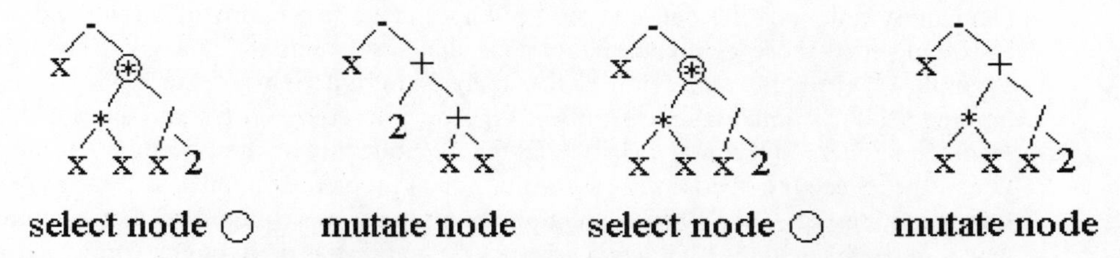

Figure 20: Node mutations. Left hand side: replacement of the node with the entire branch;, right hand side: mini-mutation that replaces the operator of the node only.

Drastic effects of node mutations may be reduced by mini-mutations that exchange the operator of a node rather than the entire branch. When all nodes correspond to operators or functions with one and the same number of arguments, the mini-mutation is easy.

Replacing an operator by another one with a different number of arguments is more difficult because it requires a modification of the tree topology.

Crossover

Since GP works with trees of variable size and architecture, one has to select two crossover points, one for each parent. Each parent is then subdivided into two parts, head and body or branch and trunk. By exchanging the two parts one obtains two offsprings, a child with head of parent one and body of parent two and another child with head of parent two and body of parent one. Figure 21 illustrates this.

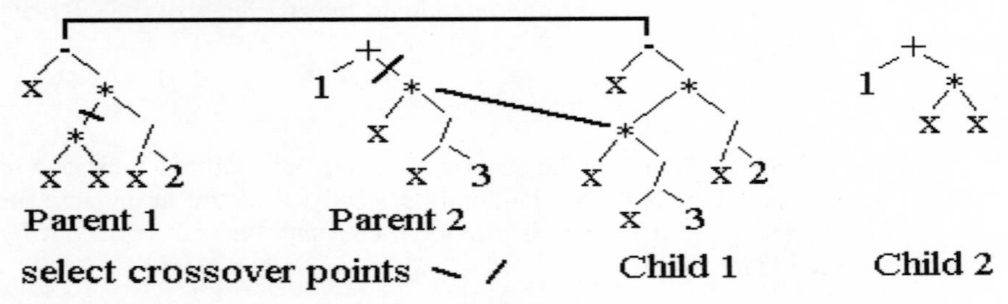

Figure 21: Crossover of two parents for generating two offsprings.

Problems of conventional genetic programming

When one analyzes the performance of conventional genetic programming codes, one can detect many more or less severe problems that make these codes very inefficient even for simple symbolic regression problems. Symbolic regression problems that are often presented for demonstrating the power of GP are well tailored for GP and of little practical value. Therefore, GP codes are often very frustrating for engineers.

It has already been mentioned that conventional GP's try to approximate parameters by more or less long and complicated branches of the formula trees. This kind of parameter optimization is neither intelligent nor efficient. Maybe it is the main source of the GP problems in engineering applications.

Another problem that is frequently observed is the fact that after a few generations some of the elementary functions die out, i.e., are no longer contained in any of the individuals. This may also happen to those elementary functions that are essential for a good solution. A typical example of symbolic regression is the analysis of functions of the type $\sin(x)/x$. For small arguments, such functions are similar to cosine. Moreover $\cos(x)$ and $\sin(x)/x$ have the same symmetry. Therefore, it may happen that the GP code throws the sine function away and favors the cosine. Since the operation / causes numerical problems for small x, it may also happen that this operation dies out. When a mutation generates a new individual that contains / and sin() again, it is very likely that the corresponding tree has a wrong architecture and that the individual has a very low fitness.

Already the simple function sin(3.14*x)/x combines the two problems mentioned above. When a conventional GP code is used for finding an approximation of this function in the interval x=0…1, one observes that many big populations are required to find solutions that are not too far away from the correct solution. To support the GP, one may reduce the set of elementary functions. Even when one only uses the elementary operations +,-,*,/ and the elementary functions sin() and cos(), the GP turns out to be extremely inefficient. In the following section, some GP extensions and new concepts are proposed that make the procedure much more promising.

Generalized Genetic Programming (GGP)

The generalization of GP codes by admitting non-linear parameters as a third terminal type corresponds to the hybrid GA combination with a non-linear parameter optimization that has already been proposed in the Hybrid GA section. From this, one can expect a much more efficient handling of real parameters. Although the idea of **parameter terminals** is very simple, its implementation requires considerable modifications of the GP code.

First of all, for the selection mechanisms of GP, a fitness value must be associated with each individual, i.e., formula. Without parameter terminals, the fitness value can be defined as a simple function of the sum of the square errors in the observation or sampling points of the given function. Note that the fitness definition can have a considerable influence on the performance of the GP code, but plausible definitions can easily be found and turn out to be useful in most cases. When parameters are involved, the fitness value of each individual becomes a function of these parameters. It seems to be natural to perform a parameter optimization for each individual and to use the error of the optimal parameter set to compute the fitness value. But non-linear parameter optimizations can be quite time consuming.

Parameter terminals increase the freedom of each individual. They allow the individual to 'learn' by adapting the parameters. Although this is a very primitive implementation of **learning**, it can be very useful and efficient and it is a step closer to natural optimization processes than traditional GP. In nature, the ability to learn is certainly important. An individual that learns quickly is probably fitter than another one that learns slowly. The learning characteristics of GGP individuals are connected to the influence of the parameter terminals on the resulting function. Some parameter terminals may have almost no influence on the genotype, whereas other ones can have a big influence. It has been found that very simple and rough non-linear parameter search algorithms combined with GP are sufficient. GGP works with a **symbolic definition of the non-linear parameter optimization**, i.e., the parameter optimization is described by a formula. Theoretically, an additional GP code could optimize this formula. Since this would be extremely time-consuming, it has not been done yet. The given default formula is simple and has been found to be sufficient in most cases. However, experienced GGP users can easily redefine it.

We have seen that **linear parameter optimization** is numerically much faster and simpler that non-linear parameter optimization. To obtain a set of linear parameters, a series expansion is required. GGP works with a generalized series expansion of the form

$$f(x) \approx \sum_{k=1}^{K} a_k f_k(p_1, p_2, \ldots, p_L, x). \qquad (3.1)$$

GGP optimizes the linear parameters a_k, the non-linear parameters p_l, and the shape of the basis functions f_k. This means that a combination of linear parameter optimization, non-linear parameter optimization, and symbolic basis function optimization is applied. For the latter, an advanced GP algorithm is used. When the number of basis functions K is equal to 1, one mainly obtains a combination of GP with non-linear parameter optimization, but there is a little difference: GGP multiplies the basis function with an **amplitude**. Although adding this amplitude is no big deal, it increases the efficiency of GGP because the construction of the formula can now be focused on the shape of the corresponding genotype, i.e., function. Assume that you have obtained the given function from measurements, then the units that have been used affect the numerical values of the function. When the units are modified, the values are scaled by a certain factor. This will only affect the amplitude. A traditional GP would have to construct an entirely new formula when the function is scaled with any factor. GGP only has to modify the amplitude, but it can keep one and the same basis.

As soon as a true series expansion with $K > 1$ is used, the implementation becomes much more complex and tricky, although the linear parameter optimization leads to a matrix equation that can be solved easily. The main problem is caused by the fitness definition. One now works with a group of individuals rather than with a single individual. From the error of the approximation of the given function, one therefore can define a fitness value for the entire group, but not for the individuals. Since this causes a quite complex extension, we postpone this generalization and focus on the GGP implementation with K=1. It should be mentioned that GGP also uses **symbolic fitness definitions**. Therefore, also the fitness definition formulae could be modified by experienced users or optimized by an additional GP algorithm.

It has already been mentioned that the extrapolation is much more difficult than the approximation of the given function. The same statement holds when one considers the construction of a theory. It is much simpler to explain known experiments than predicting how new experiments will turn out. The main benefit of good theories is the prediction of new and surprising results. Symbolic regression is a simple prototype of the theory building process. In order to obtain good extrapolations, GGP splits the known data set, i.e., the interval where the given function is known, into two parts. The first part is used for computing the basis functions, optimizing the parameter set, and for computing the error of the approximation, whereas the second one is only used for checking the quality of the extrapolation. The GGP fitness is obtained from an analysis of both approximation and **extrapolation error**.

In order to avoid time-consuming non-linear parameter optimizations, GGP performs some simple **initial checks**. Individuals that do not pass are '**repair**ed' by simple '**mini-mutations**'. Repaired individuals that still do not pass the initial test are discarded without any fitness computation. Trying to repair an individual is a typical idea of an engineer. It seems that introns play a similar role in nature. Repairing with mini-mutations corresponds to an improved neighborhood search. Mini-mutations are simple operations that leave the structure of the individual intact. Exchanging two terminals is a typical mini mutation.

The GGP user often has some idea of constants (e.g. π or e) that might play an important role. Therefore, GGP adds **user-definable constants** to the random constants of standard GP.

Mini-mutations allow GGP to convert optimized parameter values into a second type of **explicit constants**. This offers another interesting possibility. Only one non-linear parameter per individual is required. This means that (3.1) is replaced by the more simple expression

$$f(x) \approx \sum_{k=1}^{K} a_k f_k(p_k, x) \cdot \tag{3.2}$$

From time to time, the non-linear parameter is 'frozen' into a constant and a constant becomes a parameter that can be optimized. The reduction to a single non-linear parameter considerably reduces the iterations required for the non-linear parameter optimization. GGP performs some **re-evaluations of the fittest** individuals because it also uses an elitist strategy. Consequently, a very rough parameter optimization with only 2-6 iterations is sufficient in most situations.

The biggest part of the GGP code has been written in Fortran. Old-fashioned Fortran programmers allocate memory for fixed-length character strings that will contain the formula of the individuals. Thus, they would prefer a fixed formula size of the individuals. Beside this, there are also 'biological' reasons for a fixed formula size and for a fixed formula topology of all individuals. GGP can handle both **variable and fixed size individuals**. The most simple structure of a formula tree with fixed topology is obtained when all elementary functions and operations have two arguments. Let us call this a **binary tree**. A binary tree of depth 1 has one node and two terminals. A **complete binary tree** of depth 2 has 1+2=3 nodes and 2*2=4 terminals, and so on. For each depth there is exactly one complete binary tree. The restriction to complete binary trees of a fixed depth considerably reduces the search space, which can help to increase the search speed. To work with such trees, we have to find appropriate replacements for elementary functions with only one argument. This is easy. In many cases, we can replace functions with one argument by generalizations; for example, the elementary function exp(b)=e^b can be replaced by the more general function pow(a,b):=a^b. Instead, we can also introduce an arithmetic operation in the argument, for example, by defining com(a,b):=cos(a*b). In addition to **generalizations of functions** with one argument, we need elementary functions that simply pass one of the arguments: ta1(a,b):=a and ta2(a,b):=b. These are called **transfer functions**. Transfer functions make one branch of the tree inactive. The information on the inactive branches is hidden and has no influence on the phenotype. Despite this, the information is still there and can become active in another generation. It seems that this corresponds to so-called **introns** that are also observed in nature.

The default set of elementary GGP functions contains four arithmetic operations, four elementary functions with two arguments, and two transfer functions: add(a,b):=a+b, mul(a,b):=a*b, sub(a,b):=a-b, div(a,b):=a/b, com(a,b):=cos(a*b), sim(a,b):=sin(a*b), pow(a,b):=a^b, lga(a,b):=$\log_a(b)$, ta1(a,b):=a, ta2(a,b):=b. In addition, GGP knows other functions that are turned off by default. Note that most of the conventional GP implementations work with a much smaller set of elementary functions.

GGP test cases

Typical test cases for standard GP codes are functions like x-2x^3 that can easily be constructed by the three arithmetic operations +,-,* only. Note that the operation / is often turned off, because it can cause problems. GGP often finds the correct answer for such test cases already in the first generation (containing 100 individuals by default), i.e., 'by chance'. Therefore, we look for more tricky test cases.

The function sin(πx)/(πx) is a very nice function that causes severe problems for standard GP. If we compute it at 50 points in the interval x=0...1, it is hard to see from these data that

$\sin(\pi x)/(\pi x)$ is the correct solution. Since GGP takes more time to evaluate an individual with its parameter optimization, comparing the number of individuals used by GP and GGP for solving a problem would be unfair. Therefore, we consider the 'number of fitness computations' in the following. To evaluate the fitness of an individual, GGP performs n iterations in the parameter optimization loop. For each iteration, a fitness computation is made. Thus, the number of fitness computations is proportional to the program execution time when the fitness computation is time-consuming. However, traditional GP codes (with a population size of 500 individuals and a huge tree depth) were not able to solve the $\sin(\pi x)/(\pi x)$ test case within 50,000 fitness computations even when the elementary function set was well tailored. Some approximations were found, but none of them was really close to the correct solution. In fact, there is almost no chance for GP to find the correct solution even with considerably more fitness computations. Like GP, GGP has some dependence on the initial random population when a small population size is used. Therefore, several runs with different initial populations are required for obtaining some statistical description of the performance. Moreover, the success of GGP depends on the tree depth. For the test example, depth 2 would be sufficient, but with this depth, GGP never found the correct answer. Already with depth 3, GGP typically found the correct answer with 1000-5000 fitness computations. Some typical GGP solutions of $\sin(3.1416x)/(3.1416x)$ are:

$$f_1^{GP}=7.9513E3*\sin(\sin(e*1.4727E-5)*\sin(\pi)/x)$$

$$f_2^{GP}=-87.339*\sin(\sin(2*1.5702)*\sin(\pi)/(\sin(1.5702*e)*x/e))$$

Note that it is not always easy to check that a solution is correct. Also with tree depth 4 and with variable tree depth 2-5 correct solutions were found, but more fitness computations were required than for depth 3. Figure 22 and Figure 23 illustrate the GGP search.

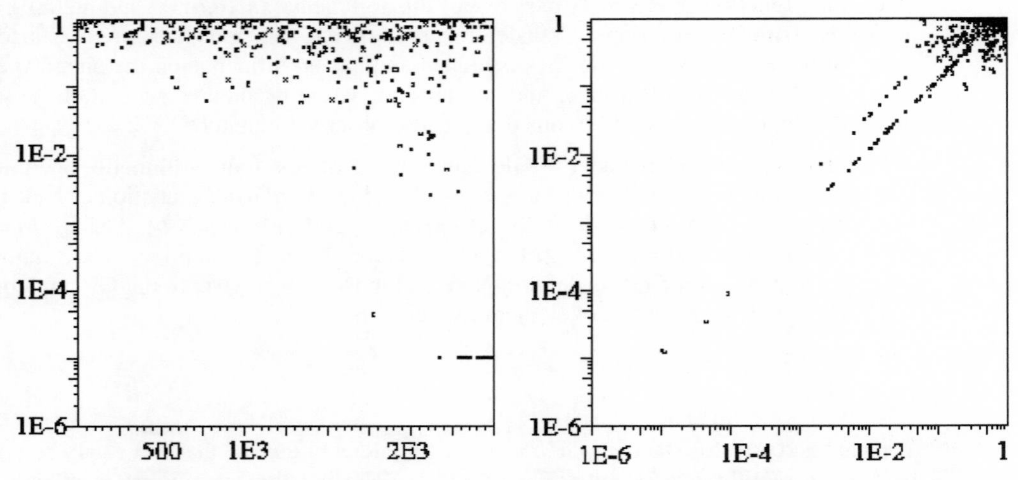

Figure 22: Typical GGP search run of the sin(3.1416x)/(3.1416x) test case with population size 100, formula tree depth 3, and GGP default settings. Approximation error as a function of the fitness computation number (left) and extrapolation error versus approximation error (right). The 'correct' solution has been found within 1800 fitness computations. Note that the GGP search is concentrated along the diagonal, where one has 'balanced' individuals.

So far, we have simulated a set of observed data by a known mathematical function evaluated at a finite number of observation points. Consequently, the accuracy of the data was very high - much higher than in real life. What happens if the accuracy of the observations is reduced? We can simulate this by adding some random values to the given function values. This considerably disturbs not only genetic, but also conventional optimization procedures. If the number of observation points is small and if the observation interval is short, there can be many completely different solutions that fit within the given accuracy and neither GGP nor any other code can decide which of them is 'correct'. However, for a sufficient number of observation points and a sufficiently large observation interval, GGP is still successful. The performance slows down with decreasing accuracy of the observations.

Although these results are very encouraging, there is still a difficult problem for GP and GGP with only one basis function. What happens when we observe, for example, the superposition of several independent resonators? In such situations, a generalized series expansion approach is promising. To tackle such problems with GGP, a combination with linear parameter optimization is required.

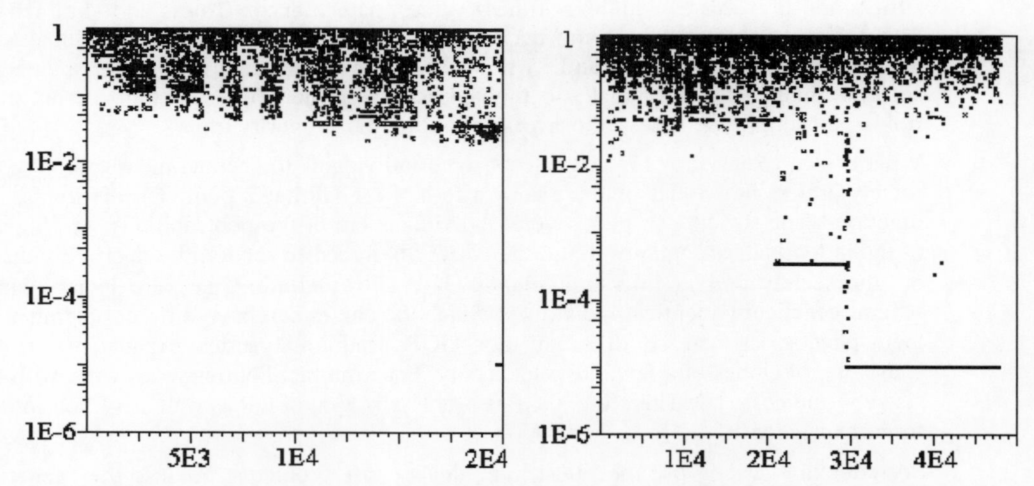

Figure 23: Same as Figure 22 for variable function tree depth 2-5 (left hand side) and for function tree depth 4. Here, the 'correct' solution has been found within 1300 fitness computations. Note that many more iterations are required here. In both cases about 20,000 fitness computations were required for finding the correct solution. A staircase behavior can be observed on the right hand side.

Further generalizations

Although one can consider GGP as a generalization of GP, its roots are completely different. GGP was obtained from a continuous generalization of Fourier series to generalized series expansions as outlined in the previous chapter. This generalization procedure was developed to illustrate the development of the MMP code for computational electromagnetics [Hafner, 1990/S], [Hafner, 1993/S]. Therefore, already the first version of GGP contained a generalized series expansion. In the following, the problems caused by this technique have to be solved.

Note that generalization always causes problems. The success of the generalizations therefore depends on the solutions of these problems.

It has already been mentioned that the definition and computation of the fitness causes the main problem. If we consider each basis function as an individual, the series expansion is a group or a society. We can evaluate the **society fitness** exactly as we have computed the **individual fitness** before, i.e., from an analysis of the approximation and the extrapolation error. But now, we need a new definition of the individual fitness. First of all, this fitness is required for the genetic selection mechanisms acting on the individuals. Individuals who live in fit societies obtain higher fitness than those who live in bad societies, but attaching the same fitness to all members of a society would certainly be wrong. How can we honor valuable members of a society? Naturally, the amplitude of a basis function gives some impression on the importance of the corresponding individual within a society, but this can be misleading. It has often been observed that two similar basis functions obtain big amplitudes of opposite sign. This corresponds to destructive work of the two individuals. Although the amount of work can be big, the resulting effect for the society can be small. Therefore, GGP **filter**s the **amplitude fitness** and it evaluates additional fitness values, named **error fitness** and **similarity fitness**. These fitness values check 1) how accurate the solution would be if the individual were to do the job of the entire society and 2) how similar the individual is to other members of the society. Finally, the individual is computed as a **combined fitness** obtained from the society fitness, the amplitude fitness, the error fitness, and the similarity fitness.

What does a society do? First, it selects some individuals to become members. Of course, the society prefers individuals that are known to be fit. GGP has a **pool of individuals**, i.e., basis functions. The society can pick several individuals out of the pool. Initially, the fitness values of the individuals are unknown and the society is forced to randomly select individuals. Later on, the society prefers fitter individuals. GGP allows **cloning**, i.e., the society can contain several genetically identical individuals. Since the clones can have different parameter values, their phenotypes can be different. For GGP, traditional series expansions are societies consisting of clones of a few individuals only. For example, Fourier series work with clones of $\sin(p*x)$ and $\cos(p*x)$. Therefore, GGP is a generalization of GP as well as of Fourier and other series expansions.

Societies that strictly use the fittest individuals cause problems, because they generate many outsiders that never get a job. Although such societies can be efficient at the beginning, they turn out to be inefficient after a while. It makes no sense to generate an individual without using it. Therefore, GGP prohibits selecting used individuals as long as there are unused individuals in the pool.

From time to time, the worst individuals in the pool are killed and replaced by new ones as in the traditional GP. GGP not only keeps the fittest individuals as in the well-known **elitist** strategy, it also saves the fittest individuals in **data files**. The society can not only select individuals from the pool, it can also select them from data files. This is advantageous if one runs GGP several times on similar problems. In these cases, GGP can take advantage of the knowledge obtained in previous runs. Data files play a role similar to libraries. With data files, GGP can clone individuals who were prominent a long time ago, i.e., GGP could generate a society consisting of clones of Einstein and Newton and it could generate children of them. Although we do not have this opportunity in real life, such features can be useful and efficient for solving problems on computers.

The individual-society structure and the various generalizations outlined in this section make the program structure of GGP much more complex than the structure of traditional GP codes. The GGP structure is even so complex that it is hard to draw a flow diagram. Moreover, there are many system parameters and symbolic system formulae that can be modified by the user for improving the GGP performance. Of course, this makes the description and handling of GGP difficult. Fortunately, default settings have been found and implemented that were useful for many different test cases. Therefore, the user can simply run GGP as it is. No doubt, the default settings might be improved and adapted to specific situations.

The symbolic definition of fitness values, fitness filters, parameter optimization procedures, etc. allows a symbolic optimization of important parts of GGP by another, external GGP code. Such a self-optimization is attractive, but Pentium based PCs are too slow for doing that.

Further GGP tests

When one runs GGP with societies with more than one individual on the test cases presented above, GGP can optimize several individuals at the same time. This is especially interesting if the society fitness evaluation is much more time-consuming than the fitness evaluations of the members of the society. With an increasing society size, GGP works more and more like conventional series expansions, i.e., excellent approximations can be found very quickly (see Figure 24). As soon as the society size exceeds some value, the extrapolations become worse and it becomes harder to find 'correct' solutions. For the test cases discussed before, single individual societies, i.e., society size 1, are sufficient.

What happens when the observed function cannot be constructed by an appropriate combination of the elementary functions implemented in GGP? To test GGP, a data file containing the data of a 'tricky' function was passed to GGP. With a society size of only 5, GGP found an excellent approximation with a quite good extrapolation. Surprisingly, all members of the best society were very similar, not clones but 'cousins'. With some analytical simplification, it was found that GGP had constructed a series expansion of the type

$$f^0(x) = \sum_{k=1}^{K} (x/a_k)^{b_k/x} + error(x) \,, \tag{3.3}$$

i.e., the 'cousins' could be replaced by pure clones. In fact, the correct answer would have been erfc(2/3x), where erfc denotes the complementary error function. GGP found a strange series expansion that seems to be very good for $x>0$. Since the computation of the powers (contained in GGP's basis) is time-consuming on usual processors, such series expansions are numerically less efficient than continuous fractions. Nevertheless, this is a big success for GGP, because series expansions of the form (3.3) could not be found in mathematical literature, i.e., GGP has the potential of finding new types of series expansions.

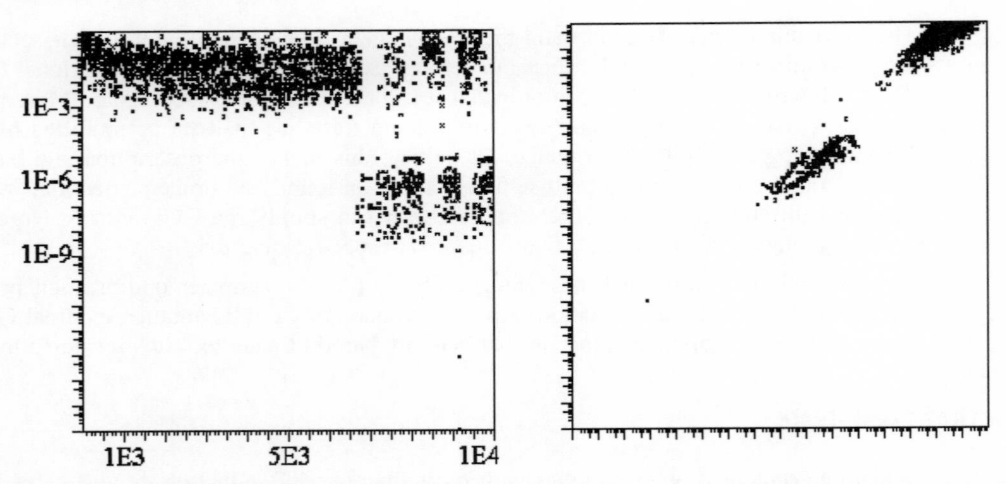

Figure 24: Same as Figure 22 for a GGP run with a society containing three individuals. Approximation error versus fitness computation number (left) and extrapolation error versus the approximation error (right). An extremely accurate solution with an error of about 1E-16 has been obtained, i.e., machine precision has been reached. The range of both axes of the right figure is extremely wide, i.e., 1E-16...1. The search is still close to the diagonal, but the extrapolation error is typically 10-100 times bigger than the approximation error, i.e., one can already observe the trends toward accurate approximations and inaccurate extrapolations that are typical for series expansions and societies with many individuals.

4 Generating matrix equations

General remarks

Matrices are the most important mathematical tool for solving linear problems on computers because matrices fit very well into computers. Computers 'like' well ordered objects, such as matrices. On modern computers, a huge variety of matrix algorithms is available. Most of these algorithms are available free or are contained in professional libraries that are widely used. Therefore, designers of codes for computational electromagnetics who use matrix formulations can usually select one of the available algorithms instead of inventing a new one.

When one wants to obtain an efficient code based on matrix equations, the selection of the matrix solver is most important, because the computer will usually spend most of the computation time for the matrix solver and often an almost negligible portion of the computation time for the matrix setup and for the post-processing. Source codes for computational electromagnetics can have thousands of lines, but the matrix solvers typically consist of ten up to a few hundred lines. Thus, it is worth optimizing the code lines that contain the matrix solver, but this is outside the scope of this book. Detailed descriptions of matrix algorithms may be found in many textbooks, for example [Datta, 1995], [Golub, 1983], [Dongarra, 1979].

The aim of this chapter is to demonstrate how matrix equations may be obtained. There are many computational methods in mathematics, physics, and engineering based on matrix equations. We have already considered the approximation of a given function by a series expansion. This class of problems is most simple and easily leads to matrix equations. In computational electromagnetics and other areas of computational physics, one deals with fields rather than with functions. This makes the procedure more complicated, but the same principles may be applied for generating matrix equations. To illustrate how matrix methods can be obtained, we assume that the following operator equation holds within a given domain D:

$$L f = g, \text{ in } D, \tag{4.1}$$

where L is a linear operator. L can be a differential operator, an integral operator or anything else. This operator acts on f, which can be a function or a field. The right hand side, i.e., the inhomogeneity g, is given function or field. When the domain D is bounded by a boundary ∂D, one may have boundary conditions that are of the form

$$\mathrm{L} f = h, \text{ on } \partial D, \tag{4.2}$$

where L is also a linear operator and h a given inhomogeneity. For the most simple case of the approximation of a function that is given in a certain interval, D is the interval, ∂D consists of the two end points of the interval, g is the given function in the interval, h corresponds to the values of the given function at the two end points, and the two operators L and L are unit operators that can be omitted. In this case, f denotes the approximation to be found and (4.2) can be omitted. In electrostatics, L might be the Laplacian, f the scalar potential. For the most simple Dirichlet boundary conditions[2] one would again set a unit operator L, and so on. In computational electromagnetics, the situation is often more complicated because the entire space is subdivided into several domains with different material properties and because one often works with a system of coupled differential or integral equations. However, the procedure can easily be adapted to more complex situations.

In the following, we call (4.1) the field equation and (4.2) the boundary condition. In the most general case, both the field equation and the boundary condition are inhomogeneous, i.e. g and h are non-zero. In this case, we can split f into two parts: $f=f_g+f_h$. Because of the linearity of the operators L and L we can write

$$L f_g = g; \quad L f_h = 0, \text{ in } D, \tag{4.3}$$

$$\mathrm{L} f_g = 0; \quad \mathrm{L} f_h = h, \text{ on } \partial D. \tag{4.4}$$

This means that either the field equation or the boundary condition is homogeneous. Both cases are formally identical. Moreover, the f_g case can be converted into the f_h case as soon as a special solution f_s is known with the following properties: $Lf_s = g$ and $\mathrm{L}f_s = h_s$, where h_s is an arbitrary function or field defined on the boundary. This means that the special solution fulfills the field equation, but not the boundary condition. One now can search for a solution f_h that fulfills $Lf_h = 0$ and $\mathrm{L}f_h = -h_s$. Then, $f_g = f_s + f_h$ is the desired solution of the f_g case. The same technique may be used to convert the f_h case into the f_g case.

Series expansions

In order to take advantage of the linearity of the operators involved in the field equations and in the boundary conditions, we now assume that the solution f can be approximated by a series expansion of the form

$$f = f^0 + error = f_0 + \sum_{k=1}^{K} a_k f_k + error, \tag{4.5}$$

where f^0 denotes the approximation of f. The basis functions f_0 and f_k can be arbitrary functions (or fields), but it is reasonable to use a basis that fulfills either the field equations or the boundary conditions. This leads to two different types of matrix methods, so-called boundary methods and domain methods.

For a good set of basis functions the following properties are desirable: 1) simple implementation, i.e., short code length, 2) short computation time, 3) accurate approximation of f with small number of basis functions K. A basis that satisfies all three conditions can only be found for very simple cases.

The simple implementation is important for those who write a code, but not for the users who apply the code - as long as the entire code remains reasonably short. Therefore, one should

---

[2] Dirichlet boundary conditions define the field along the boundary, i.e., $f = h$, where h is known along the boundary.

focus on the second and third properties when one wants to design a numerically efficient code, but most of today's codes for computational electromagnetics use a basis with a very simple implementation.

Mathematicians have found a huge number of algorithms for the accurate evaluation of almost any mathematical function that may occur in engineering. Most of the corresponding algorithms are contained in libraries or are available for free on the World Wide Web. Note that the focus of most of these algorithms is on precision and on a big argument range rather than on computation time, i.e., most of these algorithms evaluate the corresponding functions within machine precision for any argument, provided that the result is within the numeric range of the machine. These algorithms are usually based on a few well-known techniques, namely traditional, convergent series expansions, asymptotic expansions, continued fractions, etc. Within a code for computational electromagnetics, the argument range and the required accuracy of a basis function is often limited. To evaluate these functions more efficiently than with the traditional algorithms, new algorithms must be found and implemented. Generalized series expansions and symbolic regression methods have been introduced in the second chapter. These techniques can be used to find new, unconventional algorithms that speed up the computation time basis functions.

Finding a short basis (with small K) that allows one to find an accurate solution is extremely difficult, except in very simple cases. Therefore, one often encounters codes that work with huge numbers of unknowns, i.e., huge K. The size of K required for solving a problem with a given accuracy mainly depends on the selection of the basis functions f_k and on the method for computing the parameters a_k. Experienced scientists may find a basis for a given problem with relatively small K, but general purpose codes often work with much bigger K. Especially 'user-friendly' codes that do not expect that the user is experienced require a huge K. Since also a code can collect experience, one may design user-friendly codes that work with smaller K, start with a relatively low accuracy, and successively improve the results by adapting the basis using the experience obtained from previous runs. A typical example for such an adaptive improvement of the basis are the adaptive mesh generators of finite element codes.

It should be pointed out that also the method for computing the parameters a_k has a considerable influence on the size of K required for obtaining the desired accuracy. Those who write codes usually start with simple methods that result in relatively short codes. Such methods are often not very efficient and require a considerably bigger K than more sophisticated techniques. In the following sections we consider and compare several different methods.

Domain methods

When the basis functions fulfill the boundary conditions $\mathrm{L}f_0 = h$ and $\mathrm{L}f_k = 0$, the approximation f^0 automatically fulfills the given boundary condition $\mathrm{L}f^0 = h$. Consequently, the linear parameters a_k have to be optimized in such a way that also the field equations are approximately fulfilled. This is a linear optimization problem that can be reduced to a matrix equation with one of the methods presented in the following sections.

Instead of selecting a single basis function f_0 that fulfills the inhomogeneous boundary condition, one could also use a linear combination of several basis functions. This does not cause essential problems but it makes the formalism more complicated.

There are many different classes of basis functions that may be applied. Obviously, an appropriate choice is most important for the success of the entire method. Analytic methods use basis functions that fulfill the field equations in the entire domain. Such functions are difficult to find and often numerically expensive, i.e., the implementation and the numerical evaluation of such functions is time consuming. Moreover, such a basis may only be applied to a very limited class of problems. As soon as the geometry of the domain, the field equations, or the boundary conditions are modified, one must search for a new basis. Typically, such an analytic basis can only be found for very simple cases.

Since an analytic basis cannot be found for a general field solver, we focus on more simple classes of basis functions. Remember that we already have considered different classes of basis functions in the sections on series approximations for the approximation of a given function. We now may apply essentially the same types of functions (or fields). For example, when f is a complex, scalar field in the xy plane, we might expand it using a harmonic basis in the x and y directions, for example, functions of the form $\exp(ik_x x+ik_y y)$, where k_x and k_y are complex numbers. This is essentially a two-dimensional Fourier transform. Such an approach is the core of the spectral domain analysis. Since the series expansion (4.5) is truncated, we have to select a finite number of basis functions, i.e., a finite set of (k_x,k_y) pairs. Obviously, the success of such a basis depends on an appropriate selection of the (k_x,k_y) pairs.

Remember the advantages and drawbacks of orthogonal and non-orthogonal Fourier series. For obtaining a nice matrix equation, we would prefer an orthogonal basis, but with a non-orthogonal basis we have the chance of reducing the number of unknown parameters K in the series expansion (4.5). Note that orthogonality requires the definition of a scalar product. Since domain methods work in the domain D where the function or field is defined, the scalar product must be defined over D. When the geometry of D is not very simple, it is impossible to evaluate the scalar product analytically. Consequently, it is often too difficult to find an orthogonal basis. Therefore, the basis of methods for computational electromagnetics is often non-orthogonal, with an important exception. We have already seen in the second chapter that it is easy to find an orthogonal basis when the domain is subdivided in several subdomains and when each basis function is non-zero in a single subdomain only. Therefore, subdomain basis functions are often orthogonal, whereas entire domain basis functions are non-orthogonal. As we will see, subdividing a domain into subdomains is the core of the finite element method. Subdomain basis functions are also often applied in the method of moments. Moreover, one can consider finite difference methods as subdomain methods where the basis is often not explicitly defined. A subdomain basis implies the discretization of the given domain D and this is typical for all prominent domain methods. Note that one often defines more than one basis function in each subdomain. In this case, the different basis functions in one subdomain may not be orthogonal, which leads to a partially orthogonal basis.

Boundary methods

Instead of discretizing the domain D, we can also discretize its boundary ∂D. In this case, the basis functions must be selected in such a way that the field equations hold inside D, i.e., $Lf_0 = g$ and $Lf_k = 0$. In computational electromagnetics, the boundary conditions are usually much simpler than the field equations. Therefore, finding an appropriate basis for boundary methods is much more demanding than finding a basis for domain methods, but essentially, the procedure is the same.

First of all, one can work with either orthogonal or non-orthogonal basis functions. Now, the scalar product required for defining orthogonality must be defined on the boundary ∂D because the boundary method works on the boundary. Note that the basis functions are also defined in the domain D, but for the method it is unimportant whether the basis functions are orthogonal with respect to a scalar product defined in D. This will become clearer as soon as we consider the projection technique.

For domain methods we could distinguish between entire domain and subdomain basis functions. Obviously, one can also split the boundary ∂D into several parts ∂D_l and search for basis functions that are zero outside ∂D_l, i.e., one could distinguish between 'entire boundary' and 'sub-boundary' basis functions. Designers of boundary methods often consider the boundary as the working area, or the 'domain' where the scalar product is defined, and use the terms entire domain and subdomain instead of 'entire boundary' and 'sub-boundary'; this may be confusing for those who work with both domain methods and boundary methods.

The main problem of 'sub-boundary' basis functions is that the basis functions f_k must be defined in the entire domain D and must fulfill the homogeneous field equations $Lf_k = 0$. This makes it hard to find suitable 'sub-boundary' basis functions.

In computational electromagnetics, the Method of Moments (MoM) is a prominent technique that works with a discretization of the boundary ∂D in the loss-free case, but this technique is not a pure boundary method because it discretizes the domain D as soon as D is filled with lossy material. The reason for this is the following. Instead of working with the electromagnetic field, usual MoM implementations work with the sources s of the field, i.e., charges and currents. In the lossy case, one has a source distribution in the domain D and therefore D is discretized, whereas the sources are located on ∂D in the loss-free case. Therefore, one can say that MoM is a domain method that collapses to a boundary method in the loss-free case. The MoM expands the sources s and it applies a series expansion of the form (4.5) to s:

$$s = s^0 + error_s = \sum_{k=1}^{K} a_k s_k + error_s. \tag{4.6}$$

The field f is evaluated from s through integration of the form $f = L_s s$, where L_s is a linear operator. Thus, one indirectly also obtains (4.5). The domain D_s where the sources are non-zero may coincide with the domain D, with a subdomain of D or with the boundary ∂D. When D_s coincides with ∂D, one can apply basis functions s_k that are non-zero in a subdomain of $D_s = \partial D$ only. One may call this either a 'subdomain' basis with respect to D_s or a 'sub-boundary' basis with respect to ∂D. The corresponding basis of the field f is $f_k = L_s\, s_k$. In general, such a basis is non-zero in D and on ∂D. Thus, it is an entire domain basis with respect to D and an entire boundary basis with respect to ∂D.

Error minimization techniques

When we applied a series expansion to approximate a given function, we could determine the linear parameters a_k in such a way that the square error norm was minimized. We now can try the same procedure, i.e., we define the square error norm, insert the series expansion (4.5) in the field equations (4.1) and boundary conditions (4.2), and watch for a set of parameters that minimizes the square error norm. Before we do that, we decide whether we want to apply a domain method or a boundary method. Note that both methods are formally identical. Let us try a boundary method for a real valued function f. In this case, we must define the error on the boundary ∂D. The square error on the boundary can be defined as

$$e_{\partial D}^2 = \int_{\partial D} (f - f^0)^2 \, \mathrm{d}s = \int_{\partial D} (f - f_0 - \sum_{k=1}^{K} a_k f_k)^2 \, \mathrm{d}s \,, \tag{4.7}$$

where the superscript 0 denotes the numeric approximation. To minimize this error, we set all derivatives with respect to each of the parameters a_k equal to zero and obtain

$$\frac{\partial}{\partial a_i} e_{\partial D}^2 = \int_{\partial D} \frac{\partial}{\partial a_i} (f - f_0 - \sum_{k=1}^{K} a_k f_k)^2 \, \mathrm{d}s = \int_{\partial D} 2(f - f_0 - \sum_{k=1}^{K} a_k f_k) \frac{\partial}{\partial a_i} (\sum_{k=1}^{K} a_k f_k) \, \mathrm{d}s = 0 \,. \tag{4.8}$$

Here, the derivative and the integral were exchanged in the first step and the chain rule was applied in the second step. Moreover, the fact that neither f nor f_0 depends on a_i was used. The remaining derivative of the sum of basis functions is zero when i is different from k, and one obtains

$$\int_{\partial D} 2(f - f_0 - \sum_{k=1}^{K} a_k f_k) f_i \, \mathrm{d}s = 2 \int_{\partial D} (f - f_0) f_i \, \mathrm{d}s - 2 \sum_{k=1}^{K} a_k \int_{\partial D} f_k f_i \, \mathrm{d}s = 0 \,. \tag{4.9}$$

From this, one obtains the linear system of equations

$$\sum_{k=1}^{K} a_k \int_{\partial D} f_k f_i \, \mathrm{d}s = \sum_{k=1}^{K} a_k m_{ik} = \int_{\partial D} (f - f_0) f_i \, \mathrm{d}s \,. \tag{4.10}$$

Unfortunately, the right hand side of this system contains the unknown function f, which makes (4.10) useless. What can be done? Note that (4.10) has been derived without using the boundary conditions (4.2). We know that the field equations (4.1) are automatically satisfied by the basis functions, and that the parameters have to be computed in such a way that the boundary conditions hold as accurately as possible. Therefore, we now define the error of the boundary conditions (4.2) as follows:

$$e_{\partial D}^2 = \int_{\partial D} (\mathrm{L}f - \mathrm{L}f^0)^2 \, \mathrm{d}s = \int_{\partial D} (\mathrm{L}f - \mathrm{L}f_0 - \sum_{k=1}^{K} a_k \mathrm{L}f_k)^2 \, \mathrm{d}s \,. \tag{4.11}$$

Note that the terms Lf and Lf_0 are known. Lf is equal to h and Lf_0 is equal to g. Therefore, the unknown function f is eliminated and one finds the linear system of equations

$$\sum_{k=1}^{K} a_k \int_{\partial D} \mathrm{L}f_k \, \mathrm{L}f_i \, \mathrm{d}s = \sum_{k=1}^{K} a_k m_{ik} = \int_{\partial D} (h - g) \, \mathrm{L}f_i \, \mathrm{d}s = b_i \tag{4.12}$$

instead of (4.10). Note that the solution of this system of equations minimizes the square norm of the error of Lf^0 rather than of the error of the approximation f^0. In the special case of Dirichlet boundary conditions, L is the unit operator and both errors are identical. Note that this does not also minimize the square norm of the error of f^0 over the domain D.

For complex fields, the error definition (4.11) is slightly modified:

$$e_{\partial D}^2 = \int_{\partial D} (\mathrm{L}f - \mathrm{L}f^0) \cdot (\mathrm{L}f - \mathrm{L}f^0)^* \mathrm{d}s \tag{4.13}$$

where * indicates the conjugate complex. Afterwards, the procedure remains essentially the same. The same procedure may also be applied for domain methods. This leads to

$$\sum_{k=1}^{K} a_k \int_D Lf_k \, Lf_i \, \mathrm{d}s = \sum_{k=1}^{K} a_k m_{ik} = \int_D (g-h) \, Lf_i \, \mathrm{d}s = b_i \cdot \tag{4.14}$$

This system of equations minimizes the square norm of the error Lf^0 over the domain D. Note that field equations with a unit operator do not make much sense. Therefore, this error minimization never minimizes a square error norm of the approximation f^0 itself.

To obtain a linear system of equations we have modified the error definition. This can be considered as a generalization. A further generalization can be obtained by 1) inserting weights in the error definition, 2) using an arbitrary norm p instead of the square norm, and 3) replacing the operator L or L by an arbitrary operator L_e. Such a generalized error minimization technique does not necessarily lead to a linear system of equations that may be solved, as we have already seen above. The art of generalized error minimization mainly consists in finding an appropriate norm and an appropriate operator, whereas weighting does not cause problems.

Since the error minimization technique is quite difficult, methods that are more flexible and simpler are desirable.

Projection techniques

Remember that the square norm of a function may be defined through the scalar product: $\|f\|^2 = (f, f)$. The scalar product (f, g) may be interpreted as a projection of f on g. This offers a simple way to obtain scalars from functions and also from field equations and boundary conditions.

Let us start with the boundary method. First, we have to define a scalar product on the boundary ∂D. A common and simple definition of a scalar product for complex functions is

$$(f, g)_{\partial D} = \int_{\partial D} f \cdot g^* \mathrm{d}s \cdot \tag{4.15}$$

Note that * denotes the conjugate complex. This is necessary for complex functions because otherwise $\|f\|^2 = (f, f)$ would not be a correct definition of the square norm. In prominent books on the method of moments [Harrington, 1968] the conjugate complex is omitted from the product definition. Such a product is not a real scalar product, but it may also be applied to obtain matrix equations.

We now use the scalar product (4.15) to project both sides of the boundary condition (4.2) on a so-called testing function t_i:

$$(\mathrm{L}f, t_i)_{\partial D} = (h, t_i)_{\partial D} \cdot \tag{4.16}$$

Note that this equation is a scalar equation. To eliminate the unknown function f, we insert the series approximation (4.5) and obtain

$$
\begin{aligned}
&(\mathrm{L}f_0+\sum_{k=1}^{K}a_k\mathrm{L}f_k+\mathrm{L}error,t_i)_{\partial D} \\
&=(\mathrm{L}f_0,t_i)_{\partial D}+(\sum_{k=1}^{K}a_k\mathrm{L}f_k,t_i)_{\partial D}+(\mathrm{L}error,t_i)_{\partial D}=(h,t_i)_{\partial D}.
\end{aligned}
\tag{4.17}
$$

Note that *error* is the error function. When we project (4.2) on I testing functions, we obtain the following I linear equations with K unknowns:

$$
\begin{aligned}
&\sum_{k=1}^{K}a_k(\mathrm{L}f_k,t_i)_{\partial D}=\sum_{k=1}^{K}a_k m_{ik}= \\
&=(h,t_i)_{\partial D}-(g,t_i)_{\partial D}-(\mathrm{L}error,t_i)_{\partial D}=b_i+e_i\ ; \qquad i=1,2,\ldots,I.
\end{aligned}
\tag{4.18}
$$

This can also be written in the matrix form

$$
\mathrm{MA}=\mathrm{B}+\mathrm{E}\,. \tag{4.19}
$$

The matrix M contains the elements m_{ik}, the vector A the unknown parameters a_k, the vector B the known inhomogeneity b_i and the error vector E the error terms e_i that are unknown. The goal is to compute A in such a way that all elements of E are sufficiently small. It is known that a linear system of the form (4.19) can be solved with E=0 when the matrix M is square and non-singular. Therefore, one usually selects $I=K$ and omits the error vector E. Does this mean that an exact solution is obtained? The answer is certainly negative. First of all, the numeric evaluation of M and B will cause some errors. Even when these errors are negligible, i.e., when all elements of M and B are known with machine precision, the numeric solution of MA=B will have a limited accuracy depending on the algorithm that is used and depending on the properties of the matrix M. Finally, E=0 does not mean that the error term in the series expansion is zero. E=0 means that the projection of L *error* on all testing functions is zero. The square norm of the error function therefore is not necessarily minimized and the quality of the results depends on the choice of the testing functions.

What is the optimal choice of testing functions? To answer this question, we must define an error to be minimized. In the previous section we have seen that the square error norm of f on the boundary can be minimized when the boundary operator L is a unit operator. Otherwise, the square error norm of Lf is minimized when the system (4.12) is solved. When we compare (4.18) with (4.12) and keep in mind that the conjugate complex * is missing in (4.12) because real valued functions were considered there, we see that the testing functions t_i = L f_i make (4.18) identical to (4.12) when $I=K$ and $e_i = 0$ is set. When L is the unit operator, the choice $t_i = f_i$ is optimal in the sense that the square error norm is minimized. This is called Galerkin's choice of testing functions.

The scalar products contained in the elements of the matrix M and the vector B can not be evaluated analytically, especially when the basis functions and the testing functions are not very simple functions. When a very simple type of basis function is selected, it often turns out that the basis is not very well suited to accurately approximate the desired solution f. In this case, the number of unknowns K becomes very high when accurate results are to be obtained. Therefore, one may look for simple testing functions that permit analytic evaluations of the scalar products for an arbitrary basis. Dirac testing functions do this job. A Dirac delta δ function is zero everywhere, except at a single point, where its value is infinite. This is not really a function. It is a so-called distribution. Note that Dirac introduced the delta functions to simplify working with point charges and charge distributions. A point charge is a charge distribution that is zero everywhere, except at a point where the charge density is infinite. The infinite value of the charge density at this point is defined in such a way that the integral over the charge density is equal to the value of the corresponding point charge. Incidentally, this integral requires a new (Lebesgue) definition of the integral because the traditional Riemann integral definition fails. For a more detailed description, see mathematical textbooks on functional analysis, distributions, and functionals [Henrici, 1974-1986]. For our needs it is sufficient to define

$$(f(\vec{r}),\delta(\vec{r}-\vec{r}_\delta))_{\partial D} = \int_{\partial D} f(\vec{r})\cdot\delta(\vec{r}-\vec{r}_\delta)\,\mathrm{d}s = f(\vec{r}_\delta)\,. \tag{4.20}$$

The point $\vec{r}_\delta$ is the point where the Dirac function δ is singular. The Dirac function is singular when its argument is zero. Obviously, the projection of a given function on a Dirac function is the same as evaluating the value of the function at the point where δ is singular. This is nothing else than sampling the function at a given point $\vec{r}_\delta$. When we insert a set of Dirac testing functions $\delta(\vec{r}-\vec{r}_i)$ in (4.18), we can remove all scalar products, i.e., integrals contained in the elements of the matrix M and the vector B, and obtain

$$\sum_{k=1}^{K} a_k \mathrm{L}f_k|_{r_i} = \sum_{k=1}^{K} a_k m_{ik} = h(r_i) - g(r_i) - \mathrm{L}error|_{r_i} = b_i + e_i\ ; \qquad i = 1,2,\ldots,I\,. \tag{4.21}$$

The notation $\mathrm{L}f_k|_{\vec{r}_i}$ means that the operator L is applied to f_k and the result is evaluated in the point $\vec{r}_i$.

The procedure for domain methods is almost the same as for boundary methods. One defines a scalar product in the domain D,

$$(f,g)_D = \int_D f\cdot g^*\mathrm{d}s, \tag{4.22}$$

and uses it to project (4.1) on a set of testing functions t_i. After inserting the series expansion (4.5) one obtains

$$\begin{aligned}\sum_{k=1}^{K} a_k (Lf_k, t_i)_D &= \sum_{k=1}^{K} a_k m_{ik} \\ &= (g,t_i)_D - (h,t_i)_D - (Lerror,t_i)_D = b_i + e_i\ ; \qquad i = 1,2,\ldots,I.\end{aligned} \tag{4.23}$$

instead of (4.18).

Point matching, collocation

The Dirac testing functions in the previous section lead to a system of equations that can also be obtained directly by sampling the field equations (domain methods) or the boundary conditions (boundary methods). This is also called collocation or point matching. The points $\vec{r}_i$ are then called matching points.

Let us again start with the boundary method. The field equations in the domain D are fulfilled by an appropriate choice of the basis of the series expansion (4.5). To fulfill the boundary condition (4.2), we insert (4.5) in (4.2) and write down the resulting equation in a set of matching points $\vec{r}_i$. From this, we obtain (4.18) very directly without any definition of a scalar product and without introducing distributions, i.e., Dirac functions. The procedure for the domain method is analogous and needs no further explanation.

The standard point matching technique uses as many matching points as parameters, i.e., $I=K$. The resulting matrix M is square and the matrix equation (4.19) can be solved with E=0. As in the projection technique, this does not mean that the error of the solution is zero. When one applies the point matching technique to various problems, one finds that the accuracy of the results depends very much on the selection of the matching points. It is often very hard to find a good set of matching points, and with misplaced matching points one may obtain completely wrong results. A more careful analysis shows that useful results are mainly obtained when one can establish some correlation between the basis functions and the matching points. For example, when a subdomain basis is applied with a single basis function for each subdomain, a matching point should be placed near the center of each subdomain. This is a very rough rule that does not always guarantee good results.

Finding appropriate rules for setting the matching points is often very tricky and tedious. Therefore, the simple point matching technique is often replaced by a more sophisticated projection technique. An efficient alternative is obtained from the generalization of the point matching that is outlined in the following section.

Generalized point matching

Although it seems to make sense to set $I=K$ in the projection technique and in the point matching technique, this is not necessary at all because one can also solve underdetermined ($I<K$) and overdetermined ($I>K$) systems of equations. Especially overdetermined systems of equations turn out to be useful. Such systems are characterized by a matrix M with more rows than columns, i.e., more equations than unknowns. In this case, the error vector E in equation (4.19) may no longer be omitted. Note that $I>K$ can be applied to generalize projection techniques as well, but we now focus on point matching.

Since the error vector remains in the formalism, this allows one to observe what happens with it. First of all, an overdetermined system of equations of the form (4.19) has many different solutions with different error vectors. To make the solution unique, we specify that we search for the solution that minimizes the square norm of the error vector E. This is similar to what we did in the error minimization technique. There, we tried to minimize the square norm of an error function. Remember that the function L*error* was minimized when the boundary method was applied. This error function is now sampled at the matching points and the resulting values are contained in the error vector E.

The standard method to obtain a solution of (4.19) with a minimum square norm of E multiplies (4.19) with the adjoint matrix M*, i.e., the conjugate complex of the transposed matrix of M:

$$M^*MA=M^*B+M^*E\,. \tag{4.24}$$

Note that the matrix S=M*M is square and symmetric, i.e., it is of the same type as the matrix obtained from the error minimization technique. Equations (4.24) are called the normal equations of (4.19).

Let us analyze the elements of the square matrix S. Each of its elements is obtained as a scalar product of two columns of M:

$$s_{jk} = \sum_{i=1}^{I} m_{ji}^{*} m_{ik} \,. \tag{4.25}$$

In the special case of the point matched boundary method, we have

$$s_{jk} = \sum_{i=1}^{I} \mathrm{L}f_j^{*}|_{\vec{r}_i} \, \mathrm{L}f_k|_{\vec{r}_i} \,. \tag{4.26}$$

Comparing (4.26) with the matrix elements of the error minimization in (4.12), it is seen that the integral in (4.12) is replaced by a simple sum in (4.25). Note that the conjugate complex in (4.12) is missing because this equation was derived for real functions. Thus, the generalized point matching will give good results when the approximation of the integral by a sum is accurate. This is only the case when the matching points are uniformly distributed over the boundary, which is a strong constraint that is not desirable in general.

An improvement is obtained from the analysis of numerical evaluations of integrals. For reasons of simplicity, we consider a one-dimensional integral, but the procedure can easily be extended to two-dimensional integrals that are required for 3D boundary methods and for 2D domain methods as well as to three-dimensional integrals that are required for 3D domain methods. A reasonable good numeric evaluation of a one-dimensional integral is obtained from the trapezoidal algorithm. The integrand *f*(*x*) is approximated by a polygon as illustrated in Figure 25.

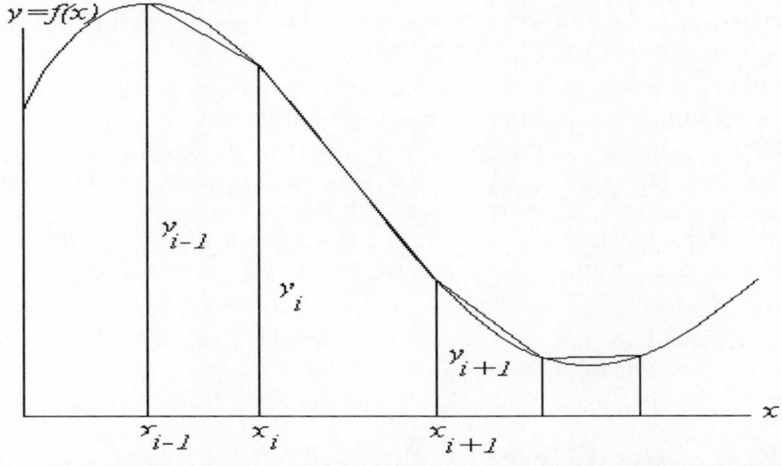

Figure 25: Trapezoidal integration of a function y=f(x) with non-equidistant sampling points.

The integral is approximated by a sum of trapezoidal areas of the size $(x_{i+1} - x_i) \cdot (y_{i+1} + y_i) / 2$. When we apply this to a closed, one-dimensional boundary integral, we obtain the approximation

$$\oint_{\partial D} f(s)\,\mathrm{d}s \approx \sum_{i=1}^{I} (f(s_{i+1}) + f(s_i)) \cdot (s_{i+1} - s_i) / 2 \cdot \tag{4.27}$$

Since the boundary is closed, one has $s_{I+1}=s_1$. The summation in (4.27) can be rearranged and one obtains

$$\oint_{\partial D} f(s)\,\mathrm{d}s \approx \sum_{i=1}^{I} f(s_i) \cdot (s_{i+1} - s_{i-1}) / 2 = \sum_{i=1}^{I} f(s_i) \cdot d_i, \tag{4.28}$$

where d_i is the average of the distances of the point s_i from its neighbors.

In order to obtain a trapezoidal approximation of the boundary integral in (4.12), we can slightly modify (4.26) by introducing a factor d_i that contains the distances between neighbor matching points:

$$s_{jk} = \sum_{i=1}^{I} \mathrm{L}f_j^*|_{\bar{r}_i} \mathrm{L}f_k|_{\bar{r}_i} \cdot d_i = \sum_{i=1}^{I} (\mathrm{L}f_j^*|_{\bar{r}_i} \cdot \sqrt{d_i})(\mathrm{L}f_k|_{\bar{r}_i} \cdot \sqrt{d_i}) \cdot \tag{4.29}$$

If the boundary integral is two-dimensional, the factor d_i obtains the size of an area, but the procedure and the formula (4.29) remain the same. Obviously, we obtain (4.29) from a generalized point-matched solution when we weight the equations (4.21) with a factor

$$w_i = \sqrt{d_i} \,. \tag{4.30}$$

Since this method also weights the components of the error vector E, i.e., the residuals, it is also called the method of weighted residuals. When the trapezoidal integration is replaced by some other algorithm, this may affect the definition of the weights w_i. With a special weighting and a locally more dense matching point distribution, one may also locally increase the accuracy of the results. As in adaptive numerical integration, the generalized point matching technique may also be implemented as an adaptive algorithm that refines the discretization, i.e., the locations of the matching points iteratively. Note that such an algorithm must also refine the basis functions because the basis has a big influence on the error distribution along the boundary.

It is very important to point out that one does not even need to set up the normal equations (4.24) for solving an overdetermined system in the least squares sense. There are direct algorithms that avoid the normal equations. Such algorithms lead to numerically much more accurate results when the matrix M has a high condition number, i.e., when it is ill-conditioned. The direct solution of overdetermined systems of equations is the key to accurate handling of ill-conditioned matrices. Since this problem is often misunderstood, it is explained in more detail below.

The generalization of the point matching technique for domain methods is straightforward and requires no explicit explanation.

Comparison of the different techniques

We have started with the error minimization technique which may be considered as a reference because obtaining results with minimum errors are desired. Unfortunately, this method turns out to be quite sophisticated and tricky and it only leads to a linear system of equations for special error definitions.

The second method, the projection technique, is more simple and always leads to a linear system of equations. The main problem with this technique is an appropriate choice of the testing functions that may have a big influence on the accuracy. We have seen that the same equations as with the error minimization technique can also be obtained from the projection technique when an appropriate choice of testing function is used. Unfortunately, matrix setup, i.e., the numeric evaluation of the elements of the matrix M and of the inhomogeneity B, becomes time-consuming when the integrals contained in these elements cannot be evaluated analytically. To avoid a time-consuming matrix setup, comfortable and reliable testing functions, e.g., Galerkin testing functions, are often replaced by more simple testing functions, especially Dirac testing functions which lead to less accurate results.

To overcome the problems of the point matching technique, i.e., the projection technique with Dirac testing functions, this method has been generalized by introducing overdetermined systems of equations and weighting of equations, which is also called the method of weighted residuals. It has been shown that this method may lead to the same equations as the error minimization technique with a numeric evaluation of the integrals involved in the matrix elements. Finding an appropriate weighting and an appropriate selection of the matching points corresponds to finding appropriate testing functions in the projection technique. The advantage of generalized point matching is that it is relatively easy to design adaptive algorithms that improve the weighting and the setting of the matching points, whereas an improvement of the testing functions in the projection technique leads quickly to scalar products that must be evaluated numerically.

If the scalar products in the elements of the matrix M of a projection technique are approximated numerically by a sum, one can proceed as in the previous section to show that M can be considered as a product of two rectangular matrices M=T*F, where T is a matrix that contains the sampled testing functions in its columns and F is a matrix that contains the sampled boundary conditions (boundary methods) or the sampled field equations (domain methods) in its columns. This means that the matrix of the projection technique can be decomposed when the integrals are replaced by summations. This is more general than what is obtained in the generalized point matching technique, where one has mainly T=F, but since this minimizes the square norm of the error vector, it is the optimal choice for most applications.

Ill-conditioned matrices

It has already been mentioned that the numerical evaluation of a matrix equation of the form (4.19) can cause numerical problems that reduce the accuracy of the results. The errors produced by the numeric matrix solver depend on both the algorithm and the properties of the matrix M. To be more precise, the condition number of the matrix M may be defined. Unfortunately, there is no unique definition of the condition number. What one would like to have is a characteristic number c for a given matrix M that indicates how accurate the results of (4.19) may be. There are some algorithms that include an estimate of such a number, but this estimate is usually very rough and rather useless. An example for such an algorithm is LINPACK's [Dongarra, 1979] Cholesky decomposition.

The most robust and accurate algorithm for solving (4.19) is the Singular Value Decomposition (SVD) [Dongarra, 1979], [Golub, 1983]. SVD algorithms transform the matrix M into a diagonal matrix that contains the so-called singular values of M. The ratio of the maximum and minimum singular value is a relatively good measure for the numerical problems caused by M, provided that the columns of M are scaled, i.e., provided that the square norms of all columns of M are identical. When M is not scaled, one should scale before SVD is applied. If we define the condition number c as the ratio of the maximum and minimum singular value of a scaled matrix M, we can apply SVD to evaluate c. Obviously, this evaluation is inaccurate when c is so high that also SVD becomes inaccurate. Since SVD is both time-consuming and memory-consuming, it makes little sense to evaluate the condition number before the matrix equation (4.19) is solved with an appropriate matrix solver that is faster and simpler than SVD. However, when one designs a new method, one should carefully analyze the matrices obtained for some typical cases and one should also evaluate the corresponding condition number.

To simplify the analysis, we return to the more simple case of the approximation and extrapolation of a given function f by a generalized series expansion of the form (4.5). This is essentially the same as a domain method with unit operator L or a boundary method with unit operator L. The advantage of this class of problems is that we can easily and accurately evaluate the error of the approximation.

The total error is caused 1) by the truncated series expansion and the selection of the basis, i.e., the inaccuracy of the model, 2) the inaccurate numerical matrix equation, i.e., the inaccuracy of the matrix solver, 3) by inaccuracies in the matrix setup, the numerical evaluation of the basis functions, etc. Since errors in the basis function evaluations, etc., are usually quite small, the total error is mainly caused by the first two parts. The goal of a numerical method is to obtain a small total error, whereas the condition number tries to estimate the loss of precision of the computation of the linear parameters a_k. Sometimes, the inaccuracy of the parameter computation does not considerably affect the error of the final result. For example, if two (or more) of the basis functions f_1 and f_2 are almost identical, c becomes very large. In this case, the matrix solver can break down completely, resulting in a total loss of accuracy. Robust algorithms like SVD and Givens [Dongarra, 1979] can still cope with such a situation. They will compute completely wrong parameters a_1 and a_2, corresponding to the almost identical basis functions f_1 and f_2, but this will not even cause an inaccurate computation of the function f, because $a_1 f_1 + a_2 f_2$ can provide an excellent approximation of $a'_1 f_1 + a'_2 f_2$, where a'_1 and a'_2 are the theoretically correct parameters. This becomes obvious in the extreme case $f_1 = f_2$, where c become infinite and identical results are obtained for an arbitrary value of a_2 when $a_1 - a_2 = a'_1$ and $a'_2 = 0$.

When implementing matrix solvers, one seems to be more interested in fast algorithms than in robust algorithms that can handle ill-conditioned matrices. For example, one of the most interesting algorithms for ill-conditioned matrices in the old LINPACK library is the QR updating procedure based on Givens plane rotations. The Givens algorithm is one of the few LINPACK algorithms that have been withdrawn when LAPACK was released. As only a few robust algorithms for ill-conditioned matrices are available, it seems to be good advice that a numerical method should be designed in such a way that c is relatively small. However, one should keep in mind that the condition number can cause inaccuracies, but these inaccuracies can be much less important than the inaccuracy of the model.

If we compare several matrix methods, we can make unexpected and very important observations. The "model accuracy" of well designed models with well designed matrix methods and larger condition numbers is often higher than that of well designed models with well designed matrix methods and smaller condition numbers. When the original method is optimally designed, techniques that can be applied for reducing the condition number reduce at the same time the "model accuracy". Improvements in matrix methods that increase the "model accuracy" typically increase also the corresponding condition number. Let us consider a few prominent examples. 1) It is well known that higher order FD or FE algorithms allow one to improve the accuracy. When one increases the order of an FE or FD scheme, one reduces the sparsity of the matrix, which increases the condition number. 2) In MoM codes, one often uses subdomain basis functions in order to reduce the condition number. At the same time, entire domain basis functions allow one to obtain more accurate results with the same number of unknowns, because the discontinuities of the derivatives of the subdomain basis functions slow down the convergence. 3) The MMP (Multiple Multipole Program) code for computational electromagnetics allows the user to compare the results of different models that can be characterized by different condition numbers. It has been shown that one usually obtains optimal results for relatively large condition numbers [Hafner, 1993]. So far, all techniques aimed at reducing the condition number of MMP matrices have caused a reduction of the accuracy as well.

Although these observations might be surprising, one can understand them by comparing models with a fixed number of unknown parameters. The restriction of the condition number by the matrix solver implies a reduction of the basis functions that can be used for the approximation of the function or field to be computed. This reduces the chance of finding an almost optimal set of basis functions for the given problem. Note that it is difficult to make any precise statement that holds in general. For example, the optimal set of basis functions would be a set containing the true solution and nothing else. This would lead to a matrix with one element only, and $c = 1$ could easily be obtained. However, in all interesting problems, we do not know the solution and can only work with basis functions that can be accurately evaluated with a known and sufficiently simple algorithm. This restricts the number of available basis functions. Usually, we choose only some of them for approximating the unknown function or field. Nonetheless, if we already have a good idea of how the result looks, we can select several basis functions that are close to this result. Since all of them will be similar, c becomes larger and the error smaller, the better we guess. Of course, we could then perform some orthogonalization of our basis to obtain $c = 1$, but this would cause the same numerical problems as the solution of the matrix equation and would lead to a set of complicated basis functions that cannot be accurately computed, i.e., we would increase the accuracy of the matrix solver and decrease the accuracy of the basis functions and of the matrix elements. Thus, we would rather shift our problems than solving them. However, this makes clear that one cannot simply say that ill-conditioned matrices allow one to obtain more accurate results, but matrix solvers that can handle ill-conditioned matrices *increase the chance* of efficiently finding more accurate results with a fixed number of parameters. This means that we have to formulate "heuristic observations" rather than exact theorems that can be proven by mathematical reasoning.

In order to demonstrate the interactions between the design of a matrix method, the condition number, and the accuracy, we now consider the orthogonal and non-orthogonal approximation and extrapolation of a given function. For example, one can approximate the most simple linear function $y=f(x)=x$ by an orthogonal Fourier series as in the second chapter. As one can see from Figure 26, Fourier series are almost useless for extrapolations.

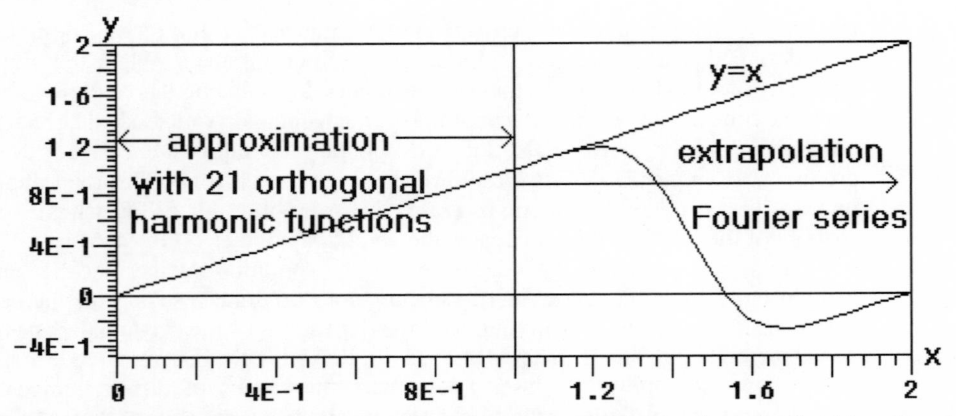

Figure 26: Approximation of the linear function $y=x$ in the interval $x=0...1$ with a Fourier series with a basis consisting of 21 orthogonal harmonic functions. Obviously the extrapolation for $x>1.2$ is useless.

Note that Fourier series even fail to extrapolate harmonic functions $y=f(x)=\cos(\omega x)$ or $y=f(x)=\sin(\omega x)$ when the frequency ω is not very close to one of the frequencies ω_k of the Fourier series. We have already seen in the second chapter that one can easily generalize Fourier series by admitting a non-orthogonal basis, i.e., arbitrary frequencies. This leads to a matrix equation for the evaluation of the linear Fourier amplitudes a_k and to a non-linear optimization problem for the frequencies ω_k. If we use a random generator to select a non-orthogonal harmonic basis, i.e., the type of each basis function (either cos or sin) and the corresponding frequency, we cannot expect to obtain a very good solution, but if we repeat the procedure and let the random generator select many sets of frequencies, we can hope that one of them is more useful than the orthogonal Fourier basis. For each set of frequencies we can easily compute the average of the error in the interval $x=0...1$ and the condition number of the corresponding matrix M. It is quite obvious that we obtain large condition numbers when two frequencies of such a set are very close to each other, whereas low condition numbers are obtained when the distances between the frequencies are large. For an orthogonal set of basis functions, the condition number is minimal, e.g., equal to one. The random search allows us to obtain some statistical probability to find accurate solutions depending on the condition number.

Analyzing Figure 27, we find that the probability of finding completely inaccurate solutions is very high and that the probability of finding accurate solutions is low, i.e., the density of solutions found with the random generator decreases rapidly with decreasing error. It seems that there are two maxima for the probability of the condition number: one near $c=100$ and the other near $c=10^8$. It is important to note here that single precision has been used throughout, and that an SVD algorithm was used for computing c. This does not allow one to accurately compute large condition numbers ($>10^7$). Large condition numbers are underestimated, i.e., replaced by values around 10^8. This explains the second maximum. Moreover, the location of the first maximum depends on the frequency range and on the number of frequencies. Looking at the low condition numbers in Figure 27 and the following figures, one can recognize that almost all solutions fulfill the condition $Error>1/c$. This means that it is extremely unlikely to find very accurate solutions with low condition numbers.

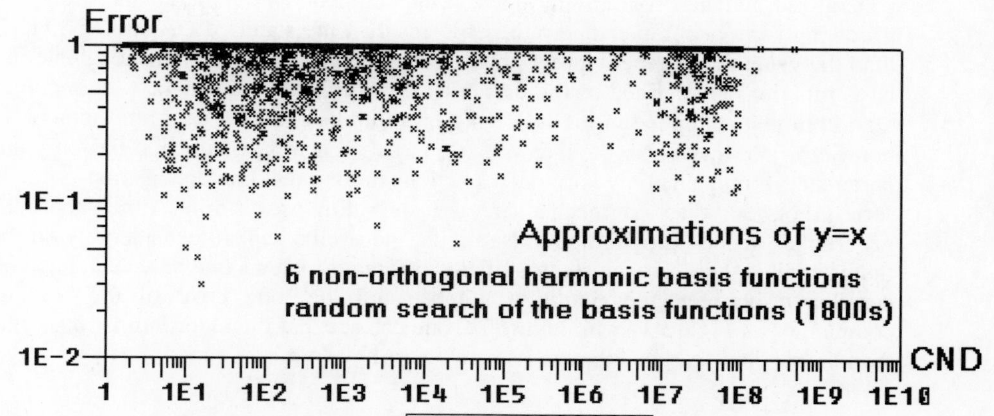

Figure 27: Averaged error $Error = \sqrt{\int_{x=0}^{x=1} (f(x) - f^0(x))^2 dx}$ *of the approximations* $f^0(x)$ *of* $y=f(x)=x$ *in the interval* $x=0...1$ *versus condition number, for various sets of six non-orthogonal harmonic basis functions with angular frequencies in the interval 0..20. The frequency spectra of the different sets have been set at random. The accuracy of the best basis found within 1800 seconds on a DEC AXP 150 personal computer is moderate (0.03).*

If either the number of frequencies is increased or the bandwidth of the frequency spectrum is reduced, the first maximum will be shifted towards bigger condition numbers. As one can see in Figure 28, a reduction of ω_{max} from 20 to 2 causes a huge shift towards higher condition numbers.

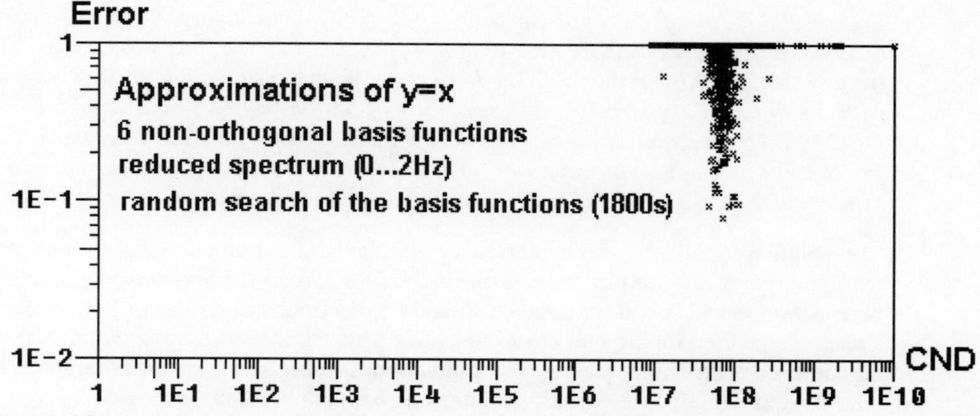

Figure 28: Error of the approximations of $y=x$ *in the interval* $x = 0...1$ *versus condition number, for various sets of six non-orthogonal harmonic basis functions with angular frequencies in the interval 0..2. The frequency spectra of the different sets have been set at random. The accuracy of the best basis found within 1800 seconds on a DEC AXP 150 personal computer is low (0.08).*

Since the random search is very slow and inefficient for finding an appropriate basis, we look for a better strategy. In the third chapter, genetic algorithms, genetic programming and generalized genetic programming have been outlined. The evolutionary strategy used in the following is more complex and more efficient than the standard GA strategy, but it is simpler than the generalized genetic programming because the basis functions available in the pool of basis functions are defined by the user and not optimized by a genetic programming code. The algorithm that is applied in the following picks up basis functions from the pool and selects a non-linear parameter p in the interval $p_{min} \dots p_{max}$. Each basis function has only one nonlinear parameter. If one puts only harmonic functions in the pool, one obtains non-orthogonal Fourier series. For the selection mechanism, the algorithm uses similar fitness definitions as the generalized genetic programming. Naturally, the results depend considerably on the algorithm that is applied, but from the study of many different problems one may get a good insight in the correlation between the condition number and the total error of the solution. From a comparison of Figure 27 with Figure 29, one can see that the algorithm is much more efficient than the random search, although its parameters have not been tuned.

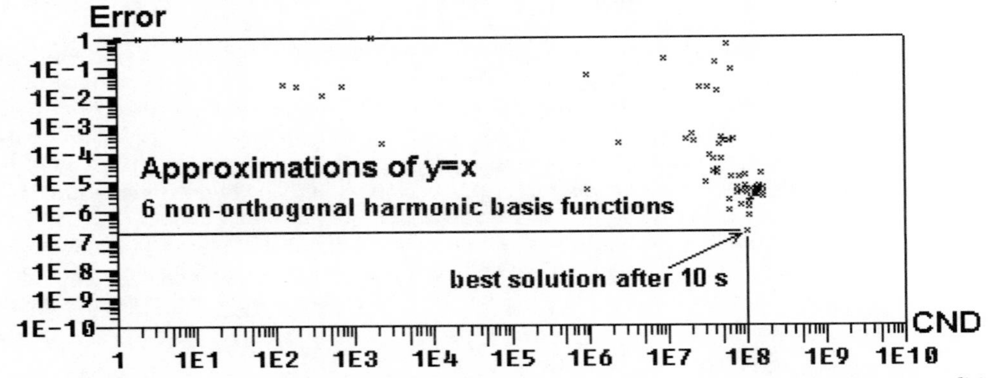

Figure 29: Error of the approximations of $y=x$ in the interval $x=0...1$ versus condition number for various sets of six non-orthogonal harmonic basis functions, with angular frequencies in the interval 0..20. The frequency spectra of the different sets have been selected with an evolutionary strategy. The accuracy of the best basis found within 10 seconds on a DEC AXP 150 personal computer is excellent, although the condition number is relatively high for single precision computations. The fitness criterion of the evolutionary search is "fitness=1/Error", which causes a search of a basis with minimum error.

The solutions found by the evolutionary strategy are extremely accurate and permit a good extrapolation. For example, for x=2 one has 0.01% error. Almost seven digits of accuracy have been obtained within a short time. (It should be mentioned that most of the computation time is "wasted" for the graphic output of the code.) This is close to the maximum accuracy that can be achieved with single precision computation. Needless to say, an orthogonal Fourier basis would require many more than six unknowns to obtain the same accuracy.

It is interesting to see that the condition number of the optimal solution is very large - especially if one keeps in mind that the SVD underestimates large condition numbers ($c>10^7$). In Figure 30 one can see that the large condition number is caused by two basis functions with very similar frequencies and that two of the six basis functions have very low amplitudes. This means that one could easily construct a set that would accurately approximate $y=x$ with only four basis functions.

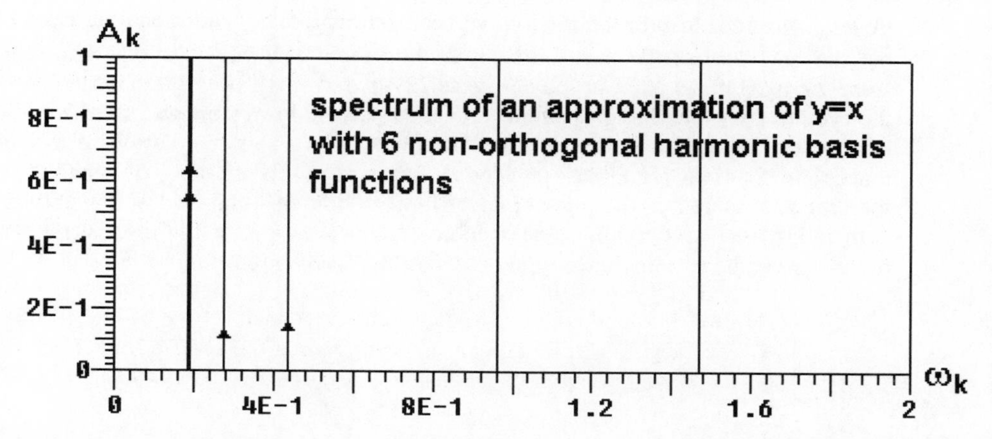

Figure 30: The spectrum of a good approximation of $y = x$ found by an evolutionary strategy with six non-orthogonal harmonic basis functions. Note that the amplitudes of the higher frequencies are so small that the corresponding basis functions have almost no influence on the solution, whereas the two lowest frequencies are almost identical, which causes a large condition number and an excellent approximation at the same time.

As in the random search, one can recognize that the inequality *Error*>1/*c* holds for most of the solutions, although the fitness definition of the individuals avoids extreme cancellations and condition numbers. In order to verify this inequality, one can increase the computation time (Figure 31).

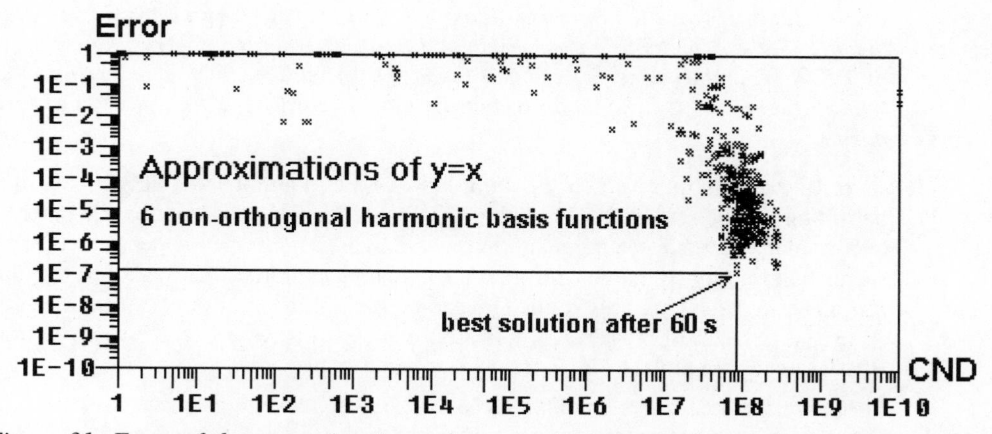

Figure 31: Error of the approximations of $y = x$ in the interval $x = 0...1$ versus condition number. Same computation as in Figure 29, after 60 seconds on a DEC AXP 150 personal computer. Note that all sets of basis functions with a low error have a relatively high condition number (single precision computation). For most of the approximations the criterion "Error > 1/CND" is valid.

If one considers the frequency spectrum of accurate solutions, one finds that there is a trend towards relatively low frequencies. The reduction of the frequency spectrum, i.e., the reduction of ω_{max} from 20 to 2 in the random search, did considerably increase the condition numbers but did not increase the probability of finding more accurate solutions as one might expect from *Error*>1/*c*. As one can see in Figure 32, the reduction of the frequency spectrum focuses the evolutionary search on a relatively small area, with low errors and high condition numbers. Obviously, the search becomes more efficient. More accurate solutions are found and very inaccurate solutions are almost avoided. It is surprising to see that - although the reduction of the spectrum increases the probability of large condition numbers - a few excellent solutions with moderate *c* and *Error*<1/*c* are found. Note that plots of the error versus the condition number describe extremely well the quality of the search algorithm.

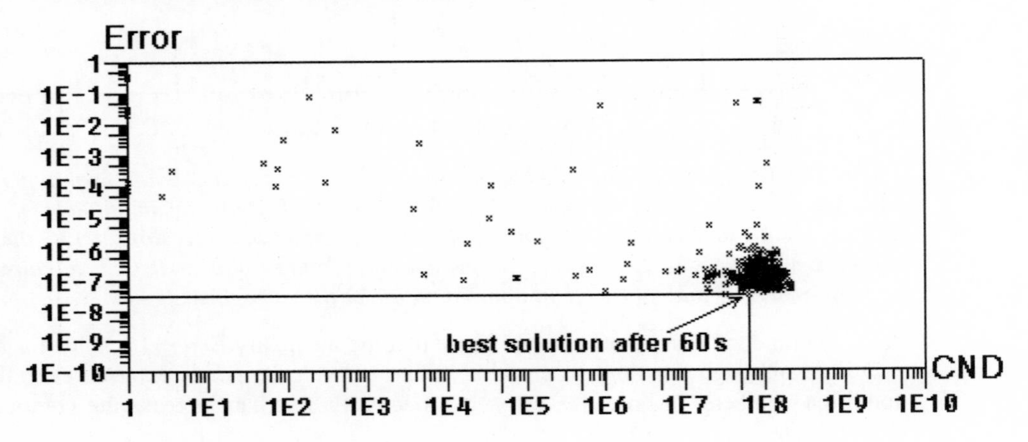

Figure 32: Error of the approximations of $y = x$ in the interval $x = 0...1$ versus condition number. Same computation as in Figure 31, with a reduced frequency spectrum (0...2 instead of 0...20). Note that some sets of basis functions with a low error have lower condition number than in the previous computations, but the most accurate solutions still have large condition numbers.

In order to find more solutions with low condition numbers, one can modify the fitness definition for the sets of basis function. Although this has an impact on the performance of the evolutionary search, accurate solutions with small condition numbers could not be found. It seems that finding solutions with *Error* < 1/CND is extremely difficult and that the probability for *Error* > 1/CND is very high. The same effects can be observed when the given function $f(x)=x$ is replaced by a more complicated one (see Figure 33) and when the non-orthogonal Fourier series are replaced by more general series expansions.

For the results that had been obtained so far, a Givens updating procedure was used to solve the overdetermined system of equations obtained from the generalized point matching procedure for computing the linear parameters of the series expansions. If we replace the Givens algorithm by the Cholesky algorithm and use the normal equations of the form (4.24), we find that the evolutionary search algorithm is heavily disturbed and performs much worse (see Figure 34). Although the computation time of the fitness has been reduced, the code has become much slower and it is no longer useful for obtaining very accurate solutions. The Cholesky algorithm breaks down for $c>10^5$ and the probability of finding solutions with *Error*

<1/c is low. Therefore, we have not found any solution with *Error* <10^{-4} within a reasonable computation time.

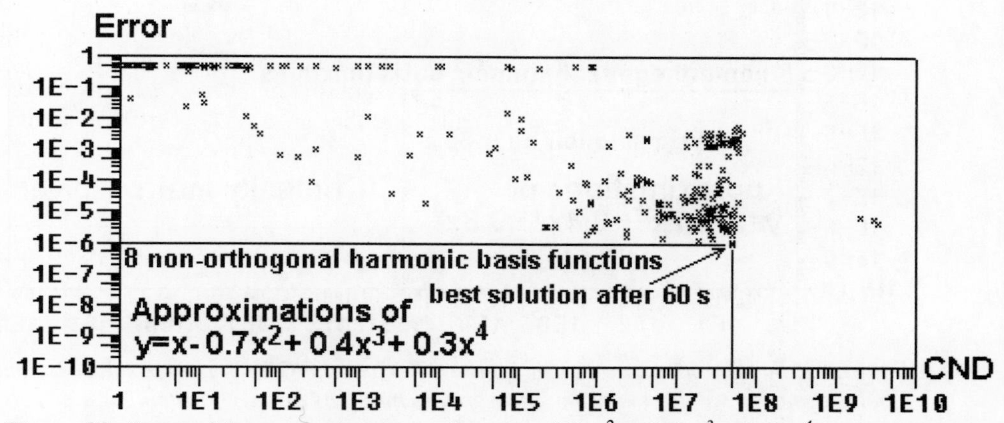

Figure 33: Error of the approximations of $y = x - 0.7\,x^2 + 0.4\,x^3 + 0.3\,x^4$ *in the interval* $x = 0...1$ *versus condition number. The same strategy as in Figure 31 has been applied. The number of non-orthogonal harmonic basis functions has been increased from six to eight.*

The loss of accuracy of a matrix solver cannot be easily estimated. Nonetheless, a comparison with a much more robust matrix solver can provide this information. In Figure 35, all matrix equations corresponding to the different sets of basis functions have been computed with both the Givens and the Cholesky algorithms. The lower end of each line in this figure indicates the error computed with the Givens algorithm, the upper end the error computed with the Cholesky algorithm. The length of each line indicates the loss of accuracy of the Cholesky algorithm. One can observe that the loss of accuracy is not a monotonic function of the condition number and that one has almost a total loss of accuracy in the area where one has a good chance of finding accurate solutions.

Although an accuracy of more than three digits is not always required, it should be pointed out that the Cholesky algorithm not only causes a loss of accuracy, it also disturbs the search procedure and causes an extreme reduction in performance. This is important not only for the specific example considered here, but also for many optimization procedures in computational physics.

For the reasons mentioned above, it is very bad advice to try to use a fast matrix solver first and to invent some technique for reducing the condition number if the matrix solver fails. It is much better to start with a robust matrix solver and to replace it by a faster algorithm only when one is sure that this does not cause any undesired loss of accuracy. Since robustness seems to be a much less attractive goal than speed for the designers of matrix solvers, much of the work in this area is going in the wrong direction. It is extremely important to fully understand that a matrix solver makes sense only in conjunction with a matrix method. Shifting the problems of large condition numbers from the matrix solver to the design of the matrix method is a bad attitude that might prevent designers of matrix methods from finding efficient techniques for solving problems in physics and engineering.

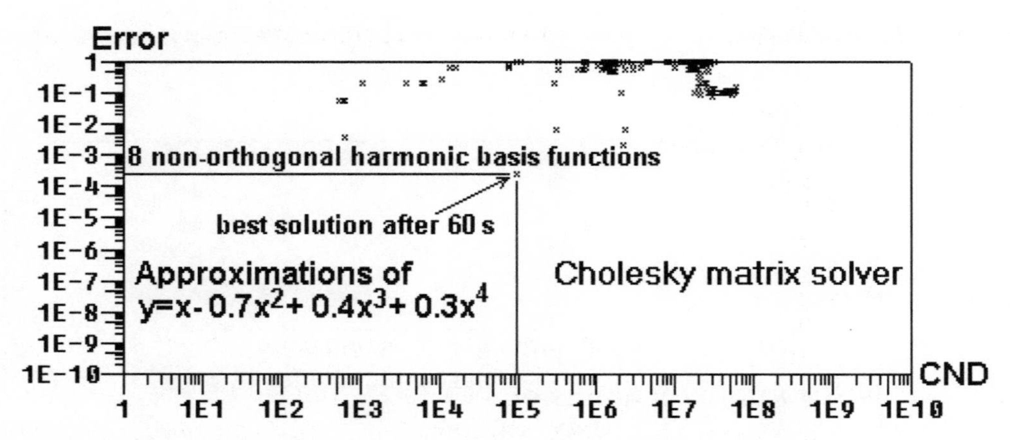

Figure 34: Same computation as in Figure 33, when the Givens matrix solver is replaced by a Cholesky matrix solver, i.e., when the normal equations are used instead of a direct solution of the overdetermined system of equations.

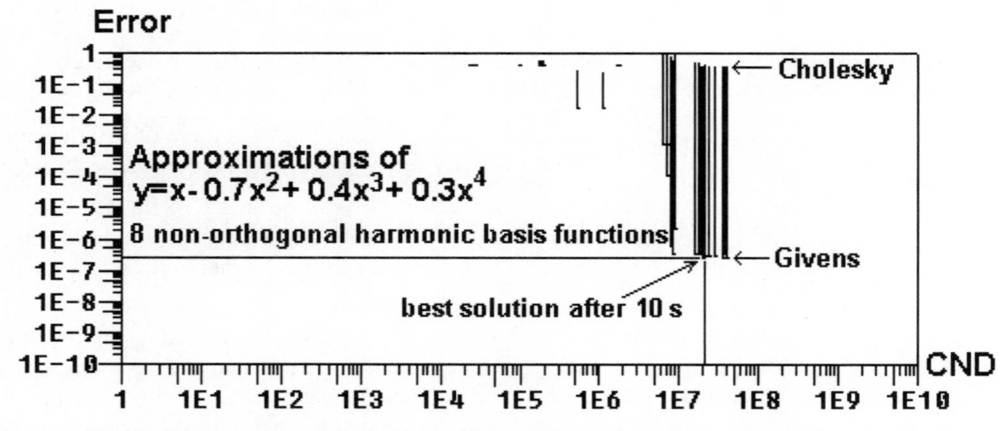

Figure 35: Same computation as in Figure 33, stopped after 10 s on a DEC AXP 150 personal computer. The vertical lines indicate the loss of accuracy when the Givens matrix solver is replaced by a Cholesky matrix solver. Note that, due to random effects in the search strategy, a higher accuracy than in Figure 33 has been achieved.

5 Maxwell theory

General remarks

Starting with Maxwell's treatise [Maxwell, 1891] there are many excellent textbooks on Maxwell theory and its applications, for example, [Jones, 1964], [Jackson, 1975], [Kong, 1990], [Misner, 1973], [Panowsky, 1975], [Stratton, 1941]. Those who are not interested in all aspects of Maxwell theory might prefer more specialized textbooks, for example, [Collin, 1991], [Haus, 1984], [Kapany, 1972], [King, 1956], [Lee, 1984], [Loewen, 1997], [Marcuse, 1991], [Ramo, 1965], [Smith, 1997], [Tamir, 1988]. This chapter presents the concepts and formulae that are most important for computational electromagnetics.

Force and field

For centuries, electric and magnetic forces were something of a topic of mysticism rather than science. Even at the time of Newton, nobody seems to have had a useful insight into the nature of these forces. Many different electrical phenomena were known, for example, lightning, electric sparks, electricity caused by friction, galvanic electricity, and animal electricity, but it was not at all obvious that all these phenomena could be explained by a single theory. So it took many years until scientists became convinced that there was a single force behind all electrical phenomena and that the laws of electricity had practically the same form as Newton's laws of gravitation. As Newton had introduced the idea of the point mass as the simplest configuration of mass, Coulomb and others introduced the point charge, which is a charge concentrated in a single point.

Electric force and field

The attraction of two point charges Q_0 and Q_1 was formulated as an 'action at a distance'. The value of the force was found to be proportional to both Q_0 and Q_1 and proportional to one over the square of the distance between Q_0 and Q_1. The direction of the force was exactly the direction from Q_0 to Q_1 (see Figure 36).

Thus, one could write for the force $\vec{F}_{01}$ produced by Q_0 and Q_1 acting on Q_0 :

$$\vec{F}_{01}(\vec{r}_0) = \frac{Q_0(\vec{r}_0) \cdot Q_1(\vec{r}_1) \cdot (\vec{r}_0 - \vec{r}_1)}{4\pi\varepsilon |\vec{r}_0 - \vec{r}_1|^3}, \tag{5.1}$$

where the factor $4\pi\varepsilon$ depends on the units. In this book, the MKSA system, i.e., the SI, is used throughout. The main difference between this law and the corresponding one for two point masses is that the charges Q_0 and Q_1 may be negative. Incidentally, this law can practically be derived by Einstein's theorem of 'simplicity'. Obviously, the simplest way of characterizing a point charge is by the use of a single scalar. If we assume the simplest Euclidean space, the force has to point in the direction from one charge to the other, for reasons of symmetry. If we suppose the action of one charge to be distributed uniformly into the surrounding space, we have the 'one over the square of the distance' law. Of course, this practically implies the idea of the electric field $\vec{E}$. The field of the charge Q_0 and Q_1 is

$$\vec{E}_1(\vec{r}_0) = \frac{Q_1(\vec{r}_1)\cdot(\vec{r}_0 - \vec{r}_1)}{4\pi\varepsilon|\vec{r}_0 - \vec{r}_1|^3} . \tag{5.2}$$

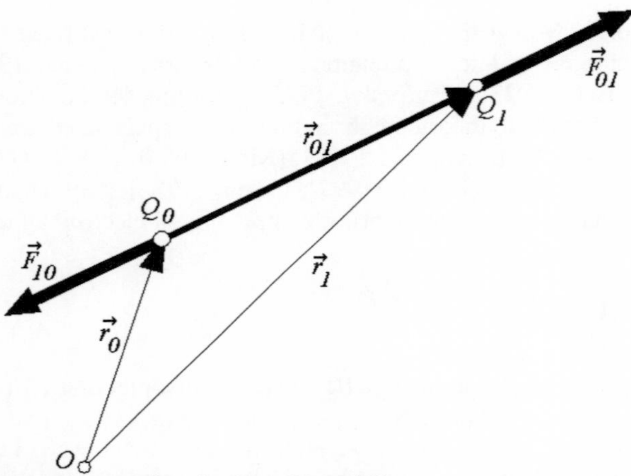

Figure 36: Interaction of two point charges Q_0 and Q_1.

For the field $\vec{E}$ produced by *N* charges $Q_1, Q_2, \ldots, Q_N$, the simplest idea is the principle of superposition:

$$\vec{E}(\vec{r}) = \sum_{n=1}^{N} \vec{E}_n(\vec{r}) = \sum_{n=1}^{N} \frac{Q_n(\vec{r}_n)\cdot(\vec{r} - \vec{r}_n)}{4\pi\varepsilon|\vec{r} - \vec{r}_n|^3} \tag{5.3}$$

and for the field produced by a continuous charge distribution ρ, one simply has to replace the summation above by an integral:

$$\vec{E}(\vec{r}) = \int \frac{\rho(\vec{r}\,')\cdot(\vec{r} - \vec{r}\,')}{4\pi\varepsilon|\vec{r} - \vec{r}\,'|^3} \mathrm{d}V' . \tag{5.4}$$

Finally, for the force on a point charge *Q*, produced by the electric field, the proportionality

$$\vec{F}(\vec{r}) = Q(\vec{r})\cdot\vec{E}(\vec{r}) \tag{5.5}$$

is very simple. Because nature behaves very kindly, these laws seem to be extremely exact, as many measurements have shown.

Einstein was probably the first one to point out very clearly that there is a difficulty in this theory: the space as well as the field and the charge distributions are continuous, but (1) the field of a point charge has a singularity and (2) a point charge cannot be considered to be a special charge distribution as this is not even a function. Although the second point seems to be worse, this difficulty can be removed by the well-known 'delta function' (δ) proposed by Dirac. Precisely speaking, δ is a distribution and not a function. To integrate δ, the usual Riemann integral has to be replaced by a Lebesgue integral. However, like Dirac, engineers do not worry about this point. We will encounter the δ function in the discussion of numerical methods.

The singularity of the electric field causes a singularity of the energy density and an infinite energy of the point charge. According to Einstein's most famous formula, $E=mc^2$ (where E is the energy and not the field), the mass m of a point charge would be infinite as well, which would cause an infinite gravitation. Thus, a point charge is physically impossible. Nonetheless, this concept is very powerful within the theory of electricity, which simply ignores gravitational forces and mathematical inconsistencies because (1) practically gravitational forces are much smaller than electric forces in many cases and (2) engineers usually have no time to wait for mathematical proofs and prefer the 'trial and error' principle. This is especially true for numerical field computations, where almost no useful mathematical theorems yet exist. As a consequence, numerical codes have to be designed as flexibly and open-mindedly as possible. 'Rigorous' algorithms may be useful at first, but later on they can show severe deficiencies which are hard to repair.

Magnetic force and field

The study of magnetism was more complicated than the study of electricity. First of all, the number of known magnetic phenomena was low and it was very hard to obtain strong magnets. The breakthrough came at the beginning of the nineteenth century, when Oersted showed experimentally that electric currents cause forces on magnets. A few months later, Ampère introduced his famous law and it suddenly became clear that electric currents are the sources of magnetism as electric charges are the sources of electricity. As currents are essentially directed one-dimensional objects, one obtains similar laws by introduction of vector sources and vector products where necessary. For a current distribution $\vec{j}$, one finds for the magnetic flux density:

$$\vec{B}(\vec{r}) = \int \vec{j}(\vec{r}\,') \times \frac{\mu \cdot (\vec{r} - \vec{r}\,')}{4\pi |\vec{r} - \vec{r}\,'|^3} \mathrm{d}V\,', \tag{5.6}$$

which appears similar to (5.4), and to obtain the force on a current element $I\,\mathrm{d}\vec{\ell}$ of infinitesimal length $\mathrm{d}\ell$, produced by the magnetic field,

$$\vec{F}(\vec{r}) = I\,\mathrm{d}\vec{\ell}(\vec{r}) \times \vec{B}(\vec{r}) \tag{5.7}$$

appears similar to (5.5). As a current consists of moving charges, one can replace $I\,\mathrm{d}\vec{\ell}$ in this equation by $Q\vec{v}$ to have the force acting on a point charge Q moving with the velocity $\vec{v}$:

$$\vec{F}(\vec{r}) = Q\vec{v}(\vec{r}) \times \vec{B}(\vec{r}). \tag{5.8}$$

Although one can have moving charges, the theory above is essentially static. In fact, most scientists at that time were not interested in the time dependence of the fields - except for Faraday. He had shown by numerous experiments that many different effects such as sparks, galvanic actions, heating of wires, and 'animal attraction' were only different faces of one and the same electricity. He was convinced that there are interactions among all other physical forces as well. Some of his experiments failed, but he was on the right track as his famous law of induction showed. This was the first dynamical law in the history of electricity. The next important step came with Maxwell, who was able to formulate mathematically Faraday's ideas of the electrodynamic field. It is well known that Maxwell's theory brought a new theory of light as well. That light is an electromagnetic phenomenon was an idea of Faraday, who showed experimentally that magnetic fields may affect the propagation of light. This effect could not be explained by Maxwell!

Faraday and Maxwell introduced a completely new concept of the electromagnetic field. Charges and currents were considered to be nothing more than terms to characterize a certain behavior of the field. The field, however, became a physical object rather than a helpful tool. The pure Maxwellians - most of them were British scientists - very soon had problems with the concept of the electron and finally with quantum mechanics. Today, Maxwell's theory is understood by most people in a 'pre-Maxwellian' sense rather than in the pure Maxwellian sense. Engineers especially prefer the idea of charge and current distributions as the 'real' things, even if they need not deal with quantum effects. Of course, most dynamic effects such as Faraday's induction can be explained by 'pre-Maxwellian' ideas as well, but the corresponding integrals are usually derived from Maxwell's equations. This will be done in the following sections.

Differential and integral forms of Maxwell equations

3D+1D: traditional time domain formalism

In most textbooks on electrodynamics for engineers, one can find the following differential form of Maxwell's equations:

$$\operatorname{curl} \vec{E}(\vec{r},t) = -\frac{\partial}{\partial t}\vec{B}(\vec{r},t), \tag{5.9}$$

$$\operatorname{curl} \vec{H}(\vec{r},t) = \vec{j}(\vec{r},t) + \frac{\partial}{\partial t}\vec{D}(\vec{r},t), \tag{5.10}$$

$$\operatorname{div} \vec{D}(\vec{r},t) = \rho(\vec{r},t), \tag{5.11}$$

$$\operatorname{div} \vec{B}(\vec{r},t) = 0. \tag{5.12}$$

This notation reflects the practical meaning of each equation rather than the known and unknown parts of the field. On the right-hand side, one has the causes, and on the left-hand side the effect:

1. The change of the $\vec{B}$ field causes a curl of the $\vec{E}$ field, which is Faraday's law of induction.

2. Both the electric current $\vec{j}$ and the change of the $\vec{D}$ field cause a curl of the $\vec{H}$ field. This is Ampère's law with an additional term $\frac{\partial}{\partial t}\vec{D}$, which has been introduced by Maxwell for reasons of consistency.

3. Electric charges are the sources of the $\vec{D}$ field. This is called Coulomb's law.

4. No magnetic object equivalent to the electric charge (i.e., no magnetic monopole) has been observed. The search for magnetic monopoles continues. Magnetic charges and the corresponding magnetic currents can be introduced for 'more beauty', i.e., a symmetric form of Maxwell's equations:

$$\operatorname{curl}\vec{E}(\vec{r},t) = \vec{j}_m - \frac{\partial}{\partial t}\vec{B}(\vec{r},t), \tag{5.13}$$

$$\operatorname{curl}\vec{H}(\vec{r},t) = \vec{j}_e(\vec{r},t) + \frac{\partial}{\partial t}\vec{D}(\vec{r},t), \tag{5.14}$$

$$\operatorname{div}\vec{D}(\vec{r},t) = \rho_e(\vec{r},t), \tag{5.15}$$

$$\operatorname{div}\vec{B}(\vec{r},t) = \rho_m(\vec{r},t). \tag{5.16}$$

Note that the introduction of magnetic currents and magnetic charges as 'fictitious' sources outside a given domain may be advantageous for the numerical modeling of the electomagnetic field inside the given domain. MMP and similar methods therefore often work with solutions of the symmetic Maxwell equations.

If everything is linear, one can simply solve the symmetric form of Maxwell's equations by superposition of two parts, one without magnetic charges and currents, and one without electric charges and currents. Formally, one has for both parts a system of partial differential equations identical to the original Maxwell equations (5.9) to (5.12).

To understand the meaning of Maxwell's equations, it is helpful to integrate them by means of the theorems of Stokes and Gauss (1.122), (1.123):

$$\int_{\partial S}\vec{E}(\vec{r},t)\,\mathrm{d}\vec{\ell} = -\int_{S}\frac{\partial}{\partial t}\vec{B}(\vec{r},t)\,\mathrm{d}\vec{S}, \tag{5.17}$$

$$\int_{\partial S}\vec{H}(\vec{r},t)\,\mathrm{d}\vec{\ell} = \int_{S}\left(\vec{j}(\vec{r},t) + \frac{\partial}{\partial t}\vec{D}(\vec{r},t)\right)\mathrm{d}\vec{S}, \tag{5.18}$$

$$\int_{\partial V}\vec{D}(\vec{r},t)\,\mathrm{d}\vec{S} = -\int_{V}\rho(\vec{r},t)\,\mathrm{d}V, \tag{5.19}$$

$$\int_{\partial V}\vec{B}(\vec{r},t)\,\mathrm{d}\vec{S} = 0. \tag{5.20}$$

As there is no difficulty even for functions that are only piecewise continuous, these integral forms are still more general than the differential forms.

From the mathematical point of view, one usually prefers to have all the unknowns on the left-hand side of an equation. If we want to do this, we have to see that the electric current density and the charge density usually consist of a known and an unknown part:

$$\vec{j} = \vec{j}_0 + \vec{j}_\sigma, \tag{5.21}$$

$$\rho = \rho_0 + \rho_\sigma. \tag{5.22}$$

$\vec{j}_0$, ρ_0 are 'impressed' currents and charges; $\vec{j}_\sigma$, ρ_σ are the unknown currents and charges in conducting media. When we rearrange the Maxwell equations in such a way that the unknown fields are on the left hand side, we have:

$$\operatorname{curl}\vec{E}(\vec{r},t) + \frac{\partial}{\partial t}\vec{B}(\vec{r},t) = 0, \tag{5.23}$$

$$\operatorname{curl}\vec{H}(\vec{r},t) - \vec{j}_\sigma(\vec{r},t) - \frac{\partial}{\partial t}\vec{D}(\vec{r},t) = \vec{j}_0(\vec{r},t), \tag{5.24}$$

$$\operatorname{div}\vec{D}(\vec{r},t) - \rho_\sigma(\vec{r},t) = \rho_0(\vec{r},t), \tag{5.25}$$

$$\operatorname{div}\vec{B}(\vec{r},t) = 0. \tag{5.26}$$

Thus, we have two homogeneous and two inhomogeneous partial differential equations. Note that, in certain cases, parts of the electromagnetic field are considered to be known and have to be moved to the right-hand side of the equations.

For a deeper insight, let us look at the schematic form of Maxwell's equations. As only 3D spatial derivatives have been used, we have to move all spatial derivatives on one side and all remaining terms on the other side to obtain:

$$\vec{E} \xrightarrow{\text{curl}} -\frac{\partial}{\partial t}\vec{B} \tag{5.27}$$

$$\vec{j} + \frac{\partial}{\partial t}\vec{D} \xleftarrow{\text{curl}} \vec{H} \tag{5.28}$$

$$\rho \xleftarrow{\text{div}} \vec{D} \tag{5.29}$$

$$\vec{B} \xrightarrow{\text{div}} 0 \tag{5.30}$$

As in Chapter 1, we 'complete' this scheme by introduction of potentials and zeros where possible:

$$\vec{E} \xrightarrow{\text{curl}} -\frac{\partial}{\partial t}\vec{B} \xrightarrow{\text{div}} 0 \tag{5.31}$$

$$0 \xleftarrow{\text{div}} \vec{j} + \frac{\partial}{\partial t}\vec{D} \xleftarrow{\text{curl}} \vec{H} \tag{5.32}$$

$$\rho \xleftarrow{\text{div}} \vec{D} \tag{5.33}$$

$$\vec{A} \xrightarrow{\text{curl}} \vec{B} \xrightarrow{\text{div}} 0 \tag{5.34}$$

If we look at these equations, we may see that the fourth can almost be derived by the first one. We have a system of loosely coupled equations with four unknown vector fields. As these equations are linear, we can have new equations by linear combinations. If we recognize that only equations with derivatives 'in the same direction' can be combined, we have few possibilities as we have two pairs of equations. As time derivatives are also linear operators, we may introduce such derivatives as well. This allows for two interesting things:

1. We can add the time derivative of the fourth equation to the first. This produces a zero at the position of $\vec{B}$. If we complete this new equation, we get the scalar potential ϕ:

$$-\phi \xrightarrow{\text{grad}} \vec{E}+\frac{\partial}{\partial t}\vec{A} \xrightarrow{\text{curl}} 0. \tag{5.35}$$

The negative sign of ϕ is usual but not necessary. The conventional form of the first part of this equation is

$$\vec{E}(\vec{r},t) = -\text{grad}\,\phi(\vec{r},t) - \frac{\partial}{\partial t}\vec{A}(\vec{r},t). \tag{5.36}$$

2. We can subtract the time derivative of the third equation from the second one and obtain the law of charge conservation:

$$-\frac{\partial}{\partial t}\rho \xleftarrow{\text{div}} \vec{j} \tag{5.37}$$

or, in the conventional form:

$$\text{div}\,\vec{j}(\vec{r},t) = -\frac{\partial}{\partial t}\rho(\vec{r},t). \tag{5.38}$$

4D formalism outlined

Up to now, we have considered the electromagnetic field as a time-dependent field in 3D space. In the differential form of Maxwell's equations we have first order derivatives with respect to both space and time. The unification of Maxwell theory [Maxwell, 1891] with classical mechanics led Einstein to the special theory of relativity [Misner, 1973]. The mathematician Minkowski presented a 4D 'space-time' formalism for the special theory of relativity that is also useful for a new formulation of Maxwell theory. The 4D formulation is useful for a profound theoretical analysis of the structure of Maxwell's equations, but most of the configurations in engineering fit much better into the 3D space + 1D time concept.

The 4D formalism starts with 4D vectors that contain both space and time. There are several ways to do this. For example, one can define a 4D space-time vector as follows:

$$(x^0,x^1,x^2,x^3) := (ct,x,y,z), \tag{5.39}$$

where c denotes the speed of light in free space. In a similar way, one can also pack the sources $\rho, \vec{j}$ of the field and the potentials $\phi, \vec{A}$ into 4D vectors **J** and **A**. Moreover, one can pack the $\vec{E}$ and $\vec{B}$ field in an antisymmetric 4×4 tensor of the form

$$\mathbf{F} := \begin{bmatrix} 0 & E_x & E_y & E_z \\ -E_x & 0 & -B_z & B_y \\ -E_y & B_z & 0 & -B_x \\ -E_z & -B_y & B_x & 0 \end{bmatrix}. \tag{5.40}$$

Similarly, one can pack the $\vec{D}$ and $\vec{H}$ field in an antisymmetric 4×4 tensor $\boldsymbol{F}$. One then can define derivatives in the 4D space. Remember that one can have 0D, 1D, 2D, and 3D objects in

3D space and that the first order spatial derivatives link n-dimensional objects with objects of dimension n+1 or n-1 (see Chapter 1). One has the three operators grad, div, and curl in 3D space. Now, one has to define four different operators in 4D space. The exterior calculus of Grassmann and Cartan [Misner, 1973] provides a generalization for n-dimensional spaces. Since this calculus has a high abstraction level, we skip it and present a schematic form of Maxwell's equations in 4D space without going into the details:

$$\mathbf{A} \rightarrow \mathbf{F} \rightarrow 0 \tag{5.41}$$

$$0 \leftarrow \mathbf{J} \leftarrow \boldsymbol{F} \tag{5.42}$$

The first equation contains the homogeneous Maxwell equations (5.9) and (5.12) and the definition of the 4D vector potential, whereas the second equation contains the inhomogeneous Maxwell equations (5.10) and (5.11) and the charge conservation law (5.37). Obviously, this notation is much more compact and its structure is also simpler than the 3D+1D notation presented in the previous section.

3D: frequency domain, t separation

There are several ways to derive the Maxwell equations for time-harmonic fields. One can simply use a time-harmonic expression for every component F of the field in the following form:

$$F(\vec{r},t) = \Re(F(\vec{r}) \cdot e^{-i\omega t}), \tag{5.43}$$

where $F(\vec{r})$ is the complex amplitude of $F(\vec{r},t)$ and $\Re$ denotes the real part. For mathematicians, it is wrong to use the same symbol for two different functions. However, this notation is convenient because one usually has a large number of different functions, and $F(\vec{r})$ and $F(\vec{r},t)$ are very closely related.

Note that the term $-i$ in the phasor convention (5.43) is sometimes replaced by the imaginary unit j. Here, the $-i$ convention is preferred, because it allows one to obtain a formalism, where most of the important complex quantities may be defined in the first quadrant of the complex plane. In textbooks that use the j convention, these quantities are either in the second or in the fourth quadrant.

The angular frequency ω may be chosen to be either real or complex. As the components of the electric and magnetic field, the components of the 'sources' ρ, $\vec{j}$, and the potentials are linked by Maxwell's and related equations, all components must have the same time dependence and especially the same frequency.

For most technical applications, the time dependence is given by a device called a 'generator'. For non-harmonic cases, there may still be good reasons to work in the frequency domain (i.e., to consider the fields to be a superposition of harmonic parts). This is especially the case if the material properties depend on the frequency. To move from the time domain to the frequency domain and *vice versa*, one can apply either the Laplace or Fourier transform [Papoulis, 1962]. There are several, slightly different forms of these transformations. For example,

$$F(\omega) = \frac{1}{\sqrt{2\pi}} \int_{-\infty}^{+\infty} f(t)\, e^{+i\omega t}, \tag{5.44}$$

$$f(t)=\frac{1}{\sqrt{2\pi}}\int_{-\infty}^{+\infty}F(\omega)\,e^{+i\omega t}, \tag{5.45}$$

where ω can be either real or complex. (5.45) is the inverse transformation.

For time-harmonic fields, the time derivatives may simply be replaced by multiplication with $-i\omega$. Thus, for Maxwell's equations, we have

$$\operatorname{curl}\vec{E}(\vec{r})=i\omega\vec{B}(\vec{r}), \tag{5.46}$$

$$\operatorname{curl}\vec{H}(\vec{r})=\vec{j}(\vec{r})-i\omega\vec{D}(\vec{r}), \tag{5.47}$$

$$\operatorname{div}\vec{D}(\vec{r})=\rho(\vec{r}), \tag{5.48}$$

$$\operatorname{div}\vec{B}(\vec{r})=0. \tag{5.49}$$

Numerical methods based on the complex Maxwell equations (5.46)-(5.49) are called *frequency-domain* methods, whereas methods working on the real, time-dependent Maxwell equations (5.9)-(5.12) are called *time-domain* methods. The Fourier transform (5.44) is the mathematical tool that links these two classes of numerical methods. It should be pointed out that the numerical evaluation of the Fourier integrals is less trivial than one might believe because of the simplicity of the formula (5.44).

2D: cylindrical structures, z separation

Applications with cylindrical geometry can be formulated completely in any transverse plane. Cylindrical geometry is especially assumed to study the propagation of guided waves on transmission lines, waveguides, optical fibers, etc. Although the fields of such structures vary along the axis, these are called two-dimensional, because the complex amplitudes of the field in a single transverse plane and a scalar, called the propagation constant γ, completely determine the wave if harmonic time dependence is assumed. Incidentally, a more rigid definition of two-dimensional fields is sometimes used. For example, R. F. Harrington [Harrington, 1968] considers only three-dimensional fields that do not vary with respect to the axis z of the cylinder (i.e., fields with $\gamma = 0$) to be two-dimensional fields. Of course, this simplifies the formalism, but, at the same time, the two-dimensional problems are restricted to the relatively simple and not very interesting case of scattering with vertically incident waves.

If waves on cylindrical structures are studied, one can separate the z-dependence like the time dependence for time-harmonic fields by the following expression:

$$F(\vec{r}_T,z,t)=\Re(F(\vec{r}_T)\cdot e^{i(\gamma z-\omega t)}), \tag{5.50}$$

where the index T indicates the transverse plane as in Chapter 1.

Usually, 2D problems are studied as if they were 3D problems. In addition, as all time derivatives for time-harmonic fields are replaced by the factor $-i\omega$, one can now replace all z-derivatives by the factor $i\gamma$. As the z-derivatives are 'inside' the spatial derivatives, this is not trivial. First, one has to write the spatial derivatives in general cylindrical coordinates and to distinguish transverse and longitudinal components of all the vectors. This leads to a longitudinal and a transverse part for the first two Maxwell equations. From the Maxwell equations (5.9)-(5.12) one obtains with (1.91) and (1.90):

$$\text{grad}_T\ E_z(\vec{r}_T) = i\gamma \vec{E}_T(\vec{r}_T) + i\omega \vec{B}_T^o(\vec{r}_T), \tag{5.51}$$

$$\text{div}_T\ \vec{E}_T^o(\vec{r}_T) = -i\omega B_z(\vec{r}_T), \tag{5.52}$$

$$\text{grad}_T\ H_z(\vec{r}_T) = i\gamma \vec{H}_T(\vec{r}_T) - i\omega \vec{D}_T^o(\vec{r}_T) + \vec{j}_T^o(\vec{r}_T), \tag{5.53}$$

$$\text{div}_T\ \vec{H}_T^o(\vec{r}_T) = i\omega D_z(\vec{r}_T) - j_z(\vec{r}_T), \tag{5.54}$$

$$\text{div}_T\ \vec{D}_T(\vec{r}_T) = -i\gamma D_z(\vec{r}_T) + \rho(\vec{r}_T), \tag{5.55}$$

$$\text{div}_T\ \vec{B}_T(\vec{r}_T) = -i\gamma B_z(\vec{r}_T), \tag{5.56}$$

where the index T denotes the part of the vector in the transverse plane, i.e., perpendicular to the z axis. Remember that the superscript o denotes a 90 degree rotation in the transverse plane as defined in (1.76).

Low frequencies and static fields

At sufficiently low frequencies the terms $\omega \vec{D}$ and $\omega \vec{B}$ become negligible. This simplifies the Maxwell equations (5.46)-(5.49) and (5.51)-(5.56), but it also has a more drastic effect. The Maxwell equations are no longer coupled as before. To understand this, one must know that several of the fields are linked through the material properties that had not been considered up to now. To proceed, we now must consider the effects of different media on the electromagnetic field.

Note that static Maxwell equations are obtained either from the limit $\omega \to 0$ or directly from the time-dependent Maxwell equations (5.9)-(5.12) by setting the time derivatives equal to zero. In (5.46)-(5.49) the fields are complex, whereas the fields in (5.9)-(5.12) are real. In the limit $\omega \to 0$ the complex fields degenerate to real ones.

Simple material properties

If we consider the Maxwell equations that were outlined in the previous sections, we see a system of coupled first order differential equations, but no way to decouple and solve them. Remember the schematic way in which similar but simpler systems of partial differential equations have been solved in the first chapter. There we combined derivatives in one direction (right or left) with derivatives in the opposite direction. In order to do so, we need equations that relate the different fields in one row of the scheme. These equations are given by the properties of the material.

For simplicity, we start with linear, homogeneous, isotropic materials with the following proportionalities:

$$\vec{D} = \varepsilon \vec{E}, \tag{5.57}$$

$$\vec{B} = \mu \vec{H}, \tag{5.58}$$

$$\vec{j}_\sigma = \sigma \vec{E}, \tag{5.59}$$

where ε is the permittivity, μ the permeability, and σ the conductivity of the material. Note that many media that are frequently used can be described quite accurately by equations (5.57)-

(5.59). Often, one can even neglect the conductivity and set $\vec{j}_\sigma = 0$. Such media are called *loss-free*. A special, loss-free medium is *free space*. It is characterized by the permittivity ε_0 and the permeability μ_0. The size of these constants depends on the system of units. In the SI or MKSA system, one has $\varepsilon_0 \approx 8.8542 \cdot 10^{-12}$ farad/m and $\mu_0 \approx 1.2566 \cdot 10^{-6}$ henry/m.

Often, free space is considered as a reference and one writes $\varepsilon = \varepsilon_r \varepsilon_0$ and $\mu = \mu_r \mu_0$, where ε_r and μ_r denote the relative permittivity and the relative permeability.

Time-dependent fields

The equations (5.57)-(5.59) operate vertically in our scheme and allow us to eliminate the $\vec{D}$, $\vec{B}$, $\vec{j}_\sigma$ terms in (5.31)-(5.34):

$$\vec{E} \xrightarrow{\text{curl}} -\frac{\partial}{\partial t}\mu\vec{H} \xrightarrow{\text{div}} 0 \tag{5.60}$$

$$0 \xleftarrow{\text{div}} \vec{j}_0 + \sigma\vec{E} + \frac{\partial}{\partial t}\varepsilon\vec{E} \xleftarrow{\text{curl}} \vec{H} \tag{5.61}$$

$$\rho \xleftarrow{\text{div}} \varepsilon\vec{E} \tag{5.62}$$

$$\vec{A} \xrightarrow{\text{curl}} \mu\vec{H} \xrightarrow{\text{div}} 0 \tag{5.63}$$

or, in the time-dependent Maxwell equations:

$$\operatorname{curl}\vec{E}(\vec{r},t) = -\frac{\partial}{\partial t}\mu\vec{H}(\vec{r},t), \tag{5.64}$$

$$\operatorname{curl}\vec{H}(\vec{r},t) = \vec{j}_0(\vec{r},t) + \sigma\vec{E}(\vec{r},t) + \frac{\partial}{\partial t}\varepsilon\vec{E}(\vec{r},t), \tag{5.65}$$

$$\operatorname{div}\varepsilon\vec{E}(\vec{r},t) = \rho(\vec{r},t), \tag{5.66}$$

$$\operatorname{div}\mu\vec{H}(\vec{r},t) = 0. \tag{5.67}$$

Note that $\frac{\partial}{\partial t}\mu\vec{H} = \mu\frac{\partial}{\partial t}\vec{H} + \vec{H}\frac{\partial}{\partial t}\mu$ holds. When the material property μ is not time-dependent, the second term is zero. Similarly, one has for a permittivity ε that is not time-dependent $\frac{\partial}{\partial t}\varepsilon\vec{E} = \varepsilon\frac{\partial}{\partial t}\vec{E}$. In most cases of practical importance, the time-dependence of the material properties is negligible. Note that ε and μ are also contained in the div operator of (5.10) and (5.12). Here, one can use the equation (1.104). Within a homogeneous domain, the gradient of ε and μ is zero and one obtains

$$\operatorname{curl}\vec{E}(\vec{r},t) = -\mu\frac{\partial}{\partial t}\vec{H}(\vec{r},t), \tag{5.68}$$

$$\operatorname{curl}\vec{H}(\vec{r},t)=\vec{j}_0(\vec{r},t)+\sigma\vec{E}(\vec{r},t)+\varepsilon\frac{\partial}{\partial t}\vec{E}(\vec{r},t), \tag{5.69}$$

$$\operatorname{div}\vec{E}(\vec{r},t)=\rho(\vec{r},t)/\varepsilon, \tag{5.70}$$

$$\operatorname{div}\vec{H}(\vec{r},t)=0, \tag{5.71}$$

for homogeneous materials that do not change their properties over time.

3D time-harmonic fields

For time-harmonic fields, we can replace the Maxwell equations (5.46)-(5.49) by

$$\operatorname{curl}\vec{E}(\vec{r})=i\omega\mu\vec{H}(\vec{r}), \tag{5.72}$$

$$\operatorname{curl}\vec{H}(\vec{r})=\vec{j}_0(\vec{r})+\sigma\vec{E}-i\omega\varepsilon\vec{E}(\vec{r})=\vec{j}_0(\vec{r})-i\omega\varepsilon'\vec{E}(\vec{r}), \tag{5.73}$$

$$\operatorname{div}\vec{E}(\vec{r})=\rho(\vec{r})/\varepsilon, \tag{5.74}$$

$$\operatorname{div}\vec{H}(\vec{r})=0, \tag{5.75}$$

for homogeneous materials that do not change their properties over time. Note that it is convenient to combine the conductivity σ and the permittivity ε into a complex permittivity ε' for describing lossy dielectrics with a single, complex number:

$$\varepsilon'=\varepsilon+\frac{i\sigma}{\omega}. \tag{5.76}$$

One can also model lossy magnetic material by a complex permeability μ', but for most of the lossy magnetic materials of practical interest this is a very rough approximation.

When we form the divergence of (5.72), we obtain (5.75). Therefore, (5.75) is implicitly solved with (5.72). When we apply divergence to (5.73), we obtain

$$\operatorname{div}\vec{E}(\vec{r})=\vec{j}_0(\vec{r})/i\omega\varepsilon'. \tag{5.77}$$

This equation should be the same as (5.74). Therefore, we obtain

$$\rho_\sigma(\vec{r})/\varepsilon=\rho(\vec{r})/\varepsilon-\rho_0(\vec{r})/\varepsilon=\vec{j}_0(\vec{r})/i\omega\varepsilon'-\rho_0(\vec{r})/\varepsilon, \tag{5.78}$$

where the index 0 denotes the impressed charge and current densities that are assumed to be known. Therefore, (5.78) allows one to compute the unknown charge density ρ_σ in a lossy medium. Thus, (5.78) and the equation (5.75) that was used to derive (5.78) are no constraints for the electromagnetic field. To compute $\vec{E}$, $\vec{H}$ in a linear, homogeneous, isotropic, time-constant medium, it is sufficient to solve the two Maxwell equations (5.72) and (5.73).

2D time-harmonic fields

When we remove the $\vec{D}$, $\vec{B}$, $\vec{j}_\sigma$ terms in the 2D Maxwell equations (5.51)-(5.56), we obtain for time-constant material properties:

$$\operatorname{grad}_T E_z(\vec{r}_T)=i\gamma\vec{E}_T(\vec{r}_T)+i\omega\mu\vec{H}^o_T(\vec{r}_T), \tag{5.79}$$

$$\mathrm{div}_T\ \vec{E}_T^o(\vec{r}_T) = -i\omega\mu H_z(\vec{r}_T), \tag{5.80}$$

$$\mathrm{grad}_T\ H_z(\vec{r}_T) = i\gamma \vec{H}_T(\vec{r}_T) - i\omega\varepsilon' E_T^o(\vec{r}_T) + \vec{j}_{0T}^{\,o}(\vec{r}_T), \tag{5.81}$$

$$\mathrm{div}_T\ \vec{H}_T^o(\vec{r}_T) = i\omega\varepsilon' E_z(\vec{r}_T) - j_{0z}(\vec{r}_T), \tag{5.82}$$

$$\mathrm{div}_T\ \vec{E}_T(\vec{r}_T) = -i\gamma E_z(\vec{r}_T) + \rho(\vec{r}_T)/\varepsilon, \tag{5.83}$$

$$\mathrm{div}_T\ \vec{H}_T(\vec{r}_T) = -i\gamma H_z(\vec{r}_T). \tag{5.84}$$

For the reasons mentioned above (see equations (5.74) and (5.75)), we can drop (5.83) and (5.84). The latter is fulfilled automatically and the former may be used to compute ρ_σ.

When one computes guided waves or propagating waves, there are no impressed charges and currents, i.e., the terms ρ_0, $\vec{j}_0$ may be omitted. In this case, the 2D Maxwell equations (5.81), (5.82) reduce to

$$\mathrm{grad}_T\ H_z(\vec{r}_T) = i\gamma \vec{H}_T(\vec{r}_T) - i\omega\varepsilon' E_T^o(\vec{r}_T), \tag{5.85}$$

$$\mathrm{div}_T\ \vec{H}_T^o(\vec{r}_T) = i\omega\varepsilon' E_z(\vec{r}_T). \tag{5.86}$$

Applying the fact that a 'derivative of a derivative in the same direction vanishes', which has been discussed in the first chapter, we can find several other formulations of first order 2D equations. Which of them we select as the set of 2D Maxwell equations is not unique. For example, we can only select the divergence equations:

$$\mathrm{div}_T\ \vec{E}_T(\vec{r}_T) = -i\gamma E_z(\vec{r}_T), \tag{5.87}$$

$$\mathrm{div}_T\ \vec{H}_T(\vec{r}_T) = -i\gamma H_z(\vec{r}_T), \tag{5.88}$$

$$\mathrm{div}_T\ \vec{E}_T^o(\vec{r}_T) = -i\omega\mu H_z(\vec{r}_T), \tag{5.89}$$

$$\mathrm{div}_T\ \vec{H}_T^o(\vec{r}_T) = -i\omega\varepsilon' E_z(\vec{r}_T). \tag{5.90}$$

Note that (5.87) and (5.88) contain 'left' derivatives, whereas (5.89) and (5.90) contain 'right' derivatives of the 2D electric and magnetic vector fields.

With 2D integrals corresponding to the theorems of Gauss and Stokes, we obtain

$$\oint_{\partial F} \vec{E}_T^0(\vec{r}_T)\,\mathrm{d}\vec{\ell} = -\int_F i\gamma E_z(\vec{r}_T)\,\mathrm{d}S, \tag{5.91}$$

$$\oint_{\partial F} \vec{H}_T^0(\vec{r}_T)\,\mathrm{d}\vec{\ell} = -\int_F i\gamma H_z(\vec{r}_T)\,\mathrm{d}S, \tag{5.92}$$

$$\oint_{\partial F} \vec{E}_T(\vec{r}_T)\,\mathrm{d}\vec{\ell} = +\int_F i\omega\mu H_z(\vec{r}_T)\,\mathrm{d}S, \tag{5.93}$$

$$\oint_{\partial F} H_T(\vec{r}_T)\,\mathrm{d}\vec{\ell} = +\int_F i\omega\mu E_z(\vec{r}_T)\,\mathrm{d}S\,\cdot \tag{5.94}$$

The 2D Maxwell equations may be 'solved' for the transverse components of the field (i.e., the transverse components can be expressed by simple derivatives of the longitudinal components), according to

$$\vec{E}_T(\vec{r}_T)=\frac{i\gamma}{\kappa^2}\operatorname{grad}_T E_z(\vec{r}_T)-\frac{i\omega\mu}{\kappa^2}\operatorname{grad}_T^o H_z(\vec{r}_T), \tag{5.95}$$

$$\vec{H}_T(\vec{r}_T)=\frac{i\gamma}{\kappa^2}\operatorname{grad}_T H_z(\vec{r}_T)+\frac{i\omega\varepsilon'}{\kappa^2}\operatorname{grad}_T^o E_z(\vec{r}_T), \tag{5.96}$$

$$\vec{E}_T^o(\vec{r}_T)=\frac{i\gamma}{\kappa^2}\operatorname{grad}_T^o E_z(\vec{r}_T)+\frac{i\omega\mu}{\kappa^2}\operatorname{grad}_T H_z(\vec{r}_T), \tag{5.97}$$

$$\vec{H}_T^o(\vec{r}_T)=\frac{i\gamma}{\kappa^2}\operatorname{grad}_T^o H_z(\vec{r}_T)-\frac{i\omega\varepsilon'}{\kappa^2}\operatorname{grad}_T E_z(\vec{r}_T). \tag{5.98}$$

where

$$\kappa=\sqrt{k^2-\gamma^2}=\sqrt{\omega^2\mu\varepsilon'-\gamma^2}=\sqrt{\omega^2\mu\varepsilon+i\omega\mu\sigma-\gamma^2} \tag{5.99}$$

is the transverse wave number. The propagation constant γ is the longitudinal wave number. One can see immediately from these equations that, for the special cases of TE- or H-waves with $E_z = 0$ and TM- or E-waves with $H_z = 0$, the electric and magnetic fields have to be orthogonal everywhere.

Obviously, the longitudinal components of the electric and magnetic fields somehow play the role of scalar potentials. Once these fields are known, the transverse components can be obtained from (5.95) and (5.96).

Static fields

It has already been mentioned that the static Maxwell equations are obtained either from the limit $\omega \to 0$ or directly from the time-dependent Maxwell equations (5.9)-(5.12) by setting the time derivatives equal to zero. When the material properties are inserted in the resulting static equations, one can observe that not all of the equations are coupled with each other. For each of the three material properties ε, μ and σ one obtains a set of two coupled Maxwell equations, namely

The electrostatic equations

$$\operatorname{curl}\vec{E}(\vec{r})=0, \tag{5.100}$$

$$\operatorname{div}\vec{D}(\vec{r})=\rho(\vec{r}), \tag{5.101}$$

$$\vec{D}(\vec{r})=\varepsilon\vec{E}(\vec{r}). \tag{5.102}$$

When (5.102) is inserted in (5.101) for a homogeneous domain, one essentially obtains the equations (1.34) and (1.35) that were discussed in Chapter 1. Therefore, the handling of these equations is straightforward. Note that the charges are assumed to be ‘impressed’, i.e., known in electrostatics.

The magnetostatic equations

$$\operatorname{curl}\vec{H}(\vec{r})=\vec{j}(\vec{r}), \tag{5.103}$$

$$\operatorname{div}\vec{B}(\vec{r})=0, \tag{5.104}$$

$$\vec{B}(\vec{r})=\varepsilon\vec{H}(\vec{r}). \tag{5.105}$$

When (5.105) is inserted in (5.104) for a homogeneous domain, one essentially obtains the equations (1.32) and (1.33) that were discussed in Chapter 1. Therefore, the handling of these equations is straightforward. Note that the currents are assumed to be 'impressed', i.e., known in magnetostatics.

The current equations

$$\operatorname{curl} \vec{E}(\vec{r}) = 0, \tag{5.106}$$

$$\operatorname{div} \vec{j}(\vec{r}) = 0, \tag{5.107}$$

$$\vec{j}(\vec{r}) = \sigma \vec{E}(\vec{r}). \tag{5.108}$$

Note that the current density in these equations is an unknown field. In the previous sections the index σ was used for this kind of current. The handling of these equations is the same as for the electrostatic equations with impressed charge density equal to zero or for the magnetostatic equations with impressed current density equal to zero.

Complex media

The description of media by three scalar numbers that was considered in the previous section is a simplification that is often but not always useful. Whether a more sophisticated description is required depends on the material and on the desired accuracy of the results. In the following, some typical extensions are outlined. For more detailed descriptions, see [Kong, 1990].

Anisotropic media

In anisotropic media, the vectors $\vec{E}$, $\vec{D}$ and $\vec{H}$, $\vec{B}$ are not parallel to each other. Consequently, the permittivity ε and the permeability μ are 3×3 tensors rather than scalars. Therefore, nine scalar numbers are required for the description of ε and μ in general. Fortunately, most of the anisotropic media may be described by simpler tensors. When the tensors are symmetrical, the medium is *reciprocal*. Moreover, either ε or μ can be a scalar. If ε is a tensor, but μ a scalar, the medium is *electrically anisotropic*. If μ is a tensor, but ε a scalar, the medium is *magnetically anisotropic*.

Crystals are well described as electrically anisotropic, reciprocal media, i.e., with a *symmetric* tensor ε. The symmetry reduces the independent tensor components to six. Moreover, a symmetric 3×3 matrix can be transformed into a diagonal one when an appropriate transform of the coordinate system is applied. In the so-called *principal system*, the matrix is diagonal, i.e., entirely described by three scalar elements. Often two of these three elements are equal, i.e., one has only two independent scalar elements. Such crystals are called *uniaxial*.

Note that anisotropic media may be either loss free or lossy. Lossy anisotropic media can easily be described in the frequency domain by complex tensors. From the frequency domain, the Fourier transform leads to the time domain.

Chiral media

Organic matter often consists of molecules that are either right handed or left handed. Typical examples are sugar and DNA. The former is a small molecule, the latter a very big one. Such molecules act on the electromagnetic fields like helical antennas, i.e. they cause some dependence between the electric and magnetic fields. Isotropic, chiral media may be described by the following material properties in the time domain:

$$\vec{D} = \varepsilon \vec{E} - \chi \frac{\partial}{\partial t} \vec{H}, \tag{5.109}$$

$$\vec{B} = \mu \vec{H} + \chi \frac{\partial}{\partial t} \vec{E}, \tag{5.110}$$

where χ is the chiral parameter. The notation of these properties is straightforward. When all molecules of a chiral medium are oriented in the same direction, the chiral medium becomes anisotropic. Moreover, the molecules may absorb electromagnetic energy. Then, one best describes the properties in the frequency domain with complex numbers.

Bi-anisotropic media

Bi-anisotropic media may be considered as a mixture and generalization of the anisotropic and chiral cases. For such media, the constitutive relations are

$$\vec{D} = \varepsilon \vec{E} - \xi \vec{H}, \tag{5.111}$$

$$\vec{B} = \mu \vec{H} + \zeta \vec{E}. \tag{5.112}$$

The analytic handling of such media is very demanding. Often, considerable simplifications can be made. For example, the matrices may be diagonal.

Inhomogeneous media

Most natural materials are macroscopically homogeneous, at least for frequencies that are not too high. This means that the material properties inside the material are constant, i.e., the gradient of each scalar component that describes the material is zero. Human technology can create inhomogeneous materials, for example, graded index fibers.

In general, each of the scalar components that describe the material properties of inhomogeneous media is a function of the location in space rather than a constant. For example, one has $\varepsilon(\vec{r})$ instead of a constant ε in a graded index fiber. As a consequence, the Maxwell equations (5.70) and (5.71) become more complicated.

It is possible to find analytic solutions in inhomogeneous media when the spatial function that describes the material properties is sufficiently simple and may be described by an analytic formula. In more difficult cases, analytic solutions cannot be found. To obtain numerical solutions, one splits inhomogeneous domains in sufficiently small subdomains and assumes that each subdomain is homogeneous. Since subdividing of domains is typical for domain methods, it is natural to apply domain methods when the media are inhomogeneous.

Time-dependent media

Time-dependent media cause slightly more difficult equations than the Maxwell equations (5.68) and (5.69) because the time derivatives of ε and μ are non-zero. This has no great consequences for time-domain methods. When frequency-domain methods are used, the Fourier transform is applied to the material properties defined in the time domain, which makes the procedure more difficult. Therefore, it is simpler to use time-domain methods when time-dependent media are present.

In practice, the time-dependence of the material properties may almost always be neglected because the time variation of the material properties is very slow compared with the time variation of the fields.

Frequency-dependent media

The electromagnetic effects that are macroscopically described by the material properties are caused by the interaction of the electromagnetic field with the molecules, atoms, and especially electrons. From this, one can easily understand that matter 'reacts' on the change of the field with some inertia and also with some resonance effects. These effects depend on the frequency of the field that is incident on the atoms. As a matter of fact, most media exhibit a pronounced frequency-dependence. For example, conducting metals lose their conductivity at sufficiently high frequencies and become transparent.

Obviously, the handling of frequency-dependent materials is easy for frequency-domain methods, but more difficult for time-domain methods. The formulation of such materials in the time domain typically leads to convolution integrals that can be numerically very demanding.

Whether the frequency-dependence of some material can be neglected or not depends on the material, on the desired accuracy, and on the frequency range of interest. Especially at high frequencies (near optic frequencies), the frequency dependence is often pronounced, whereas it can often be neglected at lower frequencies and at extremely high frequencies.

Non-linear media

Non-linear media can no longer be described by linear material properties of the form (5.57)-(5.59). For the most simple, loss-free, isotropic, homogeneous case, $\vec{D}$ is a function of the form $\vec{D}(\vec{E})$ and $\vec{B}$ is a function of the form $\vec{B}(\vec{H})$. More general cases can be much more difficult. For example, iron exhibits magnetic hysteresis effects that cannot easily be described by a function of the form $\vec{B}(\vec{H})$.

In general, no analytic solutions can be found for non-linear media, except in very special and simple situations. The main problem of handling non-linear media with numeric techniques is caused by the fact that most numerical methods work with a linear series expansion of the form (3.3). Therefore, the non-linear material properties must be linearized. Since the electromagnetic field is different at different points of space, the linearization depends on the location, which leads to the same procedure as the handling of inhomogeneous media with the additional difficulty that one first starts with a guess of the strength of the field for obtaining an appropriate linearization. When the results have been obtained with this guess, one can improve the guess and iterate the procedure. Therefore, domain methods with iterative matrix solvers are certainly the first choice for solving problems with non-linear media. Note that it may be difficult to guarantee that the iterative procedure is convergent toward the desired result.

Circuit theory and lumped elements

Electronic circuits are usually considered as systems of lumped elements connected by wires. Although electromagnetic fields act on the circuits and in the elements, one entirely describes the circuit by objects that are obtained from integrals over the electromagnetic field. First of all, the wires in such descriptions are idealized wires without any losses. It is assumed that the current flows inside the wires only and that there is an ideal dielectric outside the wires and outside the lumped elements. In this ideal dielectric, one has no currents and also no electromagnetic field – at least as a first approximation.

The lumped elements consist of a body that is considered as a black box that can be connected with other elements with at least two wires. From outside, one can only observe the currents in the wires of such a black box and the voltage between two of these wires.

Since the wires are idealized, there is no voltage along the wires. Moreover, it is assumed that the wires cannot carry any charge. This means that the entire field outside the lumped elements is described by the currents in the wires. When one connects two or several wires, one obtains a node that is also considered as an ideal object that cannot carry any charge.

In order to obtain a network, at least two lumped elements must be present and connected with each other. In such a network, one observes meshes of lumped elements. These meshes are also idealized in the sense that one assumes that no voltage is induced.

Although circuits are often used at very high frequencies, the fundamental laws and descriptions of simple lumped elements may be obtained from static Maxwell equations and from the corresponding material properties. This is outlined in the following.

Kirchhoff's laws

When we use the theorems of Stokes and Gauss, we can easily integrate the 'current' equations (5.106) and (5.107):

$$\int_{\partial S} \vec{E}(\vec{r})\, \mathrm{d}\vec{\ell} = 0, \tag{5.113}$$

$$\int_{\partial V} \vec{j}(\vec{r})\, \mathrm{d}\vec{S} = 0. \tag{5.114}$$

Now, we need the definitions of the voltage U between two wires of a lumped element and of the current I in a wire. These are simple integrals:

$$U = \int_{\ell} \vec{E}(\vec{r})\,\mathrm{d}\vec{\ell}\,, \tag{5.115}$$

$$I = \int_{S} \vec{j}(\vec{r})\,\mathrm{d}\vec{S}\,, \tag{5.116}$$

where ℓ is a path from one wire to another one and S is a cross section of a wire.

When we consider a simple mesh with n lumped elements, we see from (5.113) that the sum of all voltages over all elements is zero, when all voltages have the same orientation along the mesh. When we consider a node with n wires, we see from (5.114) that the sum of all currents that flow out of the node is zero (see Figure 37). One easily obtains:

$$\int_{\partial S} \vec{E}(\vec{r})\,\mathrm{d}\vec{\ell} = \sum_{i=1}^{n} U_i = 0, \tag{5.117}$$

$$\int_{\partial V} \vec{j}(\vec{r})\,\mathrm{d}\vec{S} = \sum_{i=1}^{n} I_i = 0. \tag{5.118}$$

These are the two fundamental laws of Kirchhoff. Note that more accurate but more complicated 'laws' could be derived from non-static Maxwell equations. For the Kirchhoff laws, it does not matter how many wires are connected to each lumped element.

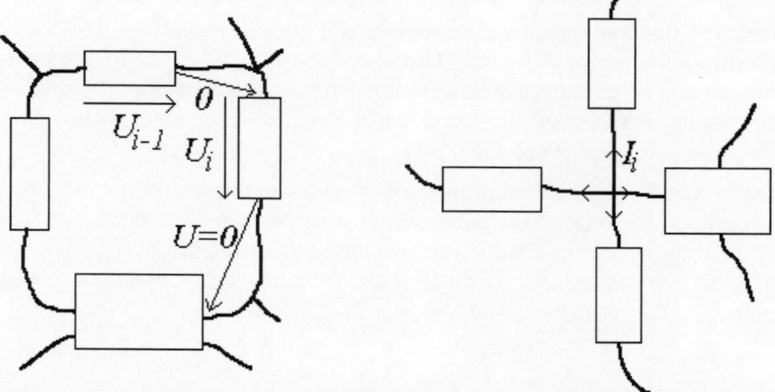

Figure 37: Mesh (left hand side) and node (right hand side) in a network of lumped elements.

Capacitance, Inductance, Resistance

The most simple lumped elements have two wires and are described by only one of the three simple material properties that were outlined above. The corresponding elements are capacitors, inductors, and resistors. This section outlines how these elements are computed in Maxwell theory. Note that also the properties of more complicated elements may be derived from Maxwell theory. Once the field in such an element is known, the currents and voltages are obtained by integration.

The electrostatic equations are used for computing the *capacitance* of capacitors. The most simple prototype of a capacitor consists of two charged electrodes. A capacitor is linked with two wires that are connected to the two electrodes, but these wires are omitted in the electrostatic model. The electrodes are considered to be conducting. As a consequence, the electric field inside the electrodes is zero and all charges are on the surface of the electrodes. Moreover, it is assumed that the total charge is zero. Consequently, the two electrodes have opposite charge $+Q$ and $-Q$. Then, the capacitance C is defined as:

$$C = \frac{Q}{U} = \frac{\int_{S_e} \sigma_e \, \mathrm{d}S}{\int_{\ell} \vec{E} \, \mathrm{d}\vec{\ell}} = \frac{\int_{S_e} \vec{D} \, \mathrm{d}\vec{S}}{\int_{\ell} \vec{E} \, \mathrm{d}\vec{\ell}}, \tag{5.119}$$

where σ_e is the surface-charge density on an electrode, S_e is the surface of the positively charged electrode, and ℓ is a path from any point on the positive electrode to any point on the negative electrode. Note that (5.101) has been used to replace the charge density by the $\vec{D}$ field.

Similarly, the magnetostatic equations are used to compute the *inductance*:

$$L = \frac{\phi}{I} = \frac{\int_{S_m} \vec{B} \, \mathrm{d}\vec{S}}{\int_{S} \vec{j} \, \mathrm{d}\vec{S}} = \frac{\int_{S_m} \vec{B} \, \mathrm{d}\vec{S}}{\oint_{\partial S} \vec{H} \, \mathrm{d}\vec{\ell}}. \tag{5.120}$$

Here, (5.103) has been used to replace the current density by the $\vec{H}$ field. Incidentally, this definition is quite problematic. The boundary of S_m is given by the currents. Thus, S_m is only defined exactly if the currents flow in wires with infinitesimally small cross sections S. This causes singularities of the field on the wires. For more detailed discussions and other definitions of L, see [Paul, 1994].

Finally, the current equations are used to compute the *resistance* of a resistor. The resistor has a configuration similar to the capacitors. Instead of two electrodes, one has two contacts that are assumed to be perfect conductors. The currents through both contacts are equal in magnitude, but have opposite signs. The wires connected to the contacts are omitted in the model for computing the resistance. The following definition is used:

$$R = 1/G = \frac{U}{I} = \frac{\int_{\ell} \vec{E} \, \mathrm{d}\vec{\ell}}{\int_{S} \vec{j} \, \mathrm{d}\vec{S}}, \tag{5.121}$$

where ℓ is a path from one contact of the resistor to the other contact and S is a surface of one of the contacts.

Note that it is easy to derive static equations also for 2D problems. In this case, the fields do not propagate along the cylinder axis z, i.e., one can set $\gamma = 0$ and $\omega = 0$ in the equations of the subsection entitled 2D: cylindrical structures, z separation. In 2D statics, one defines the capacity per unit length

$$C' = \frac{Q'}{U} = \frac{\oint_{\partial S} \vec{D}\, \mathrm{d}\vec{\ell}^{\circ}}{\int_{\ell} \vec{E}\, \mathrm{d}\vec{\ell}}, \tag{5.122}$$

where S is the cross section of the positively charged electrode. ∂S is the boundary of S, i.e., a path along this electrode. Moreover the path ℓ is a line from this electrode to the other one. Similarly, one can define the inductivity per unit length

$$L' = \frac{\phi'}{I} = \frac{\int_{\ell} \vec{B}\, \mathrm{d}\vec{\ell}^{0}}{\oint_{\partial S} \vec{H}\, \mathrm{d}\vec{\ell}}. \tag{5.123}$$

Usually, this definition is used for transmission lines, where one assumes that the current in the conductors flows in the z direction. Then, ℓ is a path from one conductor to the other and ∂S is a path around one of the conductors. Finally, one can define the resistance per unit length

$$R' = 1/G' = \frac{U}{I'} = \frac{\int_{\ell} \vec{E}\, \mathrm{d}\vec{\ell}}{\int_{S} \vec{j}\, \mathrm{d}\vec{S}}, \tag{5.124}$$

where ℓ is a path from one contact of the resistor to the other contact and S is the surface of one of the contacts.

Resistors, capacitors, and inductors are lumped electronic elements that are usually applied when time-dependent fields are present. When one applies the theorems of Gauss (1.123) and Stokes (1.122), one obtains from the charge conservation law (5.38) and from the induction law (5.9):

$$I(t) = \frac{\partial}{\partial t} Q(t), \tag{5.125}$$

$$U(t) = \frac{\partial}{\partial t} \phi(t). \tag{5.126}$$

For a capacitor, the charge is integrated over one electrode and the current is the current on the wire that connects the electrode with the exterior. Similarly, ϕ is the magnetic flux through a coil and U is the voltage between the two wires that connect the coil with the exterior. One now can insert (5.122) and (5.123) to eliminate Q and ϕ.

Obviously, the definitions above can be adapted to time-harmonic fields. Then, the fields become complex and (5.119)-(5.121) become complex as well. Moreover, the time derivatives are replaced by the factors $-i\omega$. Thus, we obtain

$$I = -i\omega Q = -i\omega CU, \tag{5.127}$$

$$U = -i\omega \quad = -i\omega LI. \tag{5.128}$$

When one compares this with the definition (5.121) of the resistance R and of the conductance G, it is natural to introduce complex properties, namely the *impedance* Z=R-$i\omega L$ or the *admittance* $Y=1/Z=G\text{-}i\omega C$. Note that $-i$ is replaced by j in many textbooks to simplify the notation.

Transmission lines

Transmission lines consist of at least two wires embedded in one or several dielectrics. Although the electromagnetic field of transmission lines is not lumped in certain areas, one can describe transmission lines very well by simple networks of capacitors, inductors, and resistors. To understand this, we best consider two cylindrical wires in a single dielectric. We assume that most of the current that flows in the direction z of the axis is inside the wires, whereas the current in the dielectric caused by losses, the electric field in the dielectric, and the magnetic field in the dielectric are mainly perpendicular to the axis. With these rather crude assumptions, we split the electromagnetic field into several parts.

Inside the wire, we compute R' from (5.124) with the assumption that both the electric field and the current density have a z component only. We obtain I from an integral over the cross section S of the first wire, and for U we integrate along the z axis along a path of unit length. Note that the computation of R' is usually refined by taking skin effect and proximity effect into account, i.e., not a pure static model is used. These effects are frequency dependent. In a wire, the current density is uniform only at zero frequency. The higher the frequency, the bigger the portion of the current that flows near the surface of the conductor, i.e., in the 'skin' of the conductor. This essentially is the skin effect. When a second conductor is near a given conductor, the magnetic field caused by the second conductor has an influence on the current distribution in the given conductor. This effect is weak, except when the second conductor is in the 'proximity' of the given conductor. Skin and proximity effects are usually estimated with very rough models.

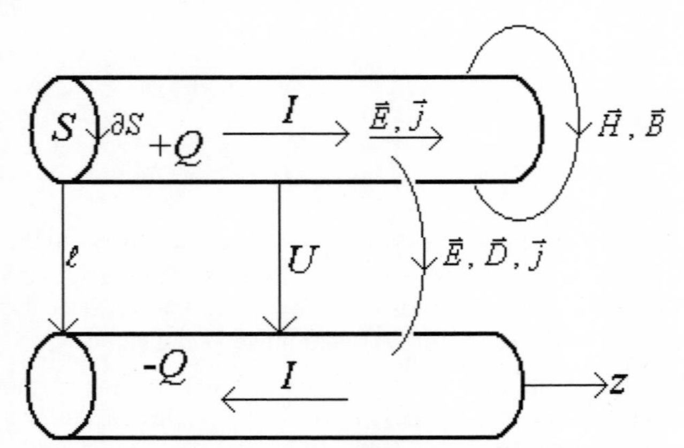

Figure 38: Short piece of a transmission line consisting of two wires in a lossy dielectric.

Outside the wire one assumes that the field is transverse, i.e., that it has no z components. One then can use (5.122)-(5.124) to compute C', L', and G', where the conductance G' is no longer the inverse of the resistance R'. R' takes the losses in the wires into account, whereas G' is responsible for the losses in the dielectric, i.e., G' and R' are now different quantities obtained from different computations of different parts of the field. Note that also the computation of L'

is usually refined. It is split into an interior and an exterior part. The former takes the field inside the conductors (skin effect and proximity effect!) into account.

Once C', L', R', and G' have been computed, one can replace the transmission line by a little network. For a piece of unit length, one can set the network outlined in Figure 39.

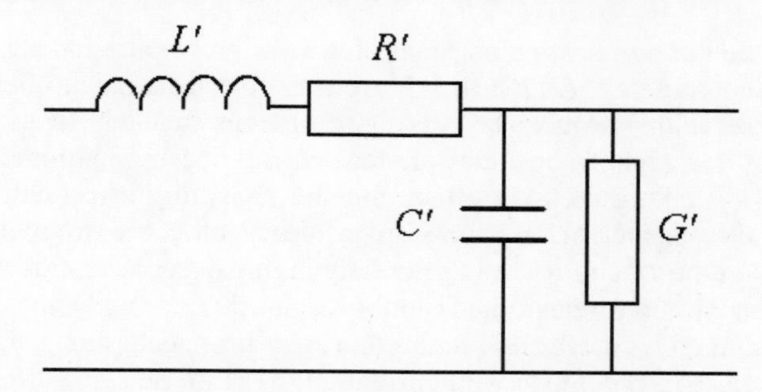

Figure 39: Network that replaces a piece of a transmission line with two wires.

Note that the same network may be used for all types of transmission lines with two wires. The geometry and the material properties of the transmission lines have an influence on the size of the lumped elements, but no influence on the structure of the network. The network shown in Figure 39 is the most simple one, but one can also use more complicated, equivalent networks.

We now can study the propagation of waves along the transmission line. For simplicity, we consider a loss-free transmission line with $R'=0$, $G'=0$. First, we consider the wires as objects of finite dimensions that can carry electric charges, i.e., we do not use the idealized wires of electronic circuits. From the charge conservation law (5.38) and with (5.122) we then obtain

$$\frac{\partial}{\partial z}I(z,t) = -\frac{\partial}{\partial t}Q'(z,t) = -C'\frac{\partial}{\partial t}U(z,t). \tag{5.129}$$

Remember that the charge conservation law is deduced from the second Maxwell equation (5.10). To study propagation effects, both the first and second Maxwell equations are important. From the induction law (5.9) and with (5.123) we obtain a second differential equation for U and I:

$$\frac{\partial}{\partial z}U(z,t) = -\frac{\partial}{\partial t}\phi'(z,t) = -L'\frac{\partial}{\partial t}I(z,t). \tag{5.130}$$

These *telegraph equations* are coupled. Uncoupled equations are obtained when we derive both equations with respect to z and t and insert the t derivative of (5.129) in the z derivative of (5.130) and vice versa:

$$\frac{\partial^2}{\partial z^2}U(z,t) = L'C'\frac{\partial^2}{\partial t^2}U(z,t), \tag{5.131}$$

$$\frac{\partial^2}{\partial z^2}I(z,t) = L'C'\frac{\partial^2}{\partial t^2}I(z,t). \tag{5.132}$$

As one can easily see, the z,t dependence (5.50) of waves on cylindrical structures solves these equations when the propagation constant is

$$\gamma = \beta + i\alpha = \omega\sqrt{L'C'} = \sqrt{-(-i\omega L')(-i\omega C')}\,. \tag{5.133}$$

Since we have assumed that the transmission line is loss-free, the damping constant α is zero. For lossy transmission lines, one obtains a complex propagation constant

$$\gamma = \beta + i\alpha = \omega\sqrt{-(R'-i\omega L')(G'-i\omega C')}\,. \tag{5.134}$$

To compute the propagation constant of a wave propagating along a transmission line, one can therefore compute C', L', R', and G' from the electromagnetic field using (5.122)-(5.124) and insert these values in (5.134). It is quite obvious that this leads to an approximation that is inaccurate for high frequencies. It has already been mentioned that the accuracy can be improved by taking the skin effect and the proximity effect into account. These effects are dynamic effects that mainly have an influence on the current distribution in the wires and therefore on the L' and R'. At sufficiently high frequencies, higher order modes can propagate along transmission lines that cannot be evaluated with the formalism outlined above. Moreover, there is a *common mode* that may propagate along transmission lines at arbitrary frequencies. This is a mode with currents flowing in the same direction in all wires. The return current flows outside the wires and is mainly a displacement current. Since displacement currents are neglected in the equations above, this mode can also not be obtained. Note that common modes are most important in EMC (electromagnetic compatibility) problems because these modes are excited when electromagnetic waves hit transmission lines. Such modes cannot be suppressed by conventional filters that consist of networks with lumped elements between the wires of a transmission line.

Wave equations

In the previous section, we have seen how the wave propagation along a transmission line can be handled. The essential steps were the decoupling of the coupled first order differential equations (5.129) and (5.130), which lead to the uncoupled second order differential equations (5.131) and (5.132). Essentially the same procedure can be applied to the Maxwell equations that are also coupled first order differential equations.

4D formalism outlined

When we look at the schematic representation (5.41) and (5.42) of the 4D formulation of Maxwell's equations, we see that the structure is almost the same as the structure (1.36)-(1.39) that had been studied in the first chapter. To obtain a 4D wave equation, we first have to replace the field tensor $\boldsymbol{F}$ with $\mathbf{F}$ by use of the material properties. Then, we have to define second order derivatives in 4D space that correspond to the Laplace operator and consist of combinations of derivatives 'to the left' with derivatives 'to the right'. This leads to 4D wave equations for $\mathbf{F}$ with an inhomogeneity that contains the sources of the field, i.e., $\mathbf{J}$.

When one analyzes the situation more precisely, one can recognize that 4D space is more similar to 2D space because both spaces have even dimension. In 2D space, the 1D object could be represented either by a vector or by a pseudovector and the operator o was introduced in equation (1.76) to link the two aspects. In 4D space, essentially the same happens with the 2D object, i.e., the field tensor $\mathbf{F}$.

In the first chapter, potentials were introduced to solve the homogeneous first order differential equations. This is also done in (5.41). Note that the derivative of the potential **A** 'to the right' is given there, whereas the derivative 'to the left' is not defined. This derivative is required for obtaining a wave equation for **A**. The derivative of **A** 'to the left' is called the *gauge*. Setting

$$0 \leftarrow \mathbf{A} \tag{5.135}$$

is most simple. This is called *Lorentz gauge*. Potentials and gauge equations will be considered in the next section.

3D+1D: traditional time domain formalism

Considering the schematic form (5.60)-(5.63) of Maxwell's equations, which includes simple material properties, we recognize that we can first make a step from $\vec{E}$ or $\vec{H}$ to either the left or right and return in a second step in the opposite direction. Introducing the vector Laplacian operator, we immediately obtain the following vector wave equations for $\vec{E}$ and $\vec{H}$:

$$\left(\Delta - \mu\sigma\frac{\partial}{\partial t} - \mu\varepsilon\frac{\partial^2}{\partial t^2}\right)\vec{E}(\vec{r},t) = \mu\frac{\partial}{\partial t}\vec{j}_0(\vec{r},t) + \operatorname{grad}\rho(\vec{r},t)/\varepsilon, \tag{5.136}$$

$$\left(\Delta - \mu\sigma\frac{\partial}{\partial t} - \mu\varepsilon\frac{\partial^2}{\partial t^2}\right)\vec{H}(\vec{r},t) = -\operatorname{curl}\vec{j}_0(\vec{r},t). \tag{5.137}$$

As the charge distribution ρ generally consists of a known and an unknown part, there may be a difficulty in (5.136) for conducting media. To express the unknown by means of the $\vec{E}$ field, one can apply the law of charge conservation for charges and currents in conducting media:

$$\frac{\partial}{\partial t}\rho_\sigma(\vec{r},t) = \operatorname{div}\vec{j}_\sigma(\vec{r},t) = \sigma\operatorname{div}\vec{E}(\vec{r},t). \tag{5.138}$$

By integrating this equation over time and introducing the result in (5.136), we have an equation more complicated than the usual wave equation. Instead, one can consider the third Maxwell equation (5.10). As both ρ_σ and $\vec{E}$ are unknowns, this equation contains no restriction for the $\vec{E}$ field and may be used to obtain ρ_σ if $\vec{E}$ has been computed. If we take only the "curl curl" part of the wave equation, we have a second-order partial differential equation for $\vec{E}$ without the Laplacian operator:

$$\operatorname{curl}\operatorname{curl}\vec{E}(\vec{r},t) = \mu\frac{\partial}{\partial t}\left(\vec{j}_0(\vec{r},t) + \sigma\vec{E}(\vec{r},t) + \varepsilon\frac{\partial}{\partial t}\vec{E}(\vec{r},t)\right). \tag{5.139}$$

These difficulties with conducting media are much less pronounced in the case of time-harmonic fields, because the time integral mentioned above can be avoided for time-harmonic fields.

3D: frequency domain, t separation

Now, one can proceed as in the section above to obtain wave equations for time-harmonic fields or, instead, one can simply replace 1) the time derivatives in the wave equations for non-harmonic fields by the factor $-i\omega$ and 2) the fields by their complex amplitudes. Thus, we have the so-called *Helmholtz equation*:

$$\left(\Delta + k^2\right) Z(\vec{r}) = Y(\vec{r}) \tag{5.140}$$

instead of the wave equation

$$\left(\Delta - \mu\sigma\frac{\partial}{\partial t} - \mu\varepsilon\frac{\partial^2}{\partial t^2}\right) Z(\vec{r},t) = Y(\vec{r},t), \tag{5.141}$$

where Z is any scalar or vector field and Y is the corresponding inhomogeneity. In (5.140), the wave number k is given by

$$k^2 = \omega^2 \mu\varepsilon' = \omega^2 \mu(\varepsilon + i\sigma / \omega). \tag{5.142}$$

2D: cylindrical structures, z separation

The z separation is similar to the t separation, but it now affects derivatives with respect to z that are contained in the Laplacian operator in (5.140), whereas the time derivatives were separated from the spatial derivatives in the 3D+1D formalism. According to (1.93), one can split the longitudinal and the transverse derivatives in the 3D Laplacian. When one now replaces a derivative with respect to z by the factor $i\gamma$, one easily obtains the 2D Helmholtz equation

$$\left(\Delta_T + \kappa^2\right) Z(\vec{r}_T) = Y(\vec{r}_T) \tag{5.143}$$

from the 3D Helmholtz equation (5.140). Here κ denotes the transverse wave number that was already defined in (5.99).

Remember that the transverse components of $\vec{E}$ and $\vec{H}$ can be derived from the longitudinal components using the equations (5.95) and (5.96). Therefore, it is sufficient when the longitudinal components E_z and H_z fulfill the Helmholtz equation (5.143) explicitly. The 'derived' transverse components will automatically fulfill the corresponding 2D Helmholtz equations. Note that the inhomogeneity Y is zero when no impressed charges and currents are present. Impressed charges and currents play a role in devices where the electromagnetic field is generated, when non-electromagnetic energy is converted into electromagnetic energy, whereas the Helmholtz equations are homogeneous when wave propagation phenomena are studied.

Potentials

In the first chapter, we have seen that potentials may be introduced to automatically solve homogeneous first order differential equations. We have already seen that the potential **A** in 4D space can easily be introduced. This allows one to derive a wave equation for **A** instead of a more complicated wave equation for the field tensor **F**. Moreover, the tensor **A** has only four independent components, whereas **F** has six of them. Since pure 4D formulation is rarely used in computational electromagnetics, we omit an explicit formulation.

We have seen that only the two longitudinal components of the electric and magnetic field must be known when one studies waves propagating along cylindrical (2D) structures. A further reduction of independent components cannot be achieved in 2D by the use of potentials. Therefore, we omit the explicit introduction of potentials for 2D problems, although this can easily be done in the usual manner.

In the following, we focus on the 3D+1D formulation which is more difficult than the 4D and 2D formulations because the separate treatment of time and 3D space is somehow unnatural for Maxwell theory.

Scalar and vector potentials

The vector potential $\vec{A}$ has already been introduced in (5.63) for completing the homogeneous Maxwell equation (5.67). The introduction of the vector potential in the 3D+1D formulation is therefore simple and natural:

$$\operatorname{curl} \vec{A}(\vec{r},t) = \vec{B}(\vec{r},t). \tag{5.144}$$

In the case of the vector (pseudovector) fields $\vec{E}$ and $\vec{H}$, we have derivatives in both directions, whereas the 'left' derivative of the vector potential $\vec{A}$ is not defined. As mentioned in Chapter 1, this derivative can be 'gauged'. At first glance, the simplest seems to be the *Coulomb gauge*:

$$\operatorname{div} \vec{A}(\vec{r},t) = 0. \tag{5.145}$$

In Maxwell theory, the gauge equation and the Coulomb gauge has no physical meaning. It simply defines the missing 'left' derivative of the vector potential in such a way that the vector potential can be computed. When another gauge equation is used, another solution for the vector potential may be found. However, this will not affect the electromagnetic field that is derived from the vector potential, i.e., the electromagnetic field is gauge invariant.

For the complete schematic form of Maxwell's equations for linear, homogeneous, isotropic media, we therefore have

$$0 \xleftarrow{\text{div}} \vec{A} \tag{5.146}$$

in addition to the Maxwell equations (5.60)-(5.63). When we add the time derivative of (5.63) to (5.60), the derivatives 'to the right' end in a zero pseudo vector. This allows us to introduce the scalar potential ϕ:

$$-\phi \xrightarrow{\text{grad}} \vec{E} + \frac{\partial}{\partial t}\vec{A} \xrightarrow{\text{curl}} 0. \tag{5.147}$$

Note that the – sign in (5.147) is conventional from electrostatics. In standard notation, we have the following definition of the scalar potential:

$$\operatorname{grad} \phi(\vec{r},t) = -\vec{E}(\vec{r},t) - \frac{\partial}{\partial t}\vec{A}(\vec{r},t). \tag{5.148}$$

If we look for steps to the right or left side and back, we easily find the following equations for the two potentials:

$$\Delta\vec{A} = -\mu\,\vec{j}_0 + \left(\mu\sigma + \mu\varepsilon\frac{\partial}{\partial t}\right)\left(\operatorname{grad}\phi + \frac{\partial}{\partial t}\vec{A}\right), \tag{5.149}$$

$$\Delta\phi = -\rho / \varepsilon. \tag{5.150}$$

Obviously, we have a static Poisson equation for the scalar potential, whereas the equation for the vector potential is more complicated than the usual wave equation and contains the scalar potential as well. If we look for a gauge, which eliminates the scalar potential in this equation, we find the *Lorentz gauge* (5.135) in 3D+1D notation:

$$\operatorname{div}\vec{A}(\vec{r},t) = -\left(\mu\sigma + \mu\varepsilon\frac{\partial}{\partial t}\right)\phi(\vec{r},t). \tag{5.151}$$

The wave equations for the potentials obtained with this gauge are:

$$\left(\Delta - \mu\sigma\frac{\partial}{\partial t} - \mu\varepsilon\frac{\partial^2}{\partial t^2}\right)\vec{A}(\vec{r},t) = -\mu\,\vec{j}_0(\vec{r},t), \tag{5.152}$$

$$\left(\Delta - \mu\sigma\frac{\partial}{\partial t} - \mu\varepsilon\frac{\partial^2}{\partial t^2}\right)\phi(\vec{r},t) = -\mu\,\rho(\vec{r},t). \tag{5.153}$$

The advantage of the potential compared with the direct solution of the equations for the $\vec{E}$ and $\vec{H}$ field is the reduction from six to four scalar components and a simplification of the inhomogeneities (i.e., the right-hand sides). This shows once more that there are dependencies between $\vec{E}$ and $\vec{H}$, although one can formulate separate wave equations for both fields. These dependencies are given by Maxwell's equations, of course. We should mention that not every combination of solutions of the wave equations for $\vec{E}$ and $\vec{H}$ would automatically be a solution of Maxwell's equations as well. If $\vec{E}$ is known, one can have $\vec{H}$ up to a constant from the second Maxwell equation, or, if $\vec{H}$ is known, one can have $\vec{E}$ up to a constant from the first Maxwell equation.

Hence, one can expect that there is still a dependence between the four scalar components of the vector and scalar potentials. As the inhomogeneities of the corresponding wave equations depend on each other by means of the law of charge conservation, it is quite clear that the scalar potential can be obtained up to a constant by integration of (5.151) when the vector potential is known. Adding any constant to the scalar potential has no effect on the electromagnetic field because only the gradient of the scalar potential is used to compute the electric field from (5.148) and because the magnetic field can be obtained from the vector potential only from equation (5.144).

Remember the difficulties with the charge density ρ that is split into a known and an unknown part. Consequently, (5.153) is not a real restriction for the field. Once the vector potential is known, the scalar potential is obtained (up to an integration constant) from (5.151). Then, (5.153) may be used to compute the unknown part of the charge density. Alternatively, a single vector, the so-called Hertz potential, can be introduced. This is not done in this book, because the Hertz potential is rarely used in computational electromagnetics.

Note that the three components of the vector potentials are not independent when no impressed charges and currents are present. In this case, one can show that the electromagnetic field in linear, homogeneous, isotropic domains can be derived from two scalar fields only. In the 2D case, we have already encountered an example: the entire field may be derived from the longitudinal components E_z and H_z. When spherical coordinates are introduced in 3D space, the radial field components can play a similar role. Selecting two appropriate scalar fields in the 3D+1D formalism without introducing special coordinates is quite difficult. This is also not done in this book because it is of no importance for most of the numerical methods.

Beside the Lorentz gauge one may use arbitrary gauge equations that define the divergence of the vector potential. This leads to more complicated formalisms and coupled differential equations for the potentials, as has been demonstrated for the Coulomb gauge that is simple and natural in statics. However, the gauge equations have no effect on the electromagnetic field that is derived from the gauged potentials.

Analytic solutions

Our main interest is to find solutions of Maxwell's equations. As these equations are coupled, first-order, partial differential equations, it is extremely difficult to find such solutions directly. In order to simplify this situation, several second-order differential equations have been derived from Maxwell's equations. Most of these equations are no longer coupled, and in all these equations the Laplacian operator is involved. The more complicated vector Laplacian operator is correlated with the scalar Laplacian operator by (1.29). This means that solutions of those vector equations which are essentially based on the vector Laplacian operator can be constructed from the solutions of formally identical scalar equations. A more general process of generating scalar equations from vector equations is the following. If an equation of the form:

$$L\vec{f}(x_1, x_2, \ldots, x_n) = \vec{g}(x_1, x_2, \ldots, x_n) \tag{5.154}$$

is given, where L is a linear operator and $\vec{f}$ and $\vec{g}$ are vectors in an n-dimensional space, we obtain for any choice of an n-dimensional vector $\vec{v}$ a scalar equation by simple projection of both sides of (5.154) on $\vec{v}$:

$$\vec{v} \cdot (L\vec{f}) = \vec{v} \cdot \vec{g} \tag{5.155}$$

where the dot indicates the dot product or scalar product. For the 3D Laplacian operator, the following equations are important:

$$\vec{e} \cdot \left(\Delta \vec{Z}(\vec{r})\right) = \Delta\left(\vec{e} \cdot \vec{Z}(\vec{r})\right), \tag{5.156}$$

$$\vec{r} \cdot \left(\Delta \vec{Z}(\vec{r})\right) = \Delta\left(\vec{r} \cdot \vec{Z}(\vec{r})\right) - 2 \cdot \operatorname{div}\left(\vec{Z}(\vec{r})\right), \tag{5.157}$$

where $\vec{e}$ is a unit vector that points in an arbitrary direction. For the 2D Laplacian operator, we have:

$$\vec{e}_T \cdot \left(\Delta_T \vec{Z}_T(\vec{r}_T)\right) = \Delta_T\left(\vec{e}_T \cdot \vec{Z}_T(\vec{r}_T)\right), \tag{5.158}$$

$$\vec{r}_T \cdot \left(\Delta_T \vec{Z}_T(\vec{r}_T)\right) = \Delta_T\left(\vec{r}_T \cdot \vec{Z}_T(\vec{r}_T)\right) - 2 \cdot \operatorname{div}_T\left(\vec{Z}_T(\vec{r}_T)\right). \tag{5.159}$$

Therefore, this is sufficient when we focus on the solution of scalar field equations in the following.

It has already been mentioned that not every solution of second-order differential equations derived from Maxwell's equations must be automatically a solution of the Maxwell equation as well. In electrodynamics, this is the case: 1) for more than three scalar field components if impressed currents and charges are present, and 2) for two scalar field components if no impressed currents and charges are present. In such situations, one has to verify that Maxwell's equations hold for the solutions obtained from second-order differential equations.

Practically, impressed charges and currents cause the electromagnetic field. They are important for computations of the field produced by 'generators' such as antennas or devices which couple mechanical, chemical, and other non-electromagnetic forces or energies. To study an electromechanical coupler, for example, one must generally know both the electrical and the mechanical equations. If the influence of the electric fields on the mechanical part of such a device can be neglected, however, the charges and currents affected by the mechanical forces can be considered to be 'impressed', and the knowledge of the mechanical equations is not necessary to compute the fields. In this case, the electromagnetic field caused by the impressed sources is obtained by the integrals outlined below.

To study the propagation of scattering and guided electromagnetic waves, one usually assumes that the electromagnetic field is produced somewhere outside the domain of interest and the source of the field is not affected by the scattered field or reflected waves. As the sources of the field are formulated by the inhomogeneities, one has homogeneous field equations in the domain of interest. In this case, one has to find a general solution of two independent scalar fields.

From a well-known mathematical theorem, one knows that the general solution of linear, inhomogeneous field equations is given by 1) the superposition of a special solution of these equations and 2) the general solution of the corresponding homogeneous equations. The first part is important for the generation, the second for the propagation, of fields. This holds for non-electromagnetic fields as well.

To obtain solutions, two techniques are most useful: Green's functions (see, for example, [Collin, 1991]) and the separation of variables (see, for example, [Ramo, 1965]).

Green's functions

It is well known that the field generated by a charge distribution can be derived from the field of a point charge by integration. Essentially, the same process can be applied for obtaining a special solution of a linear equation with any inhomogeneity g:

$$L f(\vec{r}) = g(\vec{r}). \tag{5.160}$$

A point charge essentially corresponds to a special Dirac inhomogeneity $\delta(\vec{r}-\vec{r}\,')$, which is zero everywhere, except at the point $\vec{r}=\vec{r}\,'$, where it is infinite in such a way that the integral of the δ function over the entire space is equal to one. When we replace the inhomogeneity g by δ, we obtain:

$$L G(\vec{r},\vec{r}\,') = \delta(\vec{r}-\vec{r}\,'). \tag{5.161}$$

G is called the Green's function. As mentioned earlier, strictly speaking, δ is not a function. We do not consider the rigorous mathematical background of such 'functionals' and look at δ

as a kind of generalized point charge, an inhomogeneity concentrated at one point. To fix the strength of this generalized point charge, we have to define the integral:

$$\int \delta(\vec{r}-\vec{r}')\,\mathrm{d}V'=1 \tag{5.162}$$

for any $\vec{r}$ inside the range of integration. If we integrate the product of δ with any function g, we find the value of g at the point $\vec{r}=\vec{r}'$:

$$g(\vec{r})=\int \delta(\vec{r}-\vec{r}')\,g(\vec{r}')\,\mathrm{d}V'. \tag{5.163}$$

If G is known, one can integrate both sides of (5.161) multiplied with g, and obtain

$$f(\vec{r})=\int G(\vec{r},\vec{r}')\,g(\vec{r}')\,\mathrm{d}V'. \tag{5.164}$$

For example, in electrostatics one has the Poisson equation:

$$\Delta\,\phi(\vec{r})=-\rho(\vec{r})\,/\,\varepsilon. \tag{5.165}$$

For the Green's function, the following equation holds:

$$\Delta\,G(\vec{r},\vec{r}')=\delta(\vec{r}-\vec{r}'). \tag{5.166}$$

As one looks for a special solution, one can suppose that this solution has the same symmetry as the source (i.e., spherical symmetry) with respect to the point $\vec{R}=\vec{r}-\vec{r}'=0$. If spherical coordinates are applied, one can omit all the angular derivatives because of the symmetry, and (5.166) is reduced to

$$\Delta\,G=\frac{1}{R}\frac{\partial^2}{\partial R^2}(R\,G)=\delta(R). \tag{5.167}$$

As we need only a special solution, we can immediately verify that

$$G(\vec{r},\vec{r}')=\frac{1}{-4\pi|\vec{r}-\vec{r}'|} \tag{5.168}$$

satisfies (5.167). This solution is physically meaningful as it represents the 'physically true' solution for a point charge in free space. Finally, we obtain the special, physically meaningful solution of the Poisson equation by integrating:

$$\phi(\vec{r})=\frac{1}{4\pi\varepsilon}\int\frac{\rho(\vec{r}')}{|\vec{r}-\vec{r}'|}\,\mathrm{d}V'. \tag{5.169}$$

To find an integral solution of the electric field, one has to apply the gradient. As this spatial derivative does not affect $\vec{r}'$, the Coulomb integral for the electric field is

$$\vec{E}(\vec{r})=\frac{1}{4\pi\varepsilon}\int\frac{\vec{r}-\vec{r}'}{|\vec{r}-\vec{r}'|^3}\,\rho(\vec{r}')\,\mathrm{d}V'. \tag{5.170}$$

To find a solution of the Poisson vector equation in magnetostatics, one can project this equation on three independent unit vectors $\vec{e}_i,\, i=1,2,3$ to derive three scalar Poisson equations:

$$\Delta\,A_i(\vec{r})=-\mu\, j_i(\vec{r}), \tag{5.171}$$

that are similar to (5.165) and the solutions similar to (5.169):

$$A_i(\vec{r}) = \frac{\mu}{4\pi} \int \frac{j_i(\vec{r}\,')}{|\vec{r} - \vec{r}\,'|} \mathrm{d}V\,', \tag{5.172}$$

where the index i indicates the components in the direction of the corresponding unit vector. From the components, one easily acquires the vector Coulomb integral:

$$\vec{A}(\vec{r}) = \frac{\mu}{4\pi} \int \frac{\vec{j}(\vec{r}\,')}{|\vec{r} - \vec{r}\,'|} \mathrm{d}V\,'. \tag{5.173}$$

As in electrostatics, one expresses the magnetic field by a spatial derivative which does not affect $\vec{r}\,'$. With the definition of the vector potential (5.144) we obtain:

$$\vec{B}(\vec{r}) = \frac{\mu}{4\pi} \int \vec{j}(\vec{r}\,') \times \frac{\vec{r} - \vec{r}\,'}{|\vec{r} - \vec{r}\,'|^3} \mathrm{d}V\,'. \tag{5.174}$$

For time-harmonic fields, one has the slightly more complicated Helmholtz equations. All of them essentially have the form:

$$(\Delta + k^2) f(\vec{r}) = g(\vec{r}). \tag{5.175}$$

The corresponding equation for the Green's function is

$$(\Delta + k^2) G(\vec{r}, \vec{r}\,') = \delta(\vec{r} - \vec{r}\,'). \tag{5.176}$$

If we proceed with symmetry considerations as in electrostatics, we have

$$\frac{1}{R}\left(\frac{\partial^2}{\partial R^2} + k^2\right)(RG) = \delta(R), \tag{5.177}$$

$$G(\vec{r}, \vec{r}\,') = \frac{e^{\pm ik|\vec{r} - \vec{r}\,'|}}{-4\pi|\vec{r} - \vec{r}\,'|}. \tag{5.178}$$

Together with the time-dependence, we can interpret the solution with the positive sign in the exponent as an outgoing wave with origin at the point $R = |\vec{r} - \vec{r}\,'| = 0$. The negative sign in the exponent represents an incoming wave. As we are interested in the sources of the fields, this solution is less important. By integration we find the meaningful, special solution:

$$f(\vec{r}) = \int g(\vec{r}\,') \frac{e^{ik|\vec{r} - \vec{r}\,'|}}{|\vec{r} - \vec{r}\,'|} \mathrm{d}V\,'. \tag{5.179}$$

For the scalar potential, we have

$$\phi(\vec{r}) = \frac{1}{4\pi\varepsilon} \int \rho(\vec{r}\,') \frac{e^{ik|\vec{r} - \vec{r}\,'|}}{|\vec{r} - \vec{r}\,'|} \mathrm{d}V\,'. \tag{5.180}$$

In dynamics, the electric field does not depend only on the scalar potential. For this reason, we first need a similar integral for the vector potential. For solutions of the vector Helmholtz equations, we can proceed as in statics with the vector Poisson equations or we can simply replace the scalar functions f and g by vectors. Thus, we find for the vector potential:

$$\vec{A}(\vec{r}) = \frac{1}{4\pi\varepsilon} \int \vec{j}(\vec{r}\,') \frac{e^{ik|\vec{r} - \vec{r}\,'|}}{|\vec{r} - \vec{r}\,'|} \mathrm{d}V\,'. \tag{5.181}$$

With (5.148), (5.144), and the simple material property (5.58), $\vec{E}$ and $\vec{H}$ can be computed for time-harmonic fields according to

$$\vec{E}(\vec{r}) = -\text{grad}\,\phi(\vec{r}) + i\omega\vec{A}(\vec{r}), \tag{5.182}$$

$$\vec{\text{H}}(\vec{r},t) = \frac{1}{\mu}\text{curl}\,\vec{A}(\vec{r},t). \tag{5.183}$$

By recognizing the dependence of scalar and vector potential caused by the charge conservation (5.37), we can introduce (5.37) in the integral (5.180). Instead, the scalar potential can be eliminated in (5.182) with the Lorentz gauge (5.151):

$$\vec{E}(\vec{r}) = \frac{i\omega}{k^2}\text{grad div}\,\vec{A}(\vec{r}) + i\omega\vec{A}(\vec{r}). \tag{5.184}$$

Finally, we have the following integrals for $\vec{E}$ and $\vec{H}$:

$$E(\vec{r}) = \frac{-1}{4\pi i\omega\varepsilon'}\int\left(\vec{j}(\vec{r}')\left(k^2+\frac{ik}{R}-\frac{1}{R^2}\right)-\left(\vec{j}(\vec{r}')\frac{\vec{R}}{R}\right)\frac{\vec{R}}{R}\left(k^2+\frac{3ik}{R}-\frac{3}{R^2}\right)\right)\frac{e^{ikR}}{R}\text{d}V', \tag{5.185}$$

$$\vec{H}(\vec{r}) = \frac{1}{4\pi}\int\left(\vec{j}(\vec{r}')\times\frac{\vec{R}}{R}\right)\left(ik-\frac{1}{R}\right)\frac{e^{ikR}}{R}\text{d}V' \tag{5.186}$$

where

$$\vec{R} = \vec{r} - \vec{r}', \quad R = \left|\vec{R}\right|. \tag{5.187}$$

The special solution of the wave equations for general time dependence can be found by the Fourier transform. We start with the scalar wave equation for the corresponding Green's function:

$$\left(\Delta - \mu\sigma\frac{\partial}{\partial t} - \mu\varepsilon\frac{\partial^2}{\partial t^2}\right)G(\vec{r},t,\vec{r}',t') = \delta(\vec{r}-\vec{r}')\delta(t-t') \tag{5.188}$$

with the Fourier transform (5.44), to obtain

$$(\Delta + k^2)G(\vec{r},\omega,\vec{r}',t') = \frac{1}{\sqrt{2\pi}}\delta(R)e^{i\omega t'} \tag{5.189}$$

for the Green's function in the frequency domain, which has the solutions:

$$G(\vec{r},\omega,\vec{r}',t') = \frac{e^{i\omega t'\pm ik|\vec{r}-\vec{r}'|}}{-4\pi\sqrt{2\pi}|\vec{r}-\vec{r}'|}. \tag{5.190}$$

With the inverse Fourier transform (5.45), one has the Green's function in the time domain:

$$G(\vec{r},t,\vec{r}',t') = \int\frac{e^{i\omega(t'-t)\pm ik|\vec{r}-\vec{r}'|}}{-8\pi^2|\vec{r}-\vec{r}'|}\text{d}\omega. \tag{5.191}$$

To acquire a special solution of the wave equation for a general inhomogeneity g, we must integrate over space and time:

$$f(\vec{r},t) = \int G(\vec{r},t,\vec{r}',t')\,g(\vec{r}',t')\,\text{d}V'\text{d}t'. \tag{5.192}$$

Obviously, the resulting integral is already quite complicated. Fortunately, the frequency integral (5.191) can be solved analytically in loss-free media with $k = \omega\sqrt{\mu\varepsilon}$:

$$G(\vec{r},t,\vec{r}',t') = \frac{\delta\left(t'-t \pm \sqrt{\mu\varepsilon}|\vec{r}-\vec{r}'|\right)}{-4\pi|\vec{r}-\vec{r}'|}. \tag{5.193}$$

By inserting this function in the integral (5.192), one can solve the time integral as well and obtain

$$f(\vec{r},t) = \int \frac{g(\vec{r}',t')|_{t'=t-\sqrt{\mu\varepsilon}|\vec{r}-\vec{r}'|}}{-4\pi|\vec{r}-\vec{r}'|}\,\mathrm{d}V'. \tag{5.194}$$

The solutions of the vector wave equations are again found if the scalars f and g are simply replaced by vectors. The corresponding integrals for the potentials are called retarded potentials as the values g of the sources that cause the fields f have to be computed at an earlier time $t' = t - \sqrt{\mu\varepsilon}|\vec{r}-\vec{r}'|$ than the fields. This time difference, called retardation, says that every propagation of waves takes time. If Newton's form of an 'action at a distance' is applied as in the Coulomb-Ampère theory, one has to add the retardation for dynamic effects. It is important to recognize that the simple form (5.194) cannot be applied to lossy media.

As we found two solutions with opposite signs for time-harmonic fields, one could find a second 'advanced' solution as well, which would usually not be considered because the 'advanced' solution is considered to be wrong for physical reasons.

The theory of Green's functions allows one to deduce and to generalize the integrals of the Coulomb-Ampère theory of electromagnetic fields. Moreover, these integrals hold even if the sources are not known. In this case, one obtains integral equations for the potentials and for the electromagnetic fields. As differentials are numerically more problematic than integrals, such integral equations are very interesting for numerical methods.

Superposition

The integrals in the previous subsection are not only special solutions of a given inhomogeneous field equation, they also provide solutions of the corresponding homogeneous field equations. All these integrals are solutions of homogeneous field equations in a domain D, if all the sources in the integrals are outside the domain D (i.e., if one integrates over the space outside D). As one can choose any distribution of sources outside D, one easily obtains an infinite number of solutions of the homogeneous field equations. From the analytic point of view, one has two problems. 1) It is quite clear that there will be certain dependencies between these solutions and usually one wants to have a set of linear independent solutions. 2) One wants to be sure that this set of solutions is complete (i.e., that any solution f of the field equations in D can be expanded by the given expansions). In the simplest case of a scalar, linear equation:

$$L\,f(\vec{r}) = 0 \tag{5.195}$$

one writes for every f:

$$f = \sum_{k=1}^{\infty} A_k f_k, \tag{5.196}$$

where f_k denotes solutions of (5.195) and A_k are linear parameters which have to be chosen so that the boundary conditions hold on the boundary of the domain D. The f_k are called basis functions or fields.

It is not trivial that a summation is sufficient to expand any analytic solution of the field equation, especially in the more complicated case of Maxwell's equations. From a numerical point of view, however, an infinite number of basis functions is too much anyway. If we look for numerical solutions, we look for approximations of some solutions of practical interest. This means that the series expansion (5.196) must be truncated:

$$f = \sum_{k=1}^{K} A_k f_k + \eta, \tag{5.197}$$

where the error η should be small enough. We have already studied such expansions in the previous chapters. With the method of Green's functions, we have a tool to generate appropriate basis functions for expanding the field within a domain. These basis functions can be understood as the electromagnetic field that is generated by charge and current distributions located outside the domain. Note that these charge and current distributions can be 'fictitious' rather than 'true' charge and current distributions, i.e., the sources that generate the basis functions must not have any physical meaning. This is the fundamental idea of several numerical methods, e.g., the method of fictitious sources, the Method of Auxiliary Sources (MAS), the method of Discrete Sources (DS), the Multiple Multipole Program (MMP), the Generalized Multipole Technique (GMT), the Charge Simulation (CS) method, etc., whereas standard Method of Moments (MoM) codes discretize the 'true' charge and current distributions to approximate the field.

Separation of variables

Another useful way for engineers to find solutions of homogeneous field equations is the following. First one takes any system of coordinates x_1, x_2, ..., x_K and formulates the equations in these coordinates

$$L f(x_1, x_2, \ldots, x_K) = 0. \tag{5.198}$$

Then one tries to write f as a product of functions X_1, X_2, ..., X_K which depend on only one coordinate:

$$f(x_1, x_2, \ldots, x_K) = \prod_{k=1}^{K} X_k(x_k) = X_1(x_1) \cdot X_2(x_2) \cdot \ldots \cdot X_K(x_K). \tag{5.199}$$

This so-called ansatz product is only successful if one can now rewrite (5.198) in the following form:

$$L f(x_1, x_2, \ldots, x_K) = \sum_{k=1}^{K} \left(L_k f_k(x_k) \right) = 0. \tag{5.200}$$

As all terms of this sum depend on only one variable, one can write:

$$L_k f_k(x_k) = c_k\,,\quad k = 1,2,\ldots,K \tag{5.201}$$

where the so-called separation constants c_k are generally complex. From (5.200) one has the following condition for the separation constants:

$$\sum_{k=1}^{K} c_k = 0. \tag{5.202}$$

Unfortunately, this method, called 'separation of variables,' does not work for any system of coordinates and one is usually not sure of finding a complete set of basis functions.

Cartesian coordinates - plane wave solutions

To demonstrate how the method of separation works, we consider the scalar Helmholtz equation in Cartesian coordinates x,y,z. Incidentally, to obtain the Helmholtz equation itself from the homogeneous wave equation:

$$\left(\Delta - \mu\sigma\frac{\partial}{\partial t} - \mu\varepsilon\frac{\partial^2}{\partial t^2}\right) f(\vec{r},t) = 0, \tag{5.203}$$

one can use the separation of variables as well. With the ansatz product:

$$f(\vec{r},t) = R(\vec{r})\cdot T(t), \tag{5.204}$$

one has after some simple calculations:

$$(\Delta + k^2)\,R(\vec{r}) = 0, \tag{5.205}$$

$$\left(\mu\sigma\frac{\partial}{\partial t} + \mu\varepsilon\frac{\partial^2}{\partial t^2} + k^2\right) T(t) = 0, \tag{5.206}$$

where the separation constants $-c_r = c_t = k^2$ are generally complex.

Obviously, the expression for time-harmonic fields:

$$T(t) = e^{-i\omega t} \tag{5.207}$$

is a solution of (5.206). With the ansatz product:

$$f(\vec{r}) = f(x,y,z) = X(x)\cdot Y(y)\cdot Z(z), \tag{5.208}$$

for the Helmholtz equation in Cartesian coordinates:

$$\left(\frac{\partial^2}{\partial x^2} + \frac{\partial^2}{\partial y^2} + \frac{\partial^2}{\partial z^2} + k^2\right) f(x,y,z) = f_{,xx} + f_{,yy} + f_{,zz} + k^2 f = 0, \tag{5.209}$$

we obtain three identical differential equations of the form:

$$\left(\frac{\partial^2}{\partial u^2} + k_u^2\right) U(u) = 0,\quad u = x,y,z,\quad U = X,Y,Z, \tag{5.210}$$

with the solutions:

$$U(u) = C_u^{\pm} e^{\pm i k_u u}\,,\quad u = x,y,z,\quad U = X,Y,Z, \tag{5.211}$$

where $C_u^{\pm}$ denotes complex amplitudes. One does not lose any solutions if one drops the lower sign and obtains finally the plane wave solution

$$f(\vec{r}) = Ce^{i\vec{k}\vec{r}}. \tag{5.212}$$

As every component of the electromagnetic field must fulfill a Helmholtz equation, one has such an expression for every component. This allows one to construct the simple plane wave solution of the time-harmonic electromagnetic field. As the components are coupled by the Maxwell equations, one finds 1) that one has to choose exactly the same wave vector $\vec{k}$ for all components, 2) that the vectors $\vec{E}$, $\vec{H}$, and $\vec{k}$ have to be perpendicular to each other, 3) that the amplitudes are constant everywhere, and 4) that the following holds:

$$\left|\vec{E}(\vec{r})\right| = Z_w\left|\vec{H}(\vec{r})\right|. \tag{5.213}$$

As the length of the wave vector $\vec{k}$ and the wave impedance Z_w are given by the frequency and the material properties:

$$k = \omega\sqrt{\mu\varepsilon'}, \quad Z_w = \sqrt{\frac{\mu}{\varepsilon'}}, \tag{5.214}$$

a plane wave is completely described by 1) the angular frequency ω, 2) the direction of the propagation, i.e., of the wave vector $\vec{k}$, 3) the amplitude $|\vec{E}|$, and 4) the polarization, i.e., the direction of $\vec{E}$ in the transverse plane perpendicular to $\vec{k}$:

$$\left|\vec{E}(\vec{r},t)\right| = \Re\left(\vec{E}_0 e^{i\left(\vec{k}\vec{r}-\omega t\right)}\right), \tag{5.215}$$

$$\left|\vec{H}(\vec{r},t)\right| = \Re\left(Z_w^{-1}\left(\vec{k}/k\right) \times \vec{E}_0 e^{i\left(\vec{k}\vec{r}-\omega t\right)}\right). \tag{5.216}$$

Incidentally, the characteristic data of a plane wave correspond to the following properties of an optical ray in Newton's geometric optics: 1) the color, 2) the direction of the ray, 3) the intensity, and 4) the polarization or 'sides'. For example, the intensity is proportional to the square of the amplitude. As a ray has no extension in the transverse plane, whereas a plane wave extends uniformly to infinity, these two concepts are not identical at all. However, by a spatial Fourier transform, one can superpose an infinite number of plane waves with different directions of propagation and obtain a 'ray-like' object, which seems to propagate in one direction and is essentially concentrated in an area of the transverse plane. In the limit of infinite frequency, this area can shrink to zero and one has a link to Newton's optical ray, which is important for the General Theory of Diffraction (GTD).

Spherical coordinates – 3D multipole fields

Much more important for the DS, GMT, and MMP codes are the solutions from the separation of the Helmholtz equation in spherical coordinates (r, φ, ϑ). The Helmholtz equation

$$f_{,rr} + \frac{2}{r}f_{,r} + \frac{1}{r^2}f_{,\vartheta\vartheta} + \frac{\cot\vartheta}{r^2}f_{,\vartheta} + \frac{1}{r^2\sin^2\vartheta}f_{,\varphi\varphi} + k^2 f = 0 \tag{5.217}$$

may be solved with the ansatz product:

$$f(\vec{r}) = f(r,\varphi,\vartheta) = R(r)\cdot\phi(\varphi)\cdot\theta(\vartheta). \tag{5.218}$$

From this. one obtains the three differential equations:

$$\left(\frac{\partial^2}{\partial r^2}+\frac{2}{r}\frac{\partial}{\partial r}+k^2-\frac{l(l+1)}{r^2}\right)R(r)=0\,, \tag{5.219}$$

$$\left(\frac{\partial^2}{\partial \varphi^2}+m^2\right)\phi(\varphi)=0\,, \tag{5.220}$$

$$\left(\frac{\partial^2}{\partial \vartheta^2}+\cot\vartheta\frac{\partial}{\partial \vartheta}+l(l+1)-\frac{m^2}{\sin^2\vartheta}\right)\theta(\vartheta)=0\,, \tag{5.221}$$

with the solutions:

$$R(r)=\frac{C_r^1}{\sqrt{r}}H_{l+1/2}^{(1)}(kr)+\frac{C_r^2}{\sqrt{r}}H_{l+1/2}^{(2)}(kr)\,, \tag{5.222}$$

$$\phi(\varphi)=C_\varphi^+e^{+im\varphi}+C_\varphi^-e^{-im\varphi}, \tag{5.223}$$

$$\theta(\vartheta)=C_\vartheta^P P_l^m(\cos\vartheta)+C_\vartheta^Q Q_l^m(\cos\vartheta)\,, \tag{5.224}$$

where H are the Hankel functions, and P and Q the Legendre functions. The various C terms are arbitrary constants that may be complex. Instead of (5.223) one often prefers the notation

$$\phi(\varphi)=C_\varphi^c\cos(m\varphi)+C_\varphi^s\sin(m\varphi)\,, \tag{5.225}$$

that uses real valued functions for real arguments, i.e., when the angle φ and the separation constant m are real. When the solution is complex, both notations are identical.

For the special case l=m=0, one has

$$\frac{H_{1/2}^{(1)}(kr)}{\sqrt{r}}=const\cdot\frac{e^{ikr}}{r},\quad e^{i0\varphi}=1,\quad P_0^0=1\cdot \tag{5.226}$$

From this, one recognizes that the spherical symmetric solution, which has been used in the previous subsection, is a special case of the functions found by the separation of variables with l=m=0. As the spherical coordinates have a singularity at the origin r=0, these are only regular solutions of the homogeneous Helmholtz equation outside the origin. Thus, we have the same situation for a point source at the origin. In addition to these spherical symmetric *monopoles*, we found an infinite number of slightly more complicated solutions of the Helmholtz equation outside the origin, called *multipoles*. According to the outgoing and incoming waves, we now have two different Hankel functions, which represent such waves. In this book, complex numbers such as the wave number are defined in the first quadrant of the complex plane whenever possible, whereas, in many textbooks, k is in the second or fourth quadrant. In such books, the Hankel function of the second kind $H^{(2)}$ represents outgoing waves, whereas the Hankel function of the first kind $H^{(1)}$ represents outgoing waves in this book.

In general, both l and m can be complex numbers. If the domain D for which the Helmholtz equation holds, however, allows us to go once around the origin on closed paths, l and m must be integer numbers. Moreover, the functions Q are usually not used because they are singular for $\vartheta=0$, and $\vartheta=\pi$ if m is an integer number. If D includes points of the axis $\vartheta=0$, and $\vartheta=\pi$, these functions cannot be applied to expand regular solutions. The different multipoles with integer l and m can be interpreted as the limit of configurations of several monopoles, which 'collapsed' in one single point. The best known multipole for 3D applications is the dipole, which can be considered to consist of two oscillating charges of infinite magnitude and opposite sign.

When we insert the solutions (5.222)-(5.224) in (5.218), we obtain scalar solutions for the scalar Helmholtz equation in 3D space (5.217). How can these solutions be utilized for electrodynamics? First of all, none of the components of the electric and magnetic fields in spherical coordinates fulfills (5.217). Since these fields fulfill the vector Helmholtz equation, one can use (5.157) and project the electric and magnetic vector fields on the radial vector $\vec{r}$. From this, one recognizes that $\vec{r}\cdot\vec{E}=rE_r$ and $\vec{r}\cdot\vec{H}=rH_r$ fulfill the scalar Helmholtz equation (5.217). When one omits the problematic Legendre Q functions, one obtains an electric multipole solution

$$E_r^{(lm)}=\frac{l(l+1)}{r\sqrt{r}}H_{l+1/2}^{(1)}(kr)P_l^m(\cos\vartheta)(C_{lm}^c\cos m\varphi+C_{lm}^s\sin m\varphi),\quad H_r^{(lm)}=0, \tag{5.227}$$

and a magnetic multipole solution

$$H_r^{(lm)}=\frac{l(l+1)}{r\sqrt{r}}H_{l+1/2}^{(1)}(kr)P_l^m(\cos\vartheta)(C_{lm}^c\cos m\varphi+C_{lm}^s\sin m\varphi),\quad E_r^{(lm)}=0. \tag{5.228}$$

The procedure for deriving the 'transverse' components $E_\varphi, E_\vartheta, H_\varphi, H_\vartheta$ is quite difficult. First, one uses (1.64), (1.66), and (1.67) to express the 3D time-harmonic Maxwell equations (5.72)-(5.75) in spherical coordinates. Then, one splits all components of the electromagnetic field in a radial and a 'transverse' part of the form

$$f(\vec{r})=f(r,\varphi,\vartheta)=R_f(r)\cdot Y_f(\varphi,\vartheta) \tag{5.229}$$

instead of (5.218). Then, one analyzes the dependencies between the radial and transverse parts of all components of the field caused by the Maxwell equations. Finally, one ends up with

$$E_r^{(lm)}(r,\varphi,\vartheta)=\frac{R_l(r)}{r}\cdot Y_{lm}(\varphi,\vartheta), \tag{5.230}$$

$$\vec{E}_\tau^{(lm)}(r,\varphi,\vartheta)=\frac{R_l\text{'}(r)}{l(l+1)}\cdot \operatorname{grad}_\tau Y_{lm}(\varphi,\vartheta), \tag{5.231}$$

$$H_r^{(lm)}(r,\varphi,\vartheta)=0, \tag{5.232}$$

$$\vec{H}_\tau^{(lm)}(r,\varphi,\vartheta)=\frac{i\omega\varepsilon\text{'}rR_l(r)}{l(l+1)}\cdot \operatorname{grad}_\tau^\circ Y_{lm}(\varphi,\vartheta), \tag{5.233}$$

for the electric multipole field, and

$$H_r^{(lm)}(r,\varphi,\vartheta)=\frac{R_l(r)}{r}\cdot Y_{lm}(\varphi,\vartheta), \tag{5.234}$$

$$\vec{H}_\tau^{(lm)}(r,\varphi,\vartheta)=\frac{R_l\text{'}(r)}{l(l+1)}\cdot \operatorname{grad}_\tau Y_{lm}(\varphi,\vartheta), \tag{5.235}$$

$$E_r^{(lm)}(r,\varphi,\vartheta)=0, \tag{5.236}$$

$$\vec{E}_\tau^{(lm)}(r,\varphi,\vartheta) = \frac{i\omega\mu r R_l(r)}{l(l+1)} \cdot \mathrm{grad}_\tau^\circ Y_{lm}(\varphi,\vartheta), \tag{5.237}$$

for the magnetic multipole field, where the index τ indicates the transverse part of the field. Y denotes the so-called spherical harmonics. $R'_l(r)$ is the derivative of $rR_l(r)$ with respect to r and the transverse gradient is defined as

$$\mathrm{grad}_\tau s = \frac{\partial s}{\partial \vartheta}\vec{e}_\vartheta + \frac{1}{\sin\vartheta}\frac{\partial s}{\partial \varphi}\vec{e}_\varphi, \quad \mathrm{grad}_\tau^\circ s = \frac{-1}{\sin\vartheta}\frac{\partial s}{\partial \varphi}\vec{e}_\vartheta + \frac{\partial s}{\partial \vartheta}\vec{e}_\varphi. \tag{5.238}$$

For more sophisticated and more detailed descriptions and for formulations with vector potentials, see [Stratton, 1941], [Jackson, 1975].

Static 3D multipole fields

To obtain multipoles for the Laplace equation in statics, one can use the method of separation, or the results above with the limit k=0. This affects only the radial functions R in (5.222) which have to be replaced by

$$R(r) = \frac{C_r}{r^{l+1}}. \tag{5.239}$$

For $l < 0$ one has regular functions, which are not multipoles.

In electrostatics, one usually introduces the scalar electric potential ϕ that fulfills the Laplace equation and solves the homogeneous Maxwell equation (5.100). Therefore, the procedure outlined above is used to obtain the scalar potential. From this, one derives the electric field with the equation $\vec{E} = -\mathrm{grad}\,\phi$ that defines the scalar potential. The same procedure can be used for solving the current equations (5.106)-(5.108) because (5.106) is the same as (5.100) in electrostatics.

In magnetostatics, one often uses the magnetic vector potential that solves the homogeneous Maxwell equation (5.107), but when no impressed currents are present, (5.106) is homogeneous as well and allows one to introduce a magnetic scalar potential with $\vec{H} = -\mathrm{grad}\,\phi_m$.

Polar coordinates – 2D multipole fields

In cylindrical, 2D problems, one is interested in solutions of the 2D Helmholtz equation. The Cartesian coordinates are well adapted to this type of symmetry. Therefore, one can also use plane wave solutions for modeling 2D problems. To obtain 2D multipoles, we apply polar coordinates r,φ which have a singularity at the origin r = 0 and are the 2D equivalent of the spherical coordinates. With the ansatz product:

$$f(\vec{r}) = f(r,\varphi) = R(r)\cdot\phi(\varphi), \tag{5.240}$$

we find for the 2D Helmholtz equation:

$$f_{,rr} + \frac{1}{r}f_{,r} + \frac{1}{r^2}f_{,\varphi\varphi} + \kappa^2 f = 0, \tag{5.241}$$

the following differential equations:

$$\left(\frac{\partial^2}{\partial r^2}+\frac{1}{r}\frac{\partial}{\partial r}+\kappa^2-\frac{n^2}{r^2}\right)R(r)=0\,, \tag{5.242}$$

$$\left(\frac{\partial^2}{\partial \varphi^2}+n^2\right)\phi(\varphi)=0\,, \tag{5.243}$$

with the same angular solutions as in the 3D case and similar radial functions:

$$R_n(r)=C_r^1 H_n^{(1)}(\kappa r)+C_r^2 H_n^{(2)}(\kappa r)\,, \tag{5.244}$$

$$\phi_n(\varphi)=C_\varphi^+ e^{+in\varphi}+C_\varphi^- e^{-in\varphi}=C_\varphi^c\cos(n\varphi)+C_\varphi^s\sin(n\varphi)\,. \tag{5.245}$$

Obviously, the multipoles of order n=0 are monopoles with rotational symmetry with respect to the origin r = 0. These terms are used for the special solution of the inhomogeneous 2D Helmholtz equation. Unlike the 3D case, the Hankel function of order zero cannot be replaced by a simple expression. Although 2D problems are simpler than 3D problems in general, there are some additional difficulties in the computation of 2D multipoles. Yet there are helpful recurrence relations [Abramowitz, 1970], [Gradshteyn, 1965], and the Hankel functions of the first kind still represent outgoing waves.

To obtain the electromagnetic multipole fields, one has to remember that the longitudinal components fulfill the scalar 2D Helmholtz equations of the form (5.143), whereas the transverse components can be derived using the equations (5.95) and (5.96). Therefore, one has for an n-th order electric 2D multipole

$$E_z^{(n)}(\vec{r}_T)=R_n(r)\cdot\phi_n(\varphi)\,, \tag{5.246}$$

$$\vec{E}_T^{(n)}(\vec{r}_T)=\frac{i\gamma}{\kappa^2}\,\mathrm{grad}_T\,E_z(\vec{r}_T)\,, \tag{5.247}$$

$$H_z^{(n)}(\vec{r}_T)=0, \tag{5.248}$$

$$\vec{H}_T^{(n)}(\vec{r}_T)=\frac{i\omega\varepsilon'}{\kappa^2}\,\mathrm{grad}_T^o\,E_z(\vec{r}_T)\,, \tag{5.249}$$

and for an n-th order magnetic 2D multipole

$$H_z^{(n)}(\vec{r}_T)=R_n(r)\cdot\phi_n(\varphi)\,, \tag{5.250}$$

$$\vec{H}_T^{(n)}(\vec{r}_T)=\frac{i\gamma}{\kappa^2}\,\mathrm{grad}_T\,H_z(\vec{r}_T)\,, \tag{5.251}$$

$$E_z^{(n)}(\vec{r}_T)=0, \tag{5.252}$$

$$\vec{E}_T^{(n)}(\vec{r}_T)=-\frac{i\omega\mu}{\kappa^2}\,\mathrm{grad}_T^o\,H_z(\vec{r}_T)\,. \tag{5.253}$$

Static 2D multipole fields

A difficulty occurs for the low-frequency limit, i.e., for the solution of the 2D Laplace equation. With the frequency, the 2D or transverse wave number κ tends to zero. For this limit, the Hankel functions of order n behave essentially like $1/r^n$, except for the zero order, which behaves like the logarithm ln(r). In effect, for the Laplace equation, one obtains the same angular dependence as for the Helmholtz equation and a simpler radial dependence:

$$R_0(r) = C_r \ln(r), \quad R_n(r) = \frac{C_r}{r^n}, \quad n = 1,2,\ldots \tag{5.254}$$

Note that the monopole, i.e., the multipole of order n=0, represents a 2D point charge or a homogeneous line charge parallel to the z axis for 3D cylindrical problems.

To obtain the fields in electrostatics, magnetostatics, and for the current equations, one essentially proceeds as in 3D space, where a scalar potential is introduced and the field is obtained from the gradient of the scalar potential.

Note that one can also introduce a magnetic vector potential in magnetostatics with curl $\vec{A} = \vec{B}$. Since the current on a cylindrical structure has a z component only, also the vector potential has only a z component. Therefore, one implicitly has a scalar potential A_z. From A_z one can derive the magnetic field using (1.91) for cylindrical problems. Since the derivative of the fields with respect to the z axis is zero in statics, one obtains

$$\vec{B}_T = (\text{curl}\,\vec{A})_T = -\text{grad}_\text{T}^\circ A_z. \tag{5.255}$$

Thus, one can derive the magnetic field from the magnetic vector potential in a way that is very similar to the derivative of the field from the scalar potential. Obviously, one can also introduce a vector potential for solving the current equations and in electrostatics, when there are no impressed charges.

Other types of multipole expansions - normal expansions

It is well known that one can obtain standing waves by superposition of waves propagating in opposite directions. By linear combinations of the two parts of the radial and angular dependencies, one can obtain the corresponding Bessel functions J and the Neumann functions N:

$$J_n = (H_n^{(1)} + H_n^{(2)})/2, \quad N_n = (H_n^{(1)} - H_n^{(2)})/2i. \tag{5.256}$$

This superposition can be applied to both 2D and 3D multipole fields. It is similar to the well-known superposition of harmonic functions:

$$\cos(n\varphi) = (e^{in\varphi} + e^{-in\varphi})/2, \quad \sin(n\varphi) = (e^{in\varphi} - e^{-in\varphi})/2i \tag{5.257}$$

which are real-valued for real arguments. Also the Bessel and Neumann functions are real valued for real arguments.

Since the radial dependence is most characteristic for 2D and 3D multipole fields, we call the products (5.218) and (5.240) as well as the corresponding fields (5.230)-(5.237) and (5.246)-(5.253) the Hankel expansion, Bessel expansion, or Neumann expansion depending on the function involved in the radial dependence. Instead, we also use the terms Hankel-type multipole field, Bessel-type multipole field, and Neumann-type multipole field, although the Bessel-type multipole field has no singularity for $r = 0$.

The Bessel functions are also called cylindrical functions of the first kind. They are regular for $r = 0$ (i.e., have no singularity). As these functions together with the angular dependence are not multipoles, we also call them 'normal' expansions. For large distances r, these expansions tend to infinity except for real arguments. The radial part of the corresponding static normal expansions has the form r^n, for $n>0$ and $\ln(r)$ for $n=0$, which always tends to infinity.

The Neumann functions are also called cylindrical functions of the second kind. These functions are sometimes denoted in the literature by Y and have a singularity for $r=0$ like the Hankel functions. For large arguments they behave more like the Bessel functions. Neumann functions are interesting only for real arguments because they are real-valued for real arguments.

The Hankel functions are cylindrical functions of the third kind. These multipole expansions show a 'local behavior', which is numerically very agreeable and allows use of one origin (i.e., multiple multipoles) to expand the solutions in any domain.

The Hankel functions as well as the Bessel functions are either real or imaginary valued for imaginary arguments. This allows one to introduce slightly different or 'modified' cylindrical functions I and K, which are real-valued for real arguments. This allows use of purely real expansions for loss-free problems such as guided waves on structures composed of ideal conductors and ideal dielectrics.

If the imaginary part of the argument of the cylinder functions is positive, the Hankel functions of the first kind decay exponentially for large arguments, whereas Hankel functions of the second kind grow exponentially. This means that only multipoles with Hankel functions of the first kind show a strictly local behavior. For negative imaginary parts of the argument, the Hankel functions of the second kind decay exponentially for large arguments. In electromagnetics, the argument of the Hankel functions is computed from a square root in (5.140) or (5.142). Instead of the conventional definition of the square root (where the real part of the result is positive), one can choose the sign of the square root in a way that the imaginary part of the result is positive. With this unconventional definition of the square root, the Hankel functions of the first kind represent outgoing waves, whereas the Hankel functions of the second kind represent incoming waves. Superpositions of outgoing waves are very well suited to expand, for example, the scattered electromagnetic field.

Completeness of multipole fields

As we have already found an infinite number of solutions, the question of the completeness of these functions would seem less important than the question of how to reduce this number for numerical computations. Nonetheless, there is a mathematical theory of I. N. Vekua [Vekua, 1967], which provides all the necessary theorems for 2D domains with boundaries that are continuous in the Hoelder sense, i.e., that are continuous and can be differentiated infinitely many times, except at a finite number of points where the first or higher order derivatives may be discontinuous. The situation is similar to, but much more demanding than for the approximation of a given function by a series expansion. According to Vekua, one can approximate any solution of the Helmholtz equation in a multiply connected domain with Hoelder continuous boundaries by a superposition of a Bessel expansion and K Hankel expansions of sufficiently high maximum order. The number K denotes the number of holes D_k, $k=1,2,\ldots,K$ in the domain D. To obtain a complete approximation basis, one must define K multipole expansions with one origin O_k in each of the holes D_k. The origin O_0 of the Bessel expansion is arbitrary. If the domain D has no exterior boundary, the Bessel expansion must be omitted. Figure 40 illustrates the situation for a domain with an exterior boundary ∂D_0 and K holes with the boundaries ∂D_k. All of these boundaries must be closed and continuous lines.

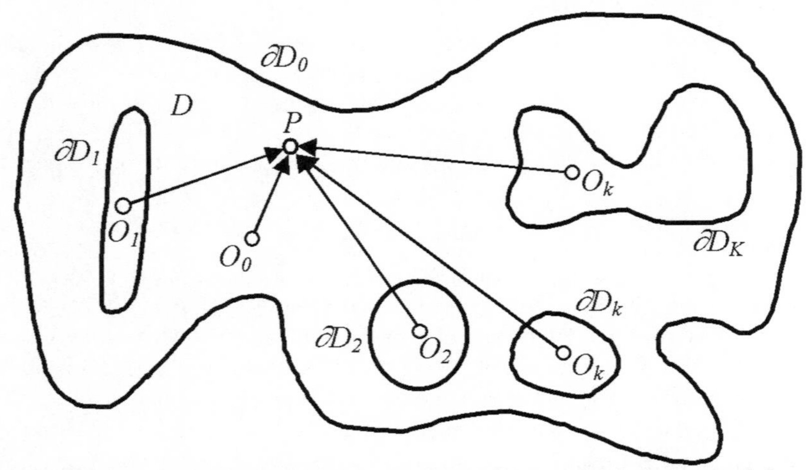

Figure 40: A complete multipole approximation basis for a multiply connected domain D with exterior boundary ∂D_0 and K interior boundaries ∂D_k consists of a normal (Bessel) expansion with an arbitrary origin O_0 and K multipole expansions with origins O_k. One multipole must be placed in each of the holes of D. P is an arbitrary point in the domain D. To evaluate the field in P, K+1 polar coordinate systems with origins O_k, k=0,1,2,...,K are required.

When one sets a normal expansion with origin O_0 and K multipole expansions with origins O_k, one has to set a couple of orders n for each of these expansions. The minimum order must always be 0 and all integer orders up to a maximum order N_k must be used for each expansion. When all maximum orders N_k are high enough, one can obtain an approximation of any regular solution of Maxwell's equations in D with an error that is below a given error bound ε. The error bound may be an arbitrarily small, positive number and the error is computed using the square norm over the domain D. This is called a *complete approximation basis.*

There is no 3D theory equivalent to Vekua's theory for 2D domains, but it is expected that a similar 3D multipole expansion would also be complete. Proving this seems to be extremely hard because there is no 3D technique that is equivalent to conformal mapping which was used for the 2D proofs by Vekua.

Although proofs of completeness are nice from the mathematical and theoretical point of view, completeness in a general sense is not required in practice because it is sufficient to have expansions that approximate the *desired* solution rather than all possible, regular solutions. Moreover, completeness is not sufficient for obtaining efficient computer codes, because the theorems do not guarantee that the numbers N_k are reasonably small.

Generalized solutions - connections

There are many ways to generalize the solutions of Maxwell's equations we have found in the previous subsections. In the following, some techniques are outlined.

First of all, Maxwell's equations and the simple material properties are linear. Therefore, any linear combination or *superposition* of known analytic solutions in a given domain D is again an analytic solution. This is also called a *connection*. A connection f_k has the form

$$f_k = \sum_{n=1}^{N} A_n f_n, \tag{5.258}$$

where f_n is any field that is an analytic solution of Maxwell's equations. The amplitudes A_n are assumed to be known numbers. Usually, N is a finite number. When the amplitudes are a known, infinite sequence, N may be *infinite*. It is obvious that a single connection can approximate the desired solution very well when an appropriate basis f_n is used, when the amplitudes are appropriately set, and when N is sufficiently high. When one finds a set of appropriate connections, one may obtain more accurate solutions by a superposition of connections, i.e., connections may be *nested*. Finally, the desired solution may be considered as a single connection. If an accurate solution of a distorted or perturbed problem is known, this solution can be embedded in a connection. The given problem may then be solved by a superposition of the connection and additional basis functions that model the effects caused by the distortion or perturbation.

Examples of connections are the superposition of fields caused by point charges, by a set of monopoles or dipoles located at different positions, etc. A further generalization of this concept leads to continuous connections that are defined by integrals rather than by summations. We have already encountered such integrals in the subsection on Green's functions. There, integrals over current and charge distributions were obtained. Since we have found other analytic solutions than the fields of electric charges and currents, we can obtain continuous connections, for example, from integrals over multipole fields along a given line. These are called *line multipoles*. Similarly, we can obtain surface multipoles, etc. Instead of multipoles, we can also use plane waves to obtain continuous connections. With such plane wave integrals, one can theoretically obtain any solution of Maxwell equations, including multipole solutions.

Continuous connections may also be considered as a combination of the Green's technique with discrete connections or as a combination of Coulomb-Ampère integrals with solutions obtained from the separation of variables.

When the given problem consists of more than a single domain D_l, one usually selects a separate set of basis functions for each domain, but one can also define basis functions that are applied to several domains. Therefore, one can also define *multi-domain connections* that are linear combinations of basis fields defined in different domains. Such connections might also be applied when sub-domain basis functions are used, i.e., in domain methods, whereas usual connections are typical for boundary methods that use basis functions that are analytic solutions of Maxwell equations within a given domain.

Connections provide an extremely high degree of freedom. Theoretically, any solution can be considered as a connection of other solutions. The high degree of freedom makes it very hard to appropriate connections for complicated problems, but experienced scientists may use patchworks of connections to obtain very efficient methods.

Energy

In Maxwell's electrodynamics, one can find the roots of the theory of relativity and quantum physics. Both the concept of fields and the finite velocity of electromagnetic waves impressed Einstein so much that he decided to rewrite the old equations of Newton's mechanics. In addition, Maxwell's theory brought a new view of energy and the law of energy conservation, which is fundamental for quantum physics and some numerical methods as well.

However, quantum concepts such as the electron and the photon, which look like a rebirth of pre-Maxwellian ideas, very soon showed the limitations of pure Maxwellian theory. For most applications of technical interest, however, the classical form of Maxwell's electrodynamics with a new (old) interpretation of charges and currents is sufficient. For this reason, only those concepts of post-Maxwellian physics that are important for numerical computations of classical electrodynamics are discussed here.

The electromagnetic fields suffer from the fact that electric and magnetic fields are not really comparable objects. It has been shown in the previous chapter that electric fields are vector fields, whereas magnetic fields are pseudovector fields. In addition, the two forms of electric fields $\vec{E}$, $\vec{D}$ and of magnetic fields $\vec{H}$, $\vec{B}$ have different dimensions. The numerical values of $\vec{E}$, $\vec{D}$, $\vec{H}$, $\vec{B}$ of a simple plane wave in free space are quite different in magnitude and depend on the units. This can cause numerical problems. Energy is a concept, which provides a link between different parts of physics, because all types of energy, such as mechanical, chemical, electric, or magnetic, can be transformed into one another. In pure electrodynamics, the concept of energy density of the electric and magnetic field is helpful in solving numerical problems such as scaling and weighting. Somehow, energy brings the idea of justice in physics, as it provides one measure for all kinds of forces. This is important for numerical computations due to their limited accuracy. Incidentally, effects of the limited accuracy of numbers appear very similar to quantum effects.

The energy concept

In classical electrodynamics, there are essentially three kinds of energy according to the three material properties of permittivity, permeability, and conductivity ((5.57)-(5.59)): 1) energy density of the electric field, 2) energy density of the magnetic field, and 3) power loss. The power loss in lossy media describes a transformation of electromagnetic energy into thermal energy and is therefore not a purely electromagnetic effect. This term is added because of its practical importance. These energy densities and the power loss are defined statically, but the law of energy conservation is essentially dynamic.

To define the electric energy density w_e, we consider a simple, ideal capacitor with two parallel, plane electrodes, with a homogeneous electric field between the electrodes. Since the field is homogeneous, also the energy density is homogeneous. When we compute the energy required for charging such a capacitor, we find for the electric energy density

$$w_e = \vec{E} \cdot \vec{D} / 2. \tag{5.259}$$

Similarly, we can analyze an ideal inductor and an ideal resistor to obtain the magnetic energy density

$$w_m = \vec{H} \cdot \vec{B} / 2, \tag{5.260}$$

and the power loss density

$$p = \vec{E} \cdot \vec{j}. \tag{5.261}$$

Note that w_e, w_m, and p are related to the three simple material properties permittivity, permeability, and conductivity.

In the pre-Maxwellian form of the law of energy conservation, only energetically closed domains or systems were considered. In such systems the total energy W is constant in time, i.e., its time derivative is zero. If we consider an energetically open domain D with boundary ∂D, the change of the total energy W in D must be compensated by a power flux P through ∂D. When energy is converted into another form of energy inside the domain, the energy conversion must be taken into account.

The idea of a power flux and energy distributed over space came with the analysis of Maxwell's equations and the concept of the electromagnetic field by Poynting. Incidentally, there were famous pioneers of electrodynamics, such as Hertz, who refuted Poynting's view of field energy density. The Poynting theorem is formally identical to the law of charge conservation. In electromagnetics, W is replaced by an integral over the energy densities w_e and w_m of the electric and the magnetic field. Since electromagnetic energy may be converted into thermal energy in lossy media, two terms reduce the electromagnetic energy contained in a domain: 1) the transport of electromagnetic energy by the power flux through ∂D, given by a surface integral, and 2) the transformation of electromagnetic energy into thermal energy within D, given by a volume integral:

$$-\frac{\partial}{\partial t} W = -\frac{\partial}{\partial t} \int_D (w_e + w_m) \mathrm{d}V = S + P = \oint_{\partial D} \vec{S} \, \mathrm{d}\vec{F} + \int_D p \, \mathrm{d}V \, . \tag{5.262}$$

If other non-electromagnetic energies are involved, additional terms have to be introduced [Landau, 1958]. The Poynting vector $\vec{S}$ indicates the oriented power flux density in space. The theorem of Gauss allows us to replace the surface integral by a volume integral and with the static definitions (5.259)-(5.261) of the energy densities and the density of power loss, we find

$$2\int_D \operatorname{div} \vec{S} \, \mathrm{d}V = -\int_D \left(\vec{E} \frac{\partial \vec{D}}{\partial t} + \vec{D} \frac{\partial \vec{E}}{\partial t} + \vec{H} \frac{\partial \vec{B}}{\partial t} + \vec{B} \frac{\partial \vec{H}}{\partial t} + 2 \vec{j}_\sigma \vec{E} \right) \mathrm{d}V \, . \tag{5.263}$$

For linear, homogeneous, isotropic materials, one has

$$\vec{D} \frac{\partial \vec{E}}{\partial t} = \vec{E} \frac{\partial \vec{D}}{\partial t} + \varepsilon \vec{D} \vec{E} \frac{\partial \varepsilon^{-1}}{\partial t}, \quad \vec{B} \frac{\partial \vec{H}}{\partial t} = \vec{H} \frac{\partial \vec{B}}{\partial t} + \mu \vec{H} \vec{B} \frac{\partial \mu^{-1}}{\partial t}. \tag{5.264}$$

If the materials are not time dependent, these equations are simplified and one can simplify (5.263) as well. Otherwise, additional energy terms are necessary because one needs energy to change the material properties. With Maxwell's equations, one can verify that the Poynting vector is the vector product of $\vec{E}$ and $\vec{H}$:

$$\vec{S}(\vec{r}, t) = \vec{E}(\vec{r}, t) \times \vec{H}(\vec{r}, t). \tag{5.265}$$

For time-harmonic fields, one can define a complex Poynting vector:

$$\vec{S}(\vec{r}) = \left(\vec{E}(\vec{r}) \times \vec{H}^*(\vec{r}) \right) / 2 \tag{5.266}$$

and find the following form of the law of energy conservation:

$$\oint_{\partial D} \vec{S}\,\mathrm{d}\vec{F} = \frac{1}{2}\int_D \mathrm{div}\left(\vec{E}\times\vec{H}^*\right)\mathrm{d}V = \frac{1}{2}\int_D \left(i\omega\mu H^2 - i\omega\varepsilon E^2 - \sigma E^2\right)\mathrm{d}V, \tag{5.267}$$

where $E^2 = \vec{E}\vec{E}^*$, $H^2 = \vec{H}\vec{H}^*$. The real part of the complex Poynting vector is the time average of the power flow density. From the real part of (5.267), we can see that the electromagnetic energy which enters the domain D in the time average is converted into thermal energy. In an energetically closed domain, one has no real part of the Poynting vector and no power losses, and thus one has from the imaginary part of (5.267) the following *resonance condition* for a *loss-free resonator*:

$$\int_D \left(\mu H^2 - \varepsilon E^2\right)\mathrm{d}V = 0. \tag{5.268}$$

Cylindrical structures

If we consider a cylindrical domain with cross section S_D and divide the complex Poynting vector into a longitudinal and a transverse part, we have

$$\int_{S_D} S_z\,\mathrm{d}F\,\Big|_{z=z_1}^{z=z_2} - \int_{z_1}^{z_2}\oint_{\partial S_D} \vec{S}_T\,\mathrm{d}\vec{\ell}^{\circ}\,\mathrm{d}z = \frac{i\omega}{2}\int_{z_1}^{z_2}\int_{S_D}\left(\mu H^2 - \varepsilon E^2 - \frac{\sigma E^2}{i\omega}\right)\mathrm{d}F\,\mathrm{d}z. \tag{5.269}$$

Note that all integrands are essentially squares of the electric or magnetic fields and therefore have the z dependence:

$$e^{i\gamma z}e^{-i\gamma^* z} = e^{i(\beta+i\alpha)z}e^{-i(\beta-i\alpha)z} = e^{-2\alpha z}. \tag{5.270}$$

Hence we have

$$\int_{z_1}^{z_2} \ldots\,\mathrm{d}z = \frac{-1}{2\alpha}\ldots\Big|_{z=z_1}^{z=z_2}, \tag{5.271}$$

and find

$$-2\alpha\int_{S_D} S_z\,\mathrm{d}F - \oint_{\partial S_D} \vec{S}_T\,\mathrm{d}\vec{\ell}^{\circ} = \frac{i\omega}{2}\int_{S_D}\left(\mu H^2 - \varepsilon E^2 - \frac{\sigma E^2}{i\omega}\right)\mathrm{d}F. \tag{5.272}$$

For the longitudinal and transverse components of the complex Poynting vector, one can write

$$S_z = \frac{1}{2}\vec{E}_T^{o}\cdot\vec{H}_T^* = -\frac{1}{2}\vec{E}_T\cdot\vec{H}_T^{o*}, \tag{5.273}$$

$$\vec{S}_T = \frac{1}{2}\left(E_z\vec{H}_T^{o*} - H_z^*\vec{E}_T^{o}\right), \tag{5.274}$$

and find the following two-dimensional form of the law of energy conservation:

$$-\alpha\int_{S_D}\vec{E}_T^{o}\cdot\vec{H}_T^*\,\mathrm{d}F - \frac{1}{2}\oint_{\partial S_D}\left(E_z\vec{H}_T^* - H_z^*\vec{E}_T\right)\mathrm{d}\vec{\ell} = \frac{i\omega}{2}\int_{S_D}\left(\mu H^2 - \varepsilon E^2 - \frac{\sigma E^2}{i\omega}\right)\mathrm{d}F. \tag{5.275}$$

As for the ideal 3D resonator, one has in the ideal 2D case (i.e., for guided waves on loss-free cylindrical structures or for ideal 2D resonators) the propagation condition or the 2D resonance condition:

$$\int_D \left(\mu H^2 - \varepsilon E^2\right) \mathrm{d}V = 0 \cdot \tag{5.276}$$

For guided waves on loss-free structures, $\gamma = \beta$ plays the role of the resonance frequency. Such problems are called eigenvalue problems. In general, they have several solutions with different eigenvalues and different field distributions. The solutions are called modes. These eigenvalue problems are nonlinear because the fields in the equations above implicitly contain the eigenvalues. Except for very special cases with additional assumptions, eigenvalue problems cannot be solved analytically. Usually, the resonance conditions given above or similar eigenvalue conditions have to be solved iteratively.

In the special case of waves on transmission lines, one can avoid the definition of an eigenvalue problem. Usually, transmission lines are replaced by electrical networks. For the computation of the network parameters, static field computations with some additional dynamic approximations such as the skin effect are applied. Unlike the 'exact' eigenvalue computations, knowledge of the propagation constant is not necessary for these field computations. The propagation constants of different modes are computed afterward from the network parameters. This technique does not allow for all modes of a transmission line. Yet it allows one to compute both the fields and the propagation constants of the most important modes quite accurately, although the law of energy conservation is not fulfilled everywhere. Theoretically, it is possible to obtain the modes of transmission lines by exact solution of an eigenvalue problem. However, if this is done numerically, one has severe numerical problems for low frequencies. For this reason, the results can be worse than the 'network approximations'.

Variational integrals

Once the importance of energy had been recognized, the idea arose that 'God created the world economically', i.e., with as little total energy as necessary. This idea has its roots in the well-known principle of Fermat, which says that light propagates along the 'optically' shortest path. In fact, the total energy:

$$W_e = \frac{1}{2} \int_D \varepsilon \left(\vec{E}(\vec{r})\right)^2 \mathrm{d}V = \frac{1}{2} \int_D \varepsilon \left(\operatorname{grad} \phi(\vec{r})\right)^2 \mathrm{d}V \tag{5.277}$$

of the electrostatic field in a charge-free domain D is minimal in the following sense. The fields are distributed in such a way that any small change of this distribution causes an increase of the total energy W_e or, mathematically speaking, the variation of the total energy vanishes:

$$\delta W_e = \frac{1}{2} \delta \int_D \varepsilon \left((\phi_{,x})^2 + (\phi_{,y})^2 + (\phi_{,z})^2 \right) \mathrm{d}V = 0 \cdot \tag{5.278}$$

For the more general variational integral:

$$\delta I = \delta \int_D F(x, y, z, u, u_{,x}, u_{,y}, u_{,z}) \, \mathrm{d}V = 0, \tag{5.279}$$

one has Euler's partial differential equation:

$$F_{,u} - \frac{\partial}{\partial x} F_{,u_x} - \frac{\partial}{\partial x} F_{,u_y} - \frac{\partial}{\partial x} F_{,u_z} = 0, \tag{5.280}$$

or in full length:

$$\begin{aligned} & F_{,u} - F_{,u_x u_x} u_{,xx} - F_{,u_y u_y} u_{,yy} - F_{,u_z u_z} u_{,zz} \\ & -2\left(F_{,u_x u_y} u_{,xy} - F_{,u_x u_z} u_{,xz} - F_{,u_y u_z} u_{,yz}\right) \\ & -F_{,u_x u} u_{,x} - F_{,u_y u} u_{,y} - F_{,u_z u} u_{,z} - F_{,u_x x} - F_{,u_y y} - F_{,u_z z} = 0 \end{aligned} \tag{5.281}$$

For the electrostatic variational integral (5.278), one finds the electrostatic Laplace equation in Cartesian coordinates as Euler's equation.

By considering the resonance and propagation condition, in the previous section, one can expect that the variation of the total energy in a dynamic system does not vanish. In fact, the total energy does not provide a correct variational integral for any dynamic system. In general, it can be quite difficult to find an appropriate variational integral for a given set of differential equations. In the case of ideal resonators and waveguides, the resonance condition and the propagation condition suggest trying the difference instead of the sum of the electric and magnetic energy. The variational formulation of electrodynamics cannot be discussed here in detail. In the following, a short outline of the simple 2D time-harmonic case without losses and sources is given. For this case, we can express the transverse components of the electric and magnetic field in terms of the longitudinal components, according to and, (8.6). Thus, one finds for the integral:

$$I = \int_{F_D} \left(\varepsilon E^2 - \mu H^2\right) \mathrm{d}F, \tag{5.282}$$

in a first step, the rather complicated integral:

$$\begin{aligned} I = & \int_{F_D} \left(\frac{\varepsilon \kappa^2}{|\kappa|^4} |\mathrm{grad}_T\, E_z|^2 - \frac{\mu \kappa^2}{|\kappa|^4} |\mathrm{grad}_T\, H_z|^2 + \varepsilon E_z^2 - \mu H_z^2 \right) \mathrm{d}F \\ & + \int_{F_D} \frac{2k^2 \beta}{\omega |\kappa|^4} \left(\mathrm{grad}_T\, E_z \cdot \mathrm{grad}_T^o\, H_z + \mathrm{grad}_T^o\, E_z \cdot \mathrm{grad}_T\, H_z \right) \mathrm{d}F\,. \end{aligned} \tag{5.283}$$

Considering the rules for the operation o, we see that the last two terms compensate each other, and we obtain

$$I = \int_{F_D} \left(\frac{\varepsilon \kappa^2}{|\kappa|^4} |\mathrm{grad}_T\, E_z|^2 - \frac{\mu \kappa^2}{|\kappa|^4} |\mathrm{grad}_T\, H_z|^2 + \varepsilon E_z^2 - \mu H_z^2 \right) \mathrm{d}F\,. \tag{5.284}$$

In Cartesian coordinates, this integral has the general form:

$$I = \int_D F(x, y, u, u_{,x}, u_{,y}, v, v_{,x}, v_{,y})\, \mathrm{d}V\,. \tag{5.285}$$

With the corresponding Euler equations:

$$F_{,u} - \frac{\partial}{\partial x} F_{,u_x} - \frac{\partial}{\partial x} F_{,u_y} = 0, \tag{5.286}$$

$$F_{,v} - \frac{\partial}{\partial x} F_{,v_x} - \frac{\partial}{\partial x} F_{,v_y} = 0, \tag{5.287}$$

one finally finds the homogeneous 2D Helmholtz equations for the longitudinal components:

$$\left(\Delta_T + \kappa^2\right) \begin{matrix} E_z(\vec{r}_T) \\ H_z(\vec{r}_T) \end{matrix} = 0 \,. \tag{5.288}$$

Although the formalism and derivation of variational integrals are quite complicated, this concept is very successful for numerical computations. For more information, see, for example, [Collin, 1991], [Landau, 1958], [Vasallo, 1991].

Perturbations

Another interesting application of the variational calculus is the computation of 'small' perturbations (see, for example, [Vasallo, 1991]). To simplify the equations and for several physical definitions, additional assumptions concerning the geometry, material properties, etc., are made. Once the 'undisturbed' fields are known from a simplified computation, one can try to improve the results by a computation of the perturbations. This is important to estimate the errors made by the additional assumptions, and to verify the stability of the solution (i.e., to verify that small deviations do not cause large effects).

To illustrate this process, we assume that the electromagnetic fields in a leaky resonator have first been computed with an idealized model. For such problems, only the first two Maxwell equations are important. From these equations, we derive the following 'disturbed' Maxwell equations in the frequency domain:

$$\operatorname{curl} \delta\vec{E} = i\vec{H}\delta(\omega\mu) + i\omega\mu\,\delta\vec{H}, \tag{5.289}$$

$$\operatorname{curl} \delta\vec{H} = i\vec{E}\delta(\omega\varepsilon) + i\omega\varepsilon\,\delta\vec{E}\,. \tag{5.290}$$

Now, we try to eliminate the unknown variations of the electric and magnetic fields in order to obtain a relation between the variation of the frequency and the material properties. To do so, we define a kind of complex Poynting vector:

$$\vec{S}_\delta = \vec{E}^* \times \delta\vec{H} + \delta\vec{E} \times \vec{H}^* \tag{5.291}$$

and we have for the divergence with (1.100) and the 'disturbed' Maxwell equations (5.289) and (5.290):

$$\operatorname{div} \vec{S}_\delta = H^2\delta(\omega\mu) + E^2\delta(\omega\varepsilon)\,. \tag{5.292}$$

By integrating over the resonator, which is energetically closed, we obtain

$$\int_D \left(H^2\delta(\omega\mu) + E^2\delta(\omega\varepsilon)\right)\mathrm{d}V = \oint_{\partial D} \vec{S}_\delta \mathrm{d}F = 0\,. \tag{5.293}$$

With

$$\delta(a\cdot b) = a\cdot\delta b + b\cdot\delta a \tag{5.294}$$

one obtains the variation of the resonance frequency in function of the variations of the material properties:

$$\delta\omega = -\frac{\omega\int_D \left(H^2\delta\mu + E^2\delta\varepsilon\right)\mathrm{d}V}{\int_D \left(\mu H^2 + \varepsilon E^2\right)\mathrm{d}V}. \tag{5.295}$$

This formula can be used, for example, to obtain the damping constant for a resonator with small losses if the fields are known for the loss-free case. With the complex frequency $\omega' = \omega - i\alpha_t$ and the complex permittivity $\varepsilon' = \varepsilon + i\sigma/\omega$ one has from the imaginary part of (5.295)

$$\alpha_t = -\frac{\int_D \sigma E^2\,\mathrm{d}V}{\int_D \left(\mu H^2 + \varepsilon E^2\right)\mathrm{d}V}. \tag{5.296}$$

In a similar manner, one finds in the 2D case for the variation of the propagation constant:

$$\delta\beta = -\frac{\int_{S_D} \left(H^2\delta(\omega\mu) + E^2\delta(\omega\varepsilon)\right)\mathrm{d}F}{2\int_{S_D} \vec{E}_T^o \cdot \vec{H}_T^*\,\mathrm{d}F} \tag{5.297}$$

and for the approximation of the damping constant of guided waves:

$$\delta\alpha \approx \frac{\int_{S_D} \sigma E^2\,\mathrm{d}F}{2\int_{S_D} \vec{E}_T^o \cdot \vec{H}_T^*\,\mathrm{d}F}. \tag{5.298}$$

Note that (5.297) is helpful for the group velocity v_g as well. With $\delta\mu = \delta\varepsilon = 0$, one obtains

$$v_g^{-1} = \frac{\delta\beta}{\delta\omega} = -\frac{\int_{S_D} \left(\mu H^2 + \varepsilon E^2\right)\mathrm{d}F}{2\int_{S_D} \vec{E}_T^o \cdot \vec{H}_T^*\,\mathrm{d}F}. \tag{5.299}$$

Boundary and continuity equations

In the previous sections, we considered the different field equations and their solutions only in linear, homogeneous, isotropic domains. For most technical applications, one has several materials (i.e., several domains). The Maxwell equations must hold everywhere, in the different domains as well as on the boundaries ∂D_{ij} between two domains, D_i and D_j. The material properties are not continuous on the boundaries. Therefore, it is difficult to apply the differential forms of the Maxwell equations but there is no problem with the corresponding integral forms (5.17) to (5.20). If we consider a sufficiently small part of a smooth boundary, the boundary looks flat and the fields are homogeneous on both sides. With a small rectangular path along this boundary (see Figure 41) and the first two Maxwell equations we get the continuity of the tangential components on ∂D_{ij}:

$$E_{it} - E_{jt} = 0\,, \qquad E_{i\tau} - E_{j\tau} = 0\,, \tag{5.300}$$

$$H_{it} - H_{jt} = \alpha\tau\,, \quad H_{i\tau} - H_{j\tau} = -\alpha_t\,, \tag{5.301}$$

where α is the density of surface currents on ∂D_{ij}. The indices t and τ are used for the tangential components. The direction of the corresponding unit vectors $\vec{e}_t, \vec{e}_\tau$ is defined as follows: 1) both are perpendicular, i.e., $\vec{e}_t \cdot \vec{e}_\tau = 0$, and 2) the vector product $\vec{e}_t \times \vec{e}_\tau$ points from domain D_j into D_i. If there is an unknown surface current as on ideal conductors, the second condition is used to compute the surface currents and does not restrict the H field.

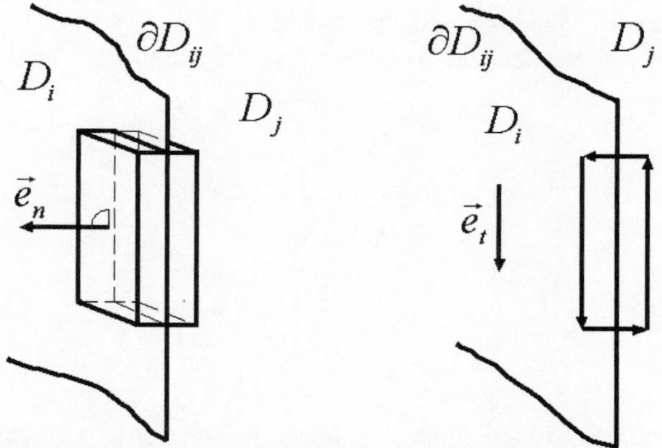

Figure 41: Paths of integration for the derivation of the continuity conditions between two domains, D_i and D_j. Left-hand side: derivation of the normal continuity conditions. Right-hand side: derivation of the tangential continuity conditions.

By integrating over a rectangular box, from the third and fourth Maxwell equations, we have the following continuity equations for the normal components on ∂D_{ij}:

$$D_{in} - D_{jn} = \varsigma\,, \tag{5.302}$$

$$B_{in} - B_{jn} = 0\,, \tag{5.303}$$

where n denotes the normal components and ς is the surface charge density on ∂D_{ij}. If there are unknown charges on ∂D_{ij} as on ideal conductors, the first condition is used to compute ς.

Perfect Electric Conductors (PEC)

Because there is no field inside a perfect electric conductor D_0, we have the following boundary conditions on the surface ∂D_{i0}:

$$E_{it} = 0\,, \qquad E_{i\tau} = 0\,, \tag{5.304}$$

$$B_{in} = 0\,, \tag{5.305}$$

and we can compute the electric surface currents and charges according to

$$\alpha_\tau = H_{it}, \qquad \alpha_t = -H_{i\tau}, \tag{5.306}$$

$$\varsigma = D_{in}. \tag{5.307}$$

Perfect Magnetic Conductors (PMC)

A PEC is an idealized material that does not exist in nature. Good electric conductors such as copper can be modeled very accurately as a PEC over a large frequency range, provided that their geometric size is sufficiently big (compared with the skin depth). On a PEC, one has electric surface currents and electric charges. In the symmetric Maxwell equations (5.13) to (5.16), we have also hypothetical magnetic currents and magnetic charges. This allows us to define a PMC as another idealized material with magnetic surface currents and magnetic surface charges. When we assume that there are no electric surface charges and currents on such a material, we have the following boundary conditions on the surface ∂D_{i0}:

$$H_{it} = 0\,, \qquad H_{i\tau} = 0\,, \tag{5.308}$$

$$D_{in} = 0\,, \tag{5.309}$$

and we can compute the magnetic surface currents and charges according to

$$\alpha_\tau = E_{it}\,, \qquad \alpha_t = -E_{i\tau}\,, \tag{5.310}$$

$$\varsigma = B_{in}\,. \tag{5.311}$$

Symmetry planes

Often the geometric structure of a device is symmetric with respect to a symmetry plane. In this case, one can decompose the electromagnetic field in two parts – at least when the materials are isotropic. The first part has an electric field that is perpendicular and a magnetic field that is tangential to the symmetry plane. This means that the symmetry plane may be replaced by a PEC plane for this part. For the second part, one has a magnetic field that is perpendicular and an electric field that is tangential to the symmetry plane. Therefore, one can replace the symmetry plane by a PMC plane for the second part.

The replacement of symmetry planes by PEC and PMC planes is often used in numeric codes to reduce the model size, because only one half of the structure must be modeled explicitly when such a plane is present. It is important to note that the symmetry decomposition into two parts is also possible when two or three symmetry planes are present in 3D space, provided that all symmetry planes are perpendicular to each other. If there are more than three symmetry planes or when the symmetry planes are not perpendicular to each other, a more sophisticated symmetry decomposition – based on the representation theory of finite symmetry groups - is required in electrodynamics [Stiefel, 1992], [Hafner, 1981].

Lossy dielectric and magnetic media

For linear, homogeneous, isotropic materials on both sides of the boundary, we can eliminate the D and B field and obtain

$$\varepsilon_i E_{in} - \varepsilon_j E_{jn} = \varsigma, \tag{5.312}$$

$$\mu_i H_{in} - \mu_j H_{jn} = 0\,, \tag{5.313}$$

instead of (5.302) and (5.303).

For the amplitudes of time-harmonic fields, one formally has the same equations. There is an additional equation of special interest: for *time-harmonic* fields one finds in the absence of impressed currents if div is applied to (5.73):

$$\varepsilon'_i E_{in} - \varepsilon'_j E_{jn} = 0. \tag{5.314}$$

This equation is a restriction for the normal component of the electric field, whereas (5.312) is used to compute unknown surface charges that are present on the boundary between lossy dielectrics. When the dielectrics on both sides of the boundary are loss-free, the left hand sides of the two equations become identical. Consequently, no surface charges may be present. When impressed surface charges would be present on the boundary between two perfect, loss-free dielectrics, these surface charges could not vary in time because no current can flow through the dielectrics to cause any change in the charge density. This means that impressed surface charges between perfect dielectrics can only be present in electrostatics.

Surface Impedance Boundary Conditions (SIBC)

When good conductors are present, it may happen that the modeling of the corresponding domains as PEC is not accurate enough, especially because the PEC modeling does not take the losses into account. Of course, one can model good conductors as lossy dielectrics with a high conductivity, but this usually leads to numerical problems. In the time-harmonic case, the field in a lossy dielectric is mainly described by the skin effect, i.e., the amplitude of the field decreases exponentially with the distance from the surface. Moreover, the field oscillates very rapidly because the wavelength inside a good conductor is high. Therefore, a very fine discretization is required inside a good conductor for modeling the field accurately. In such cases, a good approximation is obtained from the assumption that the field inside a good conductor is mainly a damped plane wave entering the conductor perpendicular to its surface. From this, one obtains the surface impedance boundary conditions on the surface ∂D_{i0}:

$$\vec{E}_i - (\vec{e}_n \cdot \vec{E}_i) \cdot \vec{e}_n = Z_0 \vec{e}_n \times \vec{H}_i\,, \quad Z_0 = \sqrt{\mu / \varepsilon'}\,, \tag{5.315}$$

where Z_0 is the wave impedance of the conducting domain D_0. Note that the electromagnetic field inside the conductor is neither contained in the SIBC (5.315), nor explicitly evaluated. Therefore, the handling of a good conductor with SIBC is very similar to the handling of a PEC, but the material properties contained in the wave impedance allow one to take losses in the conductor into account.

Periodic problems

In 3D space, periodic problems (see, for example, [Collin, 1991], [Yeh, 1988]) consist of a special geometry that is periodically repeated in one, two or three directions. This can be considered as a special type of symmetry. Periodic symmetries are described by infinite symmetry groups [Stiefel, 1992] that are quite tricky. Fortunately, a complete symmetry decomposition is not required for many periodic problems of practical interest. Instead of a rigorous analysis using representation theory, a more intuitive procedure is sufficient. Floquet theory is a well-known approach for engineers that is essentially a Fourier analysis along the axis of the periodic symmetry. In computational electromagnetics, Floquet theory may lead to heavy numerical problems, such as bad convergence. Fortunately, the periodic symmetry can be replaced by appropriate boundary conditions as the symmetry with respect to a plane could be replaced by PEC and PMC boundary conditions.

For simplicity, we consider a 2D structure with a periodic symmetry in one direction only. The structure is illuminated by a time-harmonic plane wave. This is the structure of classical *gratings*. A grating may be subdivided into cells that are separated by fictitious boundaries. Each cell has two identical neighbor cells, one to the right and one to the left side. When one considers the incident field on the two fictitious boundaries of a cell (see Figure 42), the field on the right boundary $\partial D'_p$ is equal to the field on the left boundary ∂D_p multiplied by a complex constant C.

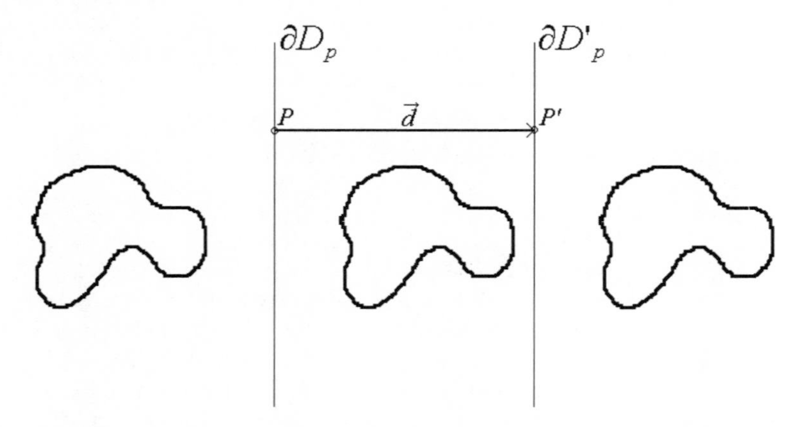

Figure 42: Grating with fictitious, periodic boundaries ∂D_p, $\partial D'_p$, displaced by the vector $\vec{d}$.

The constant C is obtained from the projection of the wave vector $\vec{k}$ on the displacement vector $\vec{d}$ of the two boundaries of the cell. When one considers the field at the two corresponding points P and P' on the two boundaries, one can easily find that the periodic boundary conditions

$$\vec{F}(\vec{r}\,') = \vec{F}(\vec{r} + \vec{d}) = C \cdot \vec{F}(\vec{r}), \quad C = e^{i\vec{k}\cdot\vec{d}}, \tag{5.316}$$

hold, where F denotes any of the fields, namely E, H, D, or B.

Independent boundary conditions

From the analytical point of view, there are dependencies between the different boundary conditions on ∂D_{ij} because the fields are linked by the Maxwell equations in the two domains, D_i and D_j. One can show, for example, that the continuity equations for the normal components are automatically satisfied, if 1) the boundary equations for the tangential components hold everywhere on the boundary and if 2) the Maxwell equations are fulfilled in the adjacent domains. For obtaining analytic solutions, one selects a set of independent boundary conditions. For example, only the tangential conditions are applied and the normal boundary conditions are omitted, i.e. fulfilled implicitly.

The same procedure is also often used in numerical methods, i.e., only a set of independent boundary conditions are explicitly used. Here, the boundary conditions or the Maxwell equations or both are not exactly fulfilled everywhere. Thus, one may encounter numerical problems if one omits, for example, the conditions for the normal components on a part of the boundary where the normal components of the field are large compared with the tangential components. Selecting an appropriate set of boundary conditions that are explicitly fulfilled by a numerical technique is not easy and may have a considerable impact on the accuracy of the results. Those boundary conditions that are not fulfilled explicitly may be used to check the accuracy of the numerical solution.

When a boundary method with overdetermined systems of equations is applied, one may also take boundary conditions into account that are theoretically dependent on each other. This allows one to obtain more reliable and balanced results, and it reduces the difficulties of finding appropriate sets of boundary conditions.

Once a boundary condition is given, one can easily obtain infinitely many derived boundary conditions either from integration or from differentiation of the boundary condition along the boundary. For example, one may impose the boundary condition

$$L(F_i, F_j) = g_{ij}\,, \tag{5.317}$$

where L is an arbitrary operator defined on the boundary ∂D_{ij}, F_i the field in the domain D_i, F_j the field in the domain D_j, and g_{ij} a given inhomogeneity. Now, the n-th order derivative of (5.317) with respect to any direction τ tangential to the boundary ∂D_{ij} must also be continuous:

$$\frac{\partial^n}{\partial \tau^n} L(F_i, F_j) = \frac{\partial^n}{\partial \tau^n} g_{ij}\,. \tag{5.318}$$

The same holds when one integrates along the boundary:

$$\int L(F_i, F_j)\,\mathrm{d}\tau = \int g_{ij}\,\mathrm{d}\tau\,. \tag{5.319}$$

When one has two or more boundary conditions, one can also obtain boundary conditions from superposition, i.e., linear combinations of the given boundary conditions:

$$\sum_{k=1}^{K} L_k(F_i, F_j) = \sum_{k=1}^{K} g_{kij}\,, \tag{5.320}$$

where the index k denotes the different 'original' boundary conditions. Obviously, derived boundary conditions are not independent from each other.

Potentials

When one is working with potentials, one can derive boundary conditions for the potentials from the boundary conditions for the electromagnetic field. To do that, one simply inserts the definition of the corresponding potential and introduces a local coordinate system that is adapted to the boundary. In 3D space, such a coordinate system may be a Cartesian coordinate system with the tangential unit vectors $\vec{e}_t, \vec{e}_\tau$ and the normal unit vector $\vec{e}_n$. To illustrate the procedure, let us consider the electrostatic scalar potential in such a local coordinate system:

$$\vec{E} = -\text{grad}\,\phi = -\frac{\partial \phi}{\partial t}\vec{e}_t - \frac{\partial \phi}{\partial \tau}\vec{e}_\tau - \frac{\partial \phi}{\partial n}\vec{e}_n. \tag{5.321}$$

When we insert this in (5.300) and (5.302), we obtain

$$\frac{\partial \phi_i}{\partial t} - \frac{\partial \phi_j}{\partial t} = 0, \quad \frac{\partial \phi_i}{\partial \tau} - \frac{\partial \phi_j}{\partial \tau} = 0, \quad \frac{\partial \phi_i}{\partial n} - \frac{\partial \phi_j}{\partial n} = \varsigma. \tag{5.322}$$

Remember that the scalar potential is defined up to a scalar constant that can be defined in such a way that a special value is given to the potential in a specific point. When we select this point somewhere on the boundary ∂D_{ij}, we have $\phi_i = \phi_j$ in this point. Now, we can integrate (5.322) from this point to any point along the boundary using (5.319). From this, we find that the continuity condition for the scalar potential

$$\phi_i = \phi_j \tag{5.323}$$

holds on the entire boundary.

Quantum theory and discrete spaces

In the formulations of electrodynamics given above and in all well-known physical theories, time and space or space-time are considered to be continuous. Of course, nobody can say that the 'true' space is continuous. There is no experimental nor any other evidence for this assumption. The only reason for the predominance of continuous spaces in physics is that such spaces are simplest if geometry and Newton's infinitesimal calculus are applied. A short time after Maxwell, it became evident that not all physical quantities are described by real numbers. The observation of such effects led to quantum mechanics. Quantum effects are not contained in Maxwell theory. For example, electric charges are multiples of an elementary charge. Therefore, the electric charge is best described by an integer number. Moreover, there are quantities such as the spin of an electron that are Boolean numbers or elements of other number spaces. Beside the presence of number spaces other than the real numbers, the most important features of quantum mechanics have to do with uncertainty and some kind of randomness. Some philosophers praised quantum theories because they are not completely deterministic as are classical mechanics, Maxwell theory, and the theory of relativity, but Einstein – who was one of the fathers of quantum mechanics – refused to accept the non-deterministic aspect of such theories. It is well known that he proposed experiments to find out whether quantum mechanics is correct or not. Today, it seems that experiments show that Einstein was wrong. In fact, Einstein saw very well that one could introduce some discrete spaces to explain quantum effects. At first, he considered this idea to be ridiculous, but later he considered it as a last resort when the experiments turned out in favor of quantum mechanics. However, he did not try to find a theory of discrete spaces, as far as we know. In fact, it is somehow strange that quantum mechanics does not use a quantum formulation of space and time. The main reason for this is certainly that it is very hard to work analytically with non-real number spaces. Moreover, space and time seem to be continuous from experience.

Today, we are very familiar with non-continuous spaces from pictures, movies, TV, and computers. That the computer with its 'pixel-worlds' can deeply affect our perception of the 'real' world and physical theories is quite clear. As the infinitesimal calculus likes infinite, continuous worlds, the computer likes finite, discrete worlds. Therefore, many successful numerical codes for computational electromagnetics work on discretized space and time, but usually this is considered to be an approximation and effects caused by the discretization are considered as numerical inaccuracies.

In the following, some of the difficulties with non-real number spaces will be outlined and it will be demonstrated that purely deterministic codes may produce effects that resemble quantum effects.

Let us start with real numbers. The information content of any real number is infinite. Therefore, we cannot exactly write down any real number, except a finite number of 'prominent' numbers for which special symbols have been created. For the same reason, no exact notation of real numbers is possible in a computer with a finite memory. Fortunately, we can approximate any real number with any desired accuracy by finite strings using an alphabet of a finite number of symbols. This requires some coding that is not unique at all. For example, we can use the decimal coding of the form +3.14E+5 with a three digit mantissa and a one digit exponent. What are the consequences of such approximations? First of all, the resulting number space is no longer complete with respect to the famous arithmetic operations +, -, *, and /. This means that the exact result of such an operation may be a number that requires longer strings, an extended alphabet, or another coding. Usually, the result of arithmetic operations is therefore an approximation with a finite accuracy, for example, +1.00E+3/+3.00E+0=+3.33E+2. The error of the result may be of the same order of magnitude as the error of the input, but it may also be considerably higher. In algorithms with a huge number of nested operations, it may be very difficult to estimate the error of the final result. Another problem is the finite size of the approximated real numbers that may lead to overflow problems, for example, +1.00E+9*+1.00E+9>+9.99E+9. However, computer arithmetic is defined by purely deterministic algorithms that associate a number of the given, discrete number space to the result of an operation.

The usual approximations of real numbers in computers are based on some binary coding that is similar to the decimal coding we use on pocket calculators and for manual calculations. Since real numbers are defined as limits of rational numbers, we now consider a rational approximation of the form

$$r=R+error=I+J/K+error\ , \tag{5.324}$$

where r is a real number, R its rational approximation, and I, J, K integer numbers. Because integer numbers may be infinite, a computer can only handle a finite set of integer numbers. Assume that M is the maximum integer number, i.e., that each integer number I is bigger than or equal to –M and smaller than or equal to +M. What about the structure of the rational numbers defined by (5.324) with a finite set of integers? Let us denote such an M-rational number as $R=(I,J,K)_M$. Obviously, $(I,J,K)_M$ and $(I,J*N,K*N)_M$ are identical. How many different M-rational numbers are there for a given M? Note that it is not easy to answer this simple question. For simplicity, we can consider the subset of M-rational numbers with I=0, J>0, K>0 only. Figure 43 shows that this number divided by M^2 is a quite noisy function of M that tends to some value close to 0.61. What is the correct limit? Also the answer to this question is not easy. The correct answer is $6/\pi^2$.

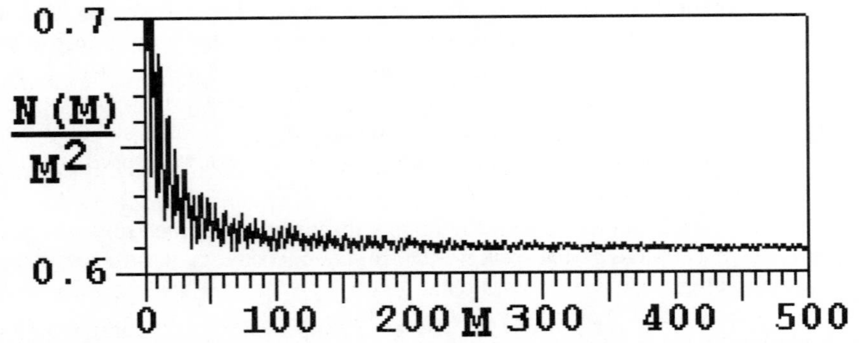

Figure 43: Number N of rational numbers of the form J/K for 0<J<M+1 and 0<K<M+1.

To get an impression of the structure of the M-rational numbers, let us consider the differences between neighbor numbers. Since we have a finite set of M-rational numbers, we can associate an integer index n in such a way that $R_{n-1}<R_n<R_{n+1}$. For a given M, the index uniquely defines the M-rational number and we can write for the differences between neighbor numbers $D_n=R_{n+1}-R_n$. These differences look quite irregular at first sight. The smallest difference for $0<R_n<1$ is 1/(M*(M-1)), whereas the biggest difference is 1/M. When M is huge (for obtaining accurate approximations of real numbers), the maximum difference becomes much bigger than the minimum difference. Figure 44 illustrates the differences between the first 1000 neighbor M-rational numbers for M=100.

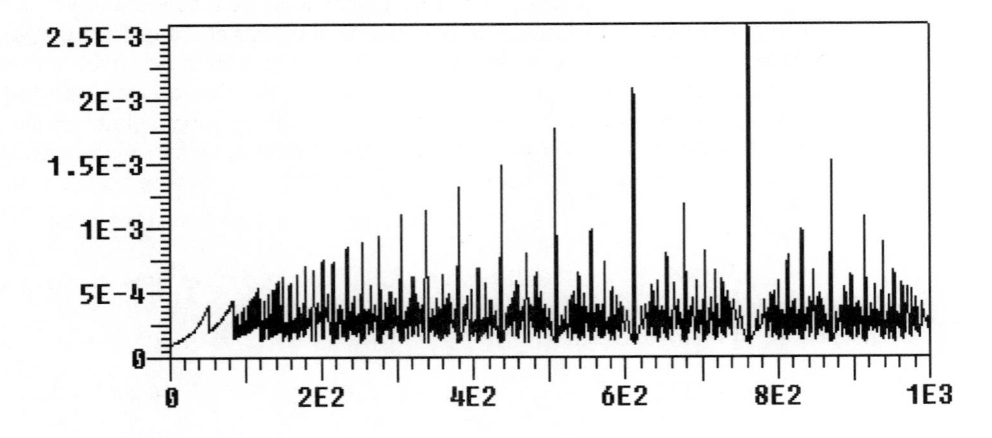

Figure 44: Distances between the first 1000 neighbor rational numbers of the form I/K with 1≤K≤M=100 as a function of the rational number.

As one can see, the pattern is very regular although there are many big jumps. Those who are familiar with fractals might guess that this is a fractal. In fact, some self-similarity (that is typical for fractals) may be observed: when one magnifies a part of the structure, one obtains a structure that is similar to the entire structure. Despite this self-similarity, the M-rational numbers are no fractal in the strong sense defined by Mandelbrot [Mandelbrot, 1977], even in the limit of M toward infinity. The reason for this is simple. M-rational numbers are finite sets and the geometric dimension of such sets is always zero. Note that fractals are often implemented on computers. Such implementations are always approximations by finite sets.

However, let us now write a computer code for simple particle mechanics based on M-rational numbers. First, we run into difficulties with the definition of derivatives, because the differences between neighbor M-rational numbers are finite. Instead of approximating Newton's formulae by introducing M-rational approximations of derivatives, we directly design algorithms working with M-rational numbers that approximate the classical behavior of particles. This is not an easy task at all and there is not a unique solution. For simplicity, let us consider a 1D model with a single particle, defined by its position x at some time t, and its mass m, where x, t, m are M-rational numbers. When we consider x, t as 2D space-time, we have a very irregular grid in 2D space-time because of the irregular differences between neighbor M-rational numbers that describe x and t. A particle moving in this M-rational space-time is now defined by a set of points on the corresponding grid. This set is the trace of the particle in space-time. What is its velocity? When there is no acceleration, the velocity v should be an M-rational number, but when we compute the ratio $q_n=(x_{n+1}-x_n)/(t_{n+1}-t_n)$ of neighbor points of the particle's trace, we see that q_n cannot be constant at all. Therefore, we can only design algorithms in such a way that the statistical average of q_n is equal to a constant velocity v. From a macroscopic view, we see the particle moving at a constant velocity, but when we zoom in or when we improve the accuracy of our observation, we see that there are fluctuations in the velocity (see Figure 45). The procedure for the acceleration is essentially the same. Since the design of such algorithms is quite tricky, we will not go into the details.

Once we have an algorithm for an accelerated particle moving in the x direction, we can model a simple experiment where a particle hits a barrier with negative acceleration, i.e., the x axis is subdivided into three parts. In the center part, the average acceleration is negative, whereas there is no acceleration in the left and right parts. The particle is incident from the left hand side with a 'constant' velocity. In classical mechanics, the particle is either reflected or transmitted with a well defined end velocity v_{end} that depends on the initial velocity v_{start}, on the acceleration in the barrier, and on the size of the barrier. In quantum mechanics, one has a statistical description, i.e., one has a certain probability that the particle 'tunnels' through the barrier, even when the classical model predicts a reflection with $v_{end}=-v_{start}$. In the M-rational model, the end velocity depends on the algorithm that is used, on the size of M, and also on the position of the barrier on the x axis and on the start time (see Figure 46).

Note that the statistical description of the M-rational models is somehow similar to quantum mechanics, but the algorithms that describe particles in an M-rational world are purely deterministic. In classical mechanics and in quantum mechanics, the result of the experiment does not depend on its location in space-time. The same result is obtained when the experiment is moved to another place or when it is repeated at a later time. Although we know that we never can exactly reproduce all initial conditions of an experiment at another location or another time, the reproduction of experiments is fundamental for classical and quantum theories. In M-rational models, there is no strict reproduction. The dependence of the results on the movement of an experiment in space-time causes the statistics of such deterministic models. Therefore, the development of such algorithms requires a fundamental change of paradigms of physics.

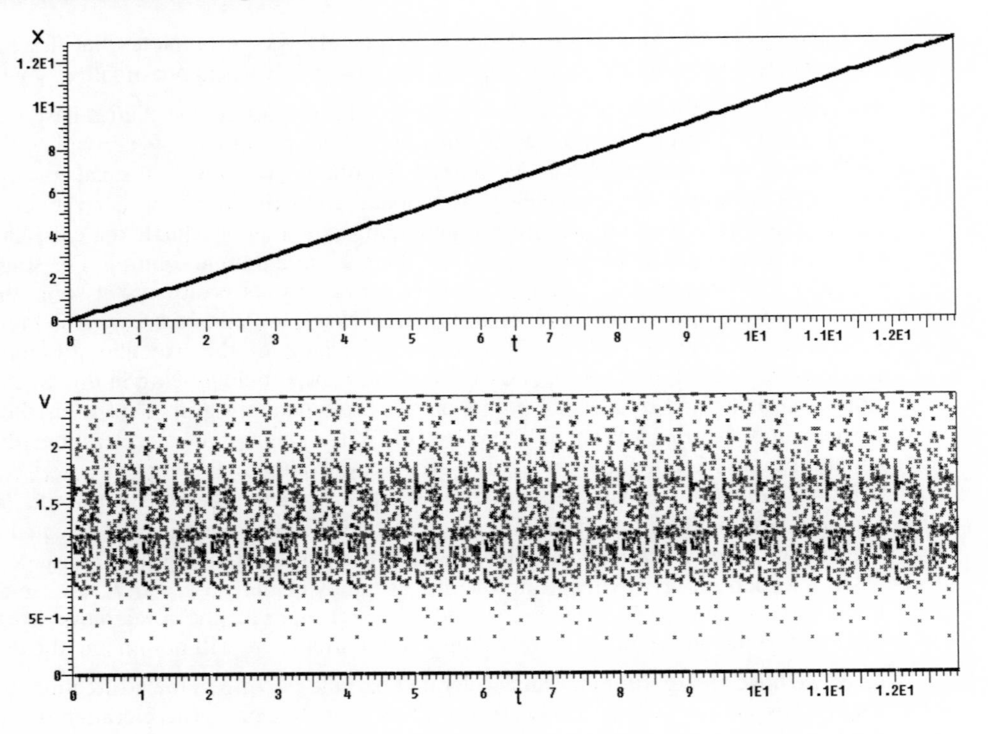

Figure 45: Particle moving with 'constant' velocity in an M-rational world. Position (top) and local velocity (bottom) as a function of time.

Statistical descriptions hide the structure of the M-rational numbers. If one were able to make microscopic observations in an M-rational world, one should be able to discover this structure. One can see this from the previous figures. From Figure 47 one can see that even the pseudo-fractal structure of M-rational numbers may become visible from results of sufficiently fine microscopic experiments.

The algorithms used for obtaining the results outlined above are 'quick and dirty' rather than sophisticated and well adapted to 'explain' experiments accurately. It is assumed that GA can be applied to tune M-rational algorithms in such a way that a much better agreement of the simulations with the observations is achieved. Note that such algorithms can also be generalized by replacing M-rational numbers by other finite number sets that approximate real numbers. It seems to be possible that future computers might be able to design new algorithms that are based on finite number spaces that are most appropriate for computers and describe nature more accurately than classical algorithms that approximate classical theories based on continuous space-time concepts. However, we now return to classical algorithms for computational electromagnetics that are based on classical Maxwell theory and that are very well suited to solving everyday problems of engineers.

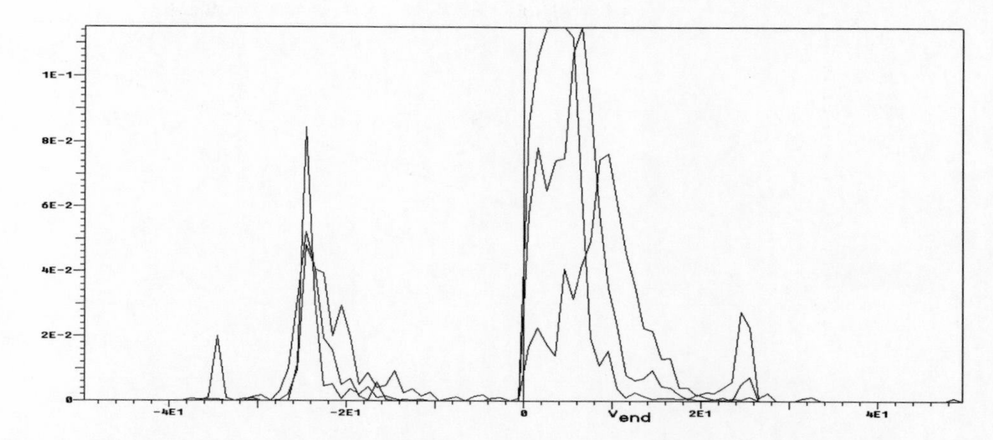

Figure 46: Statistical probability for the end velocity v_{end} *of a particle incident with* v_{start}*=25 on a barrier of negative acceleration computed with an* M*-rational algorithm. Classical prediction of the end velocity is* v_{end}*=5. The three curves show the results for different* M.

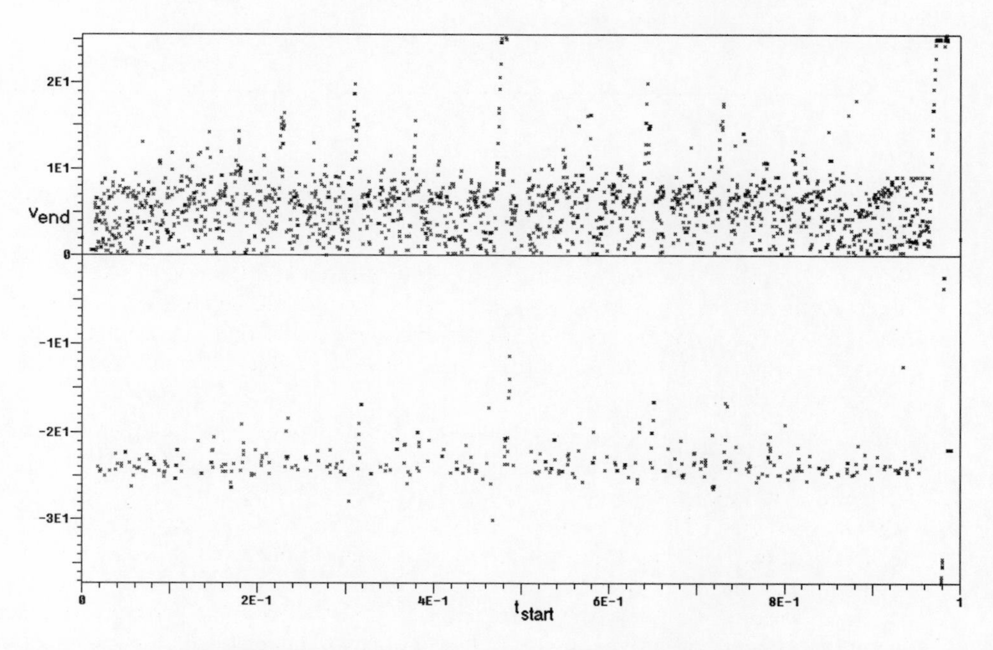

Figure 47: End velocities v_{end} *as a function of the start time for an experiment with a particle hitting a barrier with negative velocity (see Figure 46).*

6 Computational electromagnetics

General remarks

Computational electromagnetics is a complex task consisting of several steps where several decisions are made. The main steps and decisions are the following:

- Selection of the theories to be used. Beside Maxwell theory, one may work with other theories that are well suited for special cases or that extend Maxwell theory. For example, at optical frequencies, it may be reasonable to use theories based on Newton's concepts of optics. In semiconductors, quantum effects might be considered with quantum theory. In moving media, mechanics and relativistic theories might play a role. To obtain efficient codes for very specific applications one may also replace the theory by simple formulae obtained from measurements. Such codes are called *heuristic* and are often used for CAD. However, in this book we strictly focus on methods based on Maxwell theory.

- Analytic processing of the equations of the fundamental theory. Here, several decisions are made in order to simplify the equations. These decisions may have a strong impact on the range of applications of the resulting code. The main decisions concern the time dependence of the fields and the material properties: 1) no time dependence (statics) / frequency domain (time-harmonic fields) / time domain, 2) material properties are loss-free / lossy, piecewise homogeneous / inhomogeneous, isotropic / anisotropic, linear / nonlinear, frequency dependent / time dependent, and so on. For simplicity, we focus on piecewise homogeneous, isotropic, linear material properties in this book. The main equations that may be derived from Maxwell equations for such simple materials were outlined in the previous chapter.

- Discretization of the physical model. Here, one first decides whether a domain method or a boundary method shall be applied. Moreover, one often decides to restrict to a simple discretization technique rather than a general one. For example, one may discretize a given 2D domain with a regular, square grid, with an orthogonal grid with straight grid lines, by triangulation, and so on. The discretization technique affects the accuracy of the results, but also the user-friendliness of the codes. With irregular grids, one may obtain more accurate results, but the user-friendliness of such codes is usually reduced because it may be hard to define an appropriate grid.

- Most of the well-known Maxwell solvers can be considered as either explicit or implicit matrix methods. An important exception are so-called Monte Carlo simulations that are based on random exploration of the field domains and on statistics. Such simulations are outside the scope of this book. Implicit matrix methods are usually not presented as matrix methods because these methods work without the explicit notation and solution of a matrix equation. Typical examples are Finite Differences (FD) and Transmission Line Matrix (TLM). However, for comparisons one can describe implicit matrix methods as matrix methods. The procedure for matrix methods was described in the chapter on 4 Generating matrix equations.

- When a matrix method is applied, the main selections are the selection of the basis functions and the selection of the method to compute the unknown parameters. For details, see the chapter on 4 Generating matrix equations.

Most of the prominent methods were first designed for a relatively simple and restricted class of applications. Later on, these methods were generalized and extended to widen the range of applications and to overcome numerical problems. This process led to very general formulations that allow one to describe methods that were originally different as special cases of one and the same method. Therefore, one can sometimes read in textbooks on the Method of Moments (MoM) that the method of Finite Elements (FE) is a special case of MoM, whereas one can read in FE books that MoM is a special case of FE. Such considerations may be confusing but helpful for those who are familiar with a single method and would like to understand everything in the framework of this method. Although general formulations of methods are often presented in textbooks, most of the codes are much less general. Most of the commercial codes are even very simple implementations. Therefore, there are usually big differences between MoM codes and FE codes. More sophisticated implementations are often developed at universities, but these implementations are rarely user-friendly and have a very narrow range of applications.

When one considers implementations of different methods, one sometimes finds that more or less the same codes are sold under different names, which is confusing. For example, MAFIA [http://www.cst.de] is a code based on the Finite Integral (FI) method. This code is the same as standard FD codes. FI and FD are similar techniques that allow one to derive identical codes.

Most of the books on computational electromagnetics focus on one or two methods, for example, [Binns, 1963], [Binns, 1992], [Booton, 1992], [Taflove, 1995], [Hoefer, 1991]. Since there are very many different techniques, it is impossible to outline all of them. However, techniques with different names are often closely related and can easily be understood when the main ideas of the most prominent methods are known. Therefore, it is reasonable to focus on these methods.

Well-known methods outlined

In the following sections we outline the most important techniques with a wide range of applications that allow one to write general purpose codes. Special techniques, that can be added to other methods rather than being independent methods, are not considered. Typical examples are the Fast Multipole Method (FMM) [Greengard, 1987] that can be added to the MoM when large bodies are present, the Fast Fourier Transforms (FFT) that can be added to MoM [Zwamborn, 1991] and to various other methods, Conjugate Gradient (CG) methods [Choi, 1995] that are sometimes added to MoM codes, The Beam Propagation Method (BPM) [Marcuse, 1991] that is usually combined with FD or FE, etc.

For simplicity, we will consider special topics of Maxwell theory, e.g., electrostatics, although the corresponding technique is not restricted to that topic. Since these techniques are relatively general, they are not even restricted to Maxwell theory and can easily be adapted to other theories of physics. In fact, many of the numerical methods for electromagnetics were originally developed in mechanics or other disciplines of physics.

Finite Differences (FD) and Finite Integrals (FI)

Finite difference methods [Mitchell, 1980], [Taflove, 1995], [Taflove, 1998] start with a discretization of the differential operators in Maxwell's equations in differential form or in differential equations derived from Maxwell equations. Basically, the limit in the definition of the derivative of a function (2.28), is omitted, which leads directly to an approximation of the differential operators. Omitting the limit means replacing infinitesimally small differences by finite differences. To illustrate the procedure, we consider the 2D Laplace equation in Cartesian coordinates:

$$\Delta_T \phi(x,y) = \frac{\partial^2}{\partial x^2}\phi(x,y) + \frac{\partial^2}{\partial y^2}\phi(x,y) = \frac{\partial}{\partial x}\left(\frac{\partial \phi}{\partial x}\right) + \frac{\partial}{\partial y}\left(\frac{\partial \phi}{\partial y}\right) = 0 \cdot \tag{6.1}$$

Now, we approximate the first order derivatives with respect to x in two points $(x+d,y)$ and $(x-d,y)$ by the finite differences:

$$\frac{\partial}{\partial x}\phi(x+d/2,y) \approx \frac{\phi(x+d,y)-\phi(x,y)}{d}, \quad \frac{\partial}{\partial x}\phi(x-d/2,y) \approx \frac{\phi(x,y)-\phi(x-d,y)}{d}. \tag{6.2}$$

Note that this approximation is slightly different from (2.28) without the limit. Here, so-called central differences are used which are numerically more accurate. Figure 48 illustrates this.

When we repeat the procedure, we obtain central approximations for second order derivatives in the x direction:

$$\begin{aligned} \frac{\partial^2}{\partial x^2}\phi(x,y) &\approx \frac{\frac{\partial}{\partial x}\phi(x+d/2,y) - \frac{\partial}{\partial x}\phi(x-d/2,y)}{d} \\ &\approx \frac{\phi(x+d,y)+\phi(x-d,y)-2\phi(x,y)}{d^2}. \end{aligned} \tag{6.3}$$

The formula for the second order derivative in the y direction is essentially the same and we obtain for the approximation of the 2D Laplace equation:

$$0 = \Delta_T \phi(x,y) \approx \frac{\phi(x+d,y)+\phi(x-d,y)+\phi(x,y+d)+\phi(x,y-d)-4\phi(x,y)}{d^2}. \tag{6.4}$$

Although this so-called 5-point-star approximation of the Laplacian is widely used, it is not unique at all. Obviously, one can also replace the central differences by left or right differences, but this leads to a less accurate approximation that does not reduce the complexity of the formula. Taylor series and other series expansions may be applied to find more accurate approximations. The corresponding formulae are more complicated and therefore the resulting procedures are numerically more demanding.

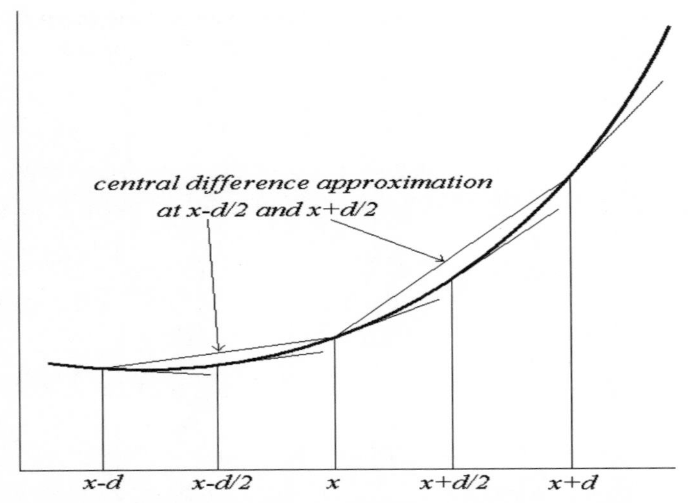

Figure 48: Approximation of the first order derivatives of a function in the two points x+d/2 and x-d/2 by central differences. The central difference in x-d/2 is equal to the right difference in x-d and equal to the left difference in x. Note that the corresponding secant is almost parallel to the tangent in x-d/2, but not parallel to the tangents in x-d and x. The same holds for x+d/2. Therefore, the central differences are much more accurate approximations of the first order derivative than the left and right differences.

When one applies series expansions to obtain an approximation of the given differential equation, one may recognize that a set of basis functions for FD is implicitly defined. These basis functions are not explicitly required as long as the field is computed on a grid. In (6.4), it is implicitly assumed that a regular grid with grid lines parallel to the x and y axis and with constant steps of length d in both directions is used. Figure 49 illustrates this. On such a grid, each grid point may be labeled with two integer indices i, j.

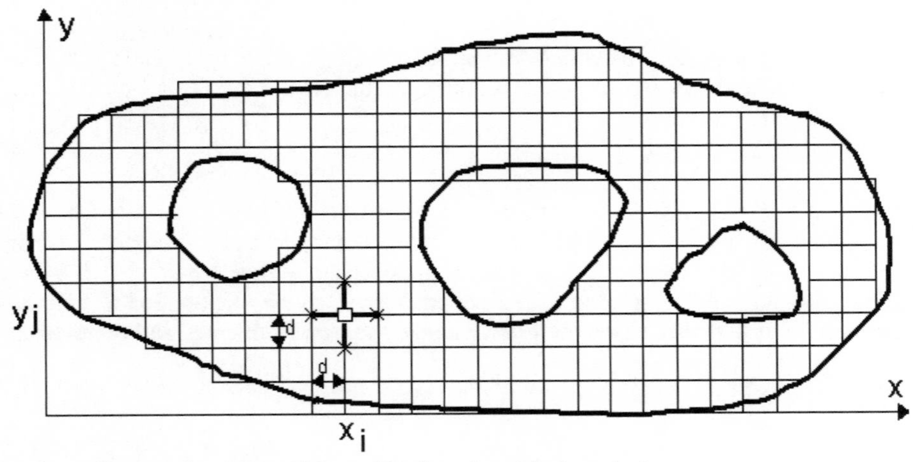

Figure 49: Regular FD grid for a 2D domain with three holes.

When we omit the errors caused by the approximation of the Laplacian and move the potential at the central grid point in (6.4) to the left hand side, we obtain:

$$\phi_{ij} = \phi(x_i, y_j) = \frac{\phi(x_{i+1}, y_j) + \phi(x_{i-1}, y_j) + \phi(x_i, y_{j+1}) + \phi(x_i, y_{j-1})}{4} = \frac{\phi_{i+1j} + \phi_{i-1j} + \phi_{ij+1} + \phi_{ij-1}}{4}. \quad (6.5)$$

This means that we could compute the potential at a grid point, when the potential at its four closest neighbor points is known. For a Dirichlet problem, the potential is known only at the grid points on the boundary of the domain. To obtain a solution, one therefore proceeds iteratively. First, one starts with an initialization that sets the field at all grid points inside the domain to some reasonable value ϕ^0_{ij} or simply to zero. Afterwards, one improves these values iteratively with a simple modification of (6.5):

$$\phi^n_{ij} = \frac{\phi^{n-1}_{i+1j} + \phi^{n-1}_{i-1j} + \phi^{n-1}_{ij+1} + \phi^{n-1}_{ij-1}}{4}, \quad (6.6)$$

where n=1,2,3,... is the number of the iteration. The interpretation of this scheme is very simple: the potential in a node is equal to the average of the potentials in the neighbor nodes.

Figure 50 shows the scalar potential in a coaxial cable, i.e., a circular domain with a circular hole after the initialization. As one can see, the potential seems to propagate from the inner boundary into the domain. In fact, the potential also propagates from the outer boundary into the domain, but this is not visible because the potential is initialized to zero, which coincides with the potential on the outer boundary. The potential propagates further into the domain until some steady state is reached (see Figure 51).

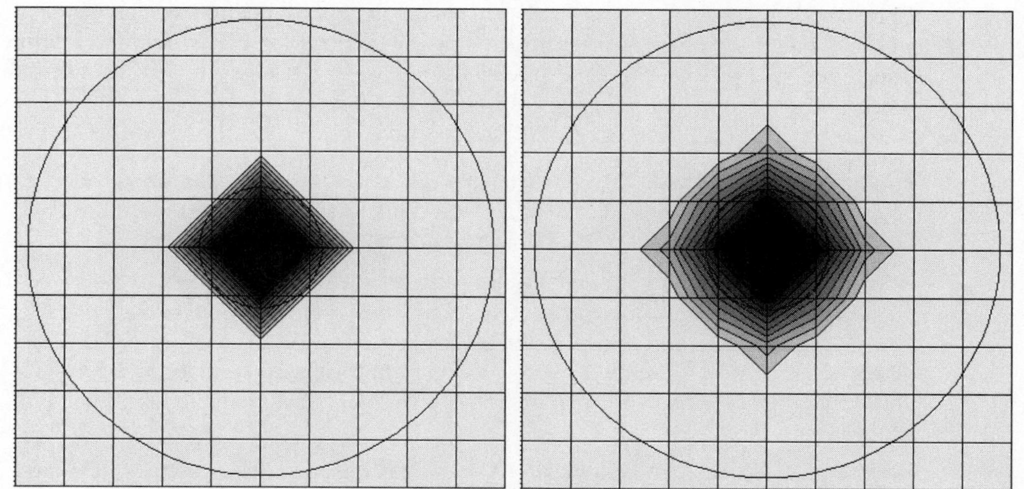

Figure 50: Electrostatic potential in a coaxial cable with circular conductors on constant potential. FD initialization (left) and first iteration (right). MaX-1 computation.

Note that several techniques have been invented to speed up the convergence and the efficiency of FD. First of all, over-relaxation and under-relaxation techniques may be applied. These techniques mainly replace the iterative scheme (6.6) by a slightly more complicated scheme of the form

$$\phi_{ij}^{n} = (1-\alpha)\frac{\phi_{i+1j}^{n-1} + \phi_{i-1j}^{n-1} + \phi_{ij+1}^{n-1} + \phi_{ij-1}^{n-1}}{4d^2} + \alpha\phi_{ij}^{n-1}, \tag{6.7}$$

where α is the relaxation factor. The convergence of this scheme depends on the choice of the relaxation factor. For a wrong choice, the procedure may diverge, i.e., become unstable. Another improvement may be expected from a better initialization. In simple situations, one may even start with the analytic solutions. Here, it is important to see that the FD scheme does not converge toward the correct, analytic solution. When one starts with the analytic solution, each iteration will modify the field until the same solution is obtained that is also obtained when the field is initially set to zero (see Figure 52).

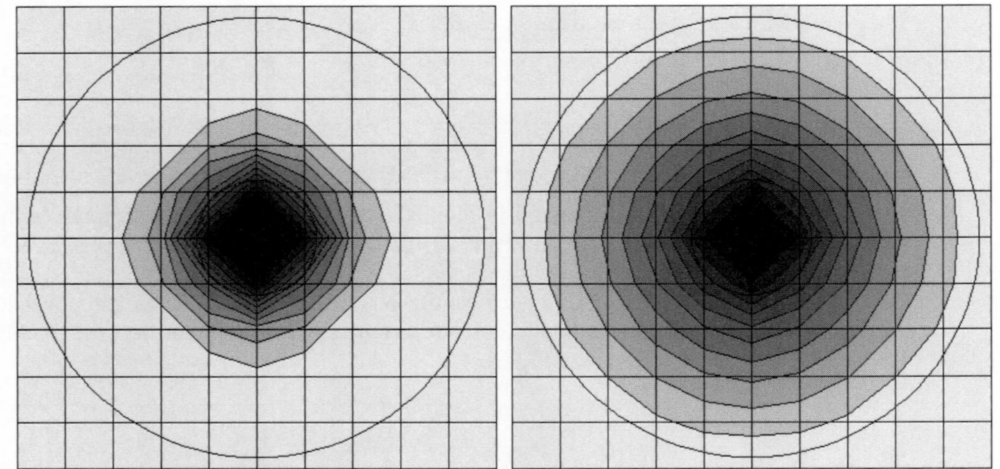

Figure 51: Electrostatic potential in a coaxial cable with circular conductors on constant potential after two iterations (left) and after 99 iterations (right). MaX-1 computation.

Further generalizations of the FD scheme allow one to use irregular grids. Such grids allow a better modeling of the boundaries of a given domain and therefore reduce the error caused by the discretization. Irregular grids will be considered in the next chapter.

Note that regular square grids seem to be optimal when the boundaries of the given domain are parallel to the grid lines. In electrostatics, a corner on an electrode produces a singularity in the electric field. Such singularities cause severe numerical problems. Usual FD schemes 'avoid' numerical problems by smoothing the field, as one can see from Figure 53. This leads to a drastic underestimation of the maximum field strengths that are most important in high voltage applications.

In computational electrodynamics, FD was first used in the frequency domain, where the Laplacian is also involved in the Helmholtz equations. Thus, only a moderate adaptation of the static FD schemes is required. As soon as the memory of modern computers became big enough, the time-domain implementation FD-TD became much more popular. The FD-TD scheme proposed by Yee as long ago as 1966 [Yee, 1966] directly starts with a FD approximation of the Maxwell equations, which makes the technique extremely simple and minimizes the analytic knowledge required for getting started. For simplicity, we consider a 2D example of an electromagnetic wave propagating in the *xy* plane in such a way that the magnetic field is perpendicular to the xy plane, i.e., the H wave propagation. In this case, the electric field has x and y components and the magnetic field has a z component. To describe the wave propagation in a simple, loss-free medium without impressed currents and charges, we need only the first two Maxwell equations, where the curl operator is involved. In Cartesian coordinates, this operator is given by (1.70). Therefore, we obtain from the Maxwell equation (5.64):

$$\frac{\partial}{\partial x}E_y - \frac{\partial}{\partial y}E_x = -\frac{\partial}{\partial t}\mu H_z. \tag{6.8}$$

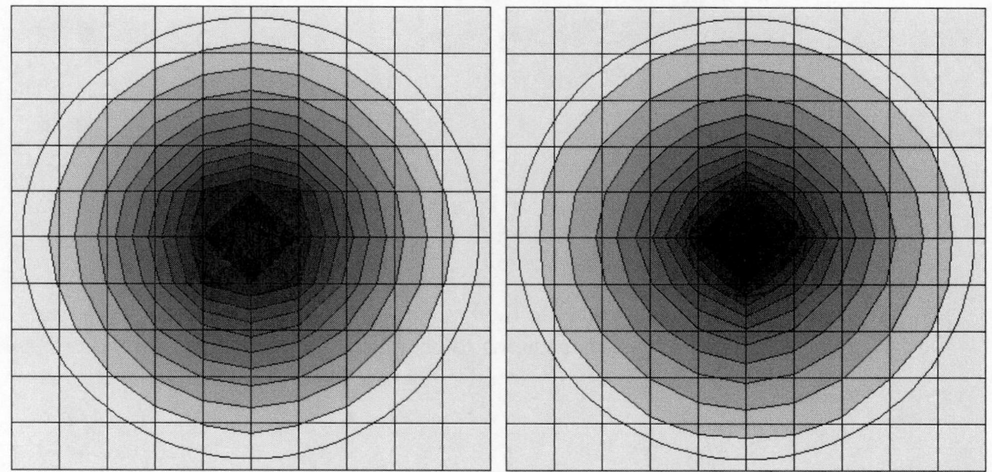

Figure 52: Electrostatic potential in a coaxial cable with circular conductors on constant potential. FD initialization with the correct, analytic solution (left) and after 99 iterations (right). MaX-1 computation.

This is a single scalar differential equation because the x and y components of this Maxwell equation are zero for H waves. From the second Maxwell equation (5.65) without impressed currents and $\sigma = 0$ we obtain two scalar equations for the x and y components:

$$\frac{\partial}{\partial y}H_z = \frac{\partial}{\partial t}\varepsilon E_x, \tag{6.9}$$

$$-\frac{\partial}{\partial x}H_z = \frac{\partial}{\partial t}\varepsilon E_y. \tag{6.10}$$

In these three differential equations for the three field components, we would like to replace the derivatives with respect to x,y,t by central finite differences to obtain a good accuracy. With a space step d in the x and y directions and a time step δ, we can approximate (6.8) by:

$$\frac{E_y(x+d/2,y,t)-E_y(x-d/2,y,t)}{d} - \frac{E_x(x,y+d/2,t)-E_x(x,y-d/2,t)}{d}$$
$$= -\frac{\mu H_z(x,y,t+\delta/2)-\mu H_z(x,y,t-\delta/2)}{\delta}. \tag{6.11}$$

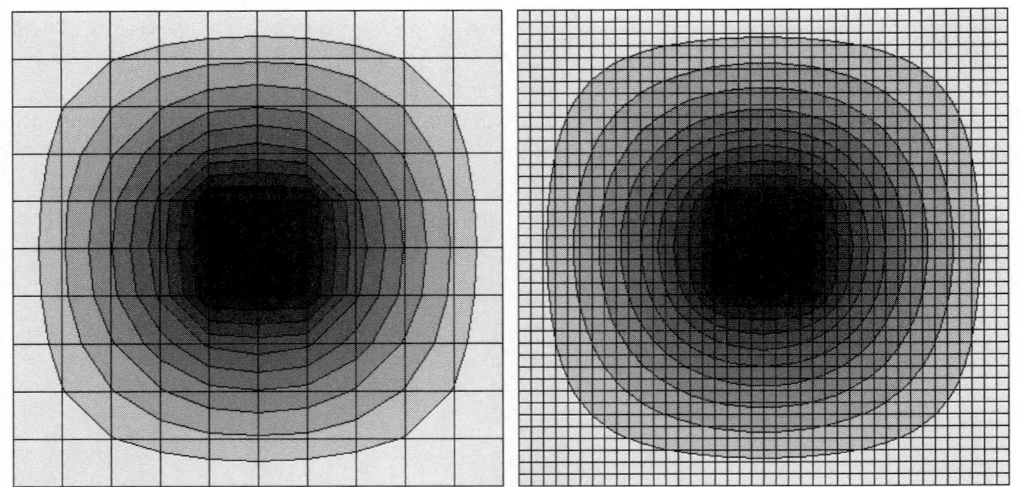

Figure 53: Electrostatic potential in a coaxial cable with square conductors on constant potential. FD solution after 99 iterations (left) and FD solution on a refined grid after 400 iterations (right). MaX-1 computation.

When we assume that the electric field is known at time t and that the magnetic field is known at time t-$\delta/2$, we can use this equation to obtain the magnetic field at time $t+\delta/2$. Now, we can approximate (6.9) and (6.10) by central differences to obtain the electric field at time $t+\delta$:

$$\frac{H_z(x,y+d,t+\delta/2)-H_z(x,y,t+\delta/2)}{d}=\frac{\varepsilon E_x(x+d/2,y,t+\delta)-\varepsilon E_x(x+d/2,y,t)}{\delta}, \quad (6.12)$$

$$\frac{H_z(x,y,t+\delta/2)-H_z(x+d,y,t+\delta/2)}{d}=\frac{\varepsilon E_y(x+d/2,y,t+\delta)-\varepsilon E_y(x+d/2,y,t)}{\delta}. \quad (6.13)$$

Finally, we can introduce a regular grid and indices as we did in the static case. It is interesting to see that each of the three field components must be evaluated at different locations of the grid (see Figure 54) and that the electric field and the magnetic field are evaluated at different time steps. This is called a leap-frog scheme. Of course, we could use three grids in space in such a way that all three field components would be evaluated at each point and we could compute both the electric and the magnetic field at each time step, but this would increase the memory required and the computation time by a factor of 6 without any improvement of the accuracy. The leap-frog scheme may be considered as a symmetry decomposition of the full scheme into six identical parts. Note that the symmetry is broken as soon as boundaries are introduced. Then, the full scheme and the leap-frog scheme may lead to different solutions. Note that the equations given above can easily be used when the material is inhomogeneous, i.e., when the permittivity ε and the permeability μ are functions of x and y. In this case, one has to insert the values of ε and μ at the position where the corresponding field is evaluated. Therefore, FD-TD can handle inhomogeneous material properties with small gradients of ε and μ more easily and accurately than piecewise homogeneous materials.

The field initialization is a minor problem in FD-TD. Usually, one assumes that the initial field is zero, which is correct when one considers the propagation of an electromagnetic pulse on a FD grid of finite size and when the start time for the first iteration is properly set.

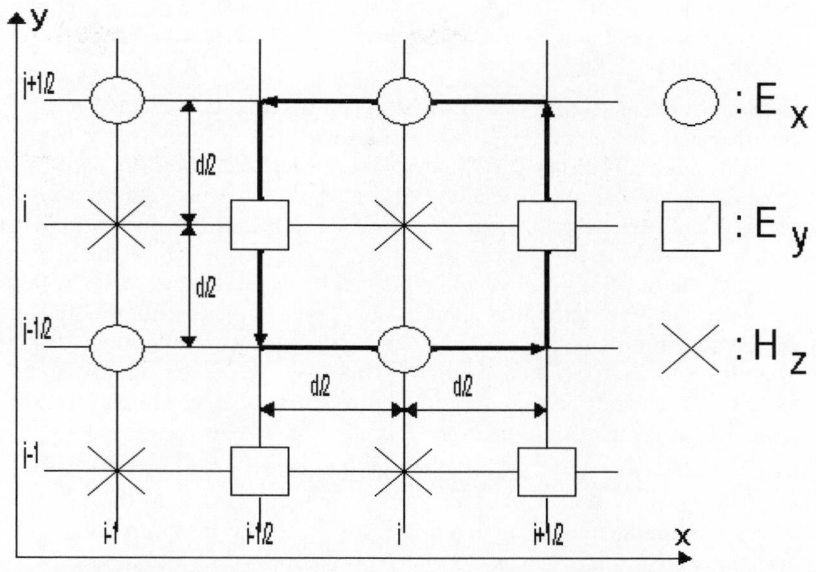

Figure 54: 2D FD-TD leap-frog grid for H waves. Only one component of the electromagnetic field is stored at each grid point.

A first problem of FD-TD schemes is the appropriate selection of the time step δ and of the space step *d*. The space step *d* depends on the desired accuracy and on the shortest wavelength that plays a role for the computation. For analyzing pulse propagation, one therefore needs a Fourier analysis of the time dependence to find the shortest wavelength. To obtain useful results, *d* should be less than a quarter of the shortest wavelength. When the time step δ is too big, the procedure becomes unstable, whereas the numerical effort increases with decreasing time steps. Stability is guaranteed when $\delta < d / (\sqrt{n}\, c)$ holds, where *n* is the number of spatial dimensions of the model and where *c* denotes the velocity of light in the corresponding medium. When the material properties are inhomogeneous or when different materials are present, *c* is the maximum velocity.

One of the most important problems of FD (and all domain methods) is the fact that one often has open structures in computational electromagnetics, whereas FD can only be implemented on grids with finite size, i.e., grids with a boundary. In the points on the boundary of the grid, one cannot apply the usual FD operators because these operators require information on the field in neighbor points that are outside the grid. There are several techniques to overcome these difficulties. First of all, one can use the information in the neighbor points on the grid to extrapolate the missing information on points outside the grid. As we have seen in the chapter on 2 Function analysis, extrapolation is quite tricky. Therefore, such extrapolations are either inaccurate or numerically expensive. When one is studying the propagation of waves, one has a similar situation to the measurement of such waves in a closed room, where absorbers are placed on all walls to simulate open space. Thus, one can invent special FD operators that simulate absorbers along the boundaries of a FD grid, i.e., one implements Absorbing Boundary Conditions (ABC). The development of good ABC's is much more demanding than the development of good FD schemes on regular grids. A relatively new technique is

Berenger's Perfectly Matched Layers (PML) that work with hypothetical absorbing material that does not exist in nature. For more information see textbooks on FD, for example, [Taflove, 1995], [Taflove, 1998].

Especially for analyzing complicated structures, one is interested in FD-TD schemes working on irregular grids. The development of such grids is very tricky when the grid lines do not coincide with Cartesian coordinates. Therefore, one often encounters numerical codes with irregular grids that are only slightly irregular in the sense that all grid lines are parallel to the axes of a Cartesian coordinate systems. The main problems of more irregular grids are the following: 1) Much additional information on the location of the grid points must be stored, whereas the location of the grid points can be easily evaluated for regular or slightly irregular grids. 2) The development of appropriate FD operators is difficult. 3) The numerical evaluation of the resulting schemes requires much more computation time than the evaluation of the simple schemes on regular grids. 4) Because of the central differences formulation, even simple FD schemes on regular grids are of second order, whereas the simplest FD schemes on irregular grids are of first order and therefore have reduced accuracy.

The simplest way to reduce the difficulties of developing FD operators for irregular grids consists of the replacement of the differential form of the Maxwell equations by the corresponding integral form. To derive a FD scheme for an irregular grid from the differential form, one has to select an appropriate coordinate system and to insert the corresponding metric coefficients in the definition of the spatial derivatives. This leads to lengthy and difficult differential equations. When one works with the integral representation, one selects the paths of the integral along the grid lines, which is relatively easy. This method is also called the Finite Integral (FI) technique. Finally, FD and FI lead to identical schemes. Therefore, FI and FD codes may be identical. Incidentally, MAFIA [http://www.cst.de], the most prominent commercial FI code, works with slightly irregular grids that may easily be derived from the differential formulation in Cartesian coordinates. For this reason, the term FI is rarely used in the literature and many authors who use the integral form of Maxwell equations call the resulting operators FD operators.

To illustrate the FI procedure, we derive (6.11) from the integral form of the first Maxwell equation (5.68):

$$\int_{\partial S} \vec{E}(\vec{r},t)\,\mathrm{d}\vec{\ell} = -\int_{S} \frac{\partial}{\partial t} \mu \vec{H}(\vec{r},t)\,\mathrm{d}\vec{S}\cdot \tag{6.14}$$

We use the square path indicated in Figure 54 for the first integral and the square area bounded by this path for the second integral. Moreover, we assume that the electric field along each side of the square is constant and equal to the value at the center point and that the magnetic field inside the square is constant and equal to the value at the center point. Thus, the first integral is simply a sum of four terms, whereas the second integral is reduced to a sum with a single term. The resulting FI scheme is identical to the FD scheme (6.11), when the central difference approximation is applied for the approximation of the time derivative.

Finite Elements (FE)

Finite Elements (FE) [Bathe, 1982], [Beer, 1992], [Chari, 1980], [Silvester, 1983], [Whiteman, 1997] were used in mechanics before they were applied to electromagnetics. The initial idea in the simulation of the statics of buildings was to replace floors and walls by a grid of elements of finite size, first of all by a grid of rods. Later on, more complicated elements were introduced. Today, the elements are usually selected in such a way that the entire space is covered by finite elements.

Lumped elements for statics

When we apply the initial FE idea to 2D electrostatics, we replace the field domain by a network of capacitors as illustrated in Figure 55. Such networks were used before computers became available for experimentally measuring the potential of a given structure. Usually, the capacitors were replaced by resistors because a network of resistors is equivalent to a network of capacitors.

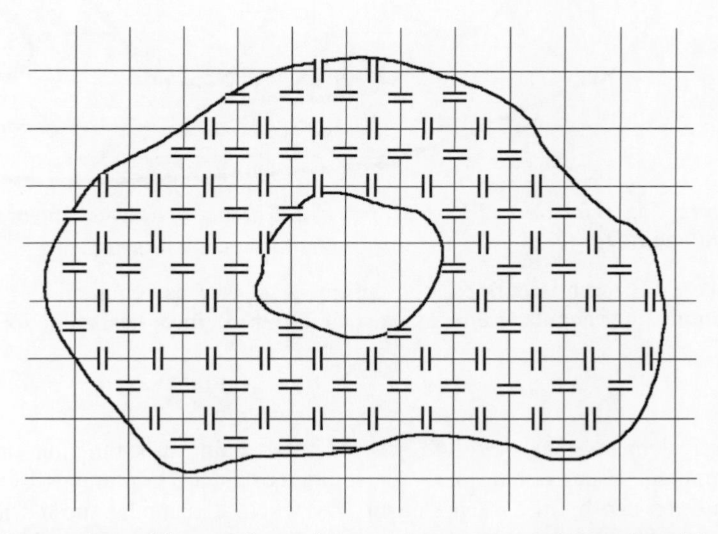

Figure 55: 2D capacitor with two electrodes on constant potential replaced by a network of 'finite' capacitors. First, a regular grid is defined. Then all grid lines inside the domain are replaced by capacitors and all grid lines in the electrodes are replaced by wires. When the boundary of the domain crosses a grid line, a capacitor is set when the longer part of the grid line is in the domain.

When one looks at a single node in such a network, one may recognize that such a node 'works' exactly like the 5-point-star operator that was introduced for static FD simulations. Therefore, this type of FE leads to codes identical with standard FD codes for statics.

Triangular elements for statics

To improve the poor approximation of the given boundary of the domain, one can introduce other types of elements. Instead of a network of lumped elements, one usually uses elements of the same dimension as the domain and one covers the entire domain by such finite elements. This is nothing else than subdividing a domain in subdomains, i.e., FE may be considered as a synonym for matrix methods with subdomain basis functions. For simplicity, triangular elements in 2D domains and tetrahedral elements in 3D domains are most frequently used. As one can see from Figure 56, even simple triangular elements approximate the given boundaries much more accurately than the FD grid or the 'lumped' FE grid.

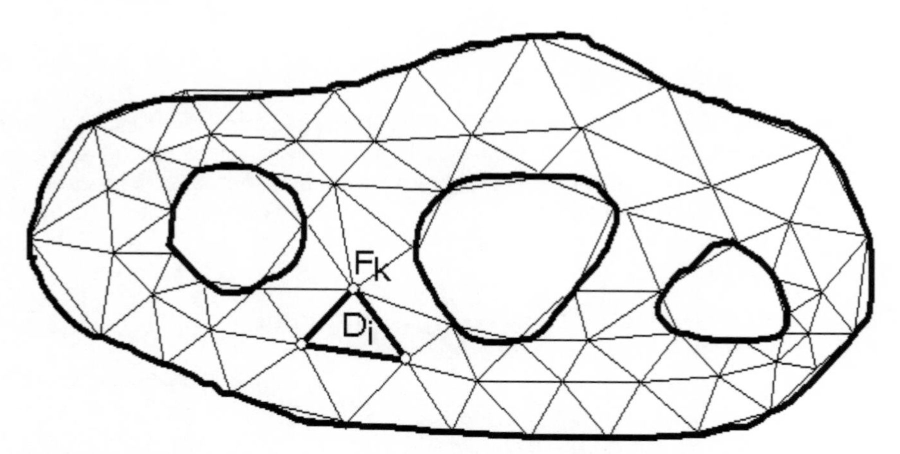

Figure 56: Triangular FE mesh for a 2D domain. Each element may be considered as a subdomain D_i.

Once a FE mesh is defined, one defines a set of J basis functions f_{ij} in each element D_i. For example, in triangular elements one can introduce linear basis functions:

$$\phi_i(x,y) = \sum_{j=1}^{J} A_{ij} f_{ij}(x,y) = A_{i1} + A_{i2}x + A_{i3}y. \tag{6.15}$$

For I elements, one therefore has $I*J$ subdomain basis functions and $I*J$ parameters to be computed. Usually, one looks for solutions that are continuous everywhere. The continuity condition can be met in an elegant way when a triangular mesh with a linear basis is used. Assume that the potentials F_k are known at the nodes of a FE mesh, i.e., at the corners of each triangle. Then one can compute the parameters A_{ij} from the F_k. In fact, the potential is only known at the nodes on the boundaries of the given domain when the boundary values are known. The potential at the nodes inside the domain is unknown. Note that the number K of unknown potential values at the nodes is considerably smaller than $I*J$. Therefore, working with the values at the nodes allows one to reduce the number of unknowns and guarantees the continuity of the solution over the entire domain. Remember that a smaller square error integral over the entire domain might be obtained without the continuity condition (see the section on Sub-domain basis functions). The main drawback of this method is that one must know which of the nodes of the entire mesh correspond to each corner of each triangle. Since FE meshes are usually irregular, several incidence matrices or incidence lists are required that store these correlations. FE codes often resemble companies with a huge administration. The mesh generation and definition of the corresponding list is often more time consuming than the solution of the resulting matrix.

Another problem arises from the use of linear subdomain basis functions. The second order derivatives of these functions are zero everywhere, except on the borders of the elements where the derivatives are undefined because the basis functions are discontinuous there. Therefore, the Laplace equation cannot be used directly to compute the unknown parameters in static applications. The following are two prominent methods to overcome these difficulties.

Variational approach

The variational formulation (5.278) is used. In the corresponding variational integral one has only first order derivatives that can be obtained very easily from (6.15). As soon as the subdomain basis functions mentioned above are inserted to approximate the field, the variational integral can be replaced by a sum over all elements. The requirement that the derivatives of this integral with respect to all unknowns is zero leads to a matrix equation that can be solved on a computer.

Weighted residuals

An error or residual r is introduced in the Laplace equation, because this equation may not be solved analytically. Then the resulting equation is weighted:

$$w \cdot \Delta_T \phi = w \cdot r . \tag{6.16}$$

Now, the weighted equation is integrated over the entire domain and the first theorem of Green (1.128) is applied:

$$\int_D (w \Delta_T \phi + \mathrm{grad}_T\, w \cdot \mathrm{grad}_T\, \phi)\, \mathrm{d}F = -\oint_{\partial D} w (\mathrm{grad}_T\, \phi) \cdot \mathrm{d}\vec{\ell}^{\circ} , \tag{6.17}$$

where the boundary integral on the right hand side is usually set equal to zero. When the Dirichlet boundary condition holds, the gradient of the potential on the boundary is unknown. In this case, one selects the weight w in such a way that it is zero on the boundary and the Dirichlet boundary condition is met by appropriately setting the values of the potentials in the nodes on the boundary. As a consequence, (6.17) allows one to replace the integral over the weighted Laplace equation by an integral over a product of gradients.

Note that the method of weighted residuals has more degrees of freedom than the variational method. Although this may be desirable, it introduces the chance of obtaining useless results when the weighting function w is not properly selected. For Galerkin's choice $w=\phi$, both methods lead to the same results.

Extensions of FE

Figure 57 shows a triangular FE grid for a coaxial cable with square wires and the resulting lines of constant potential. As one can see, these lines are less smooth than those obtained with a FD in the previous section. The main reason for this is the fact that FD with central difference operators corresponds to second order elements, whereas triangular elements with linear basis are of first order.

Of course, the accuracy of the results may be improved by a refinement of the mesh as in Figure 58, but the structure of the potential lines remains essentially the same. An improvement may be obtained from the introduction of higher order elements with higher order basis functions in each element and by the replacement of triangular elements by elements of a more general geometric shape. This also leads to a more complicated formalism and finally to matrix equations that are less sparse and have a higher condition number.

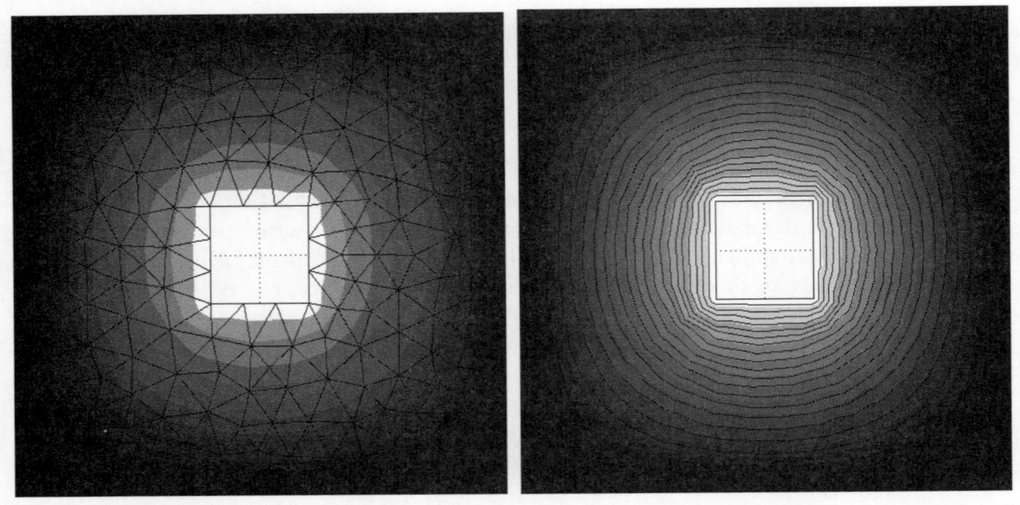

Figure 57: Rough FE mesh (left) and lines of constant potential (right) for an electrostatic computation of a coaxial cable with square conductors. Matlab PDE tool computation.

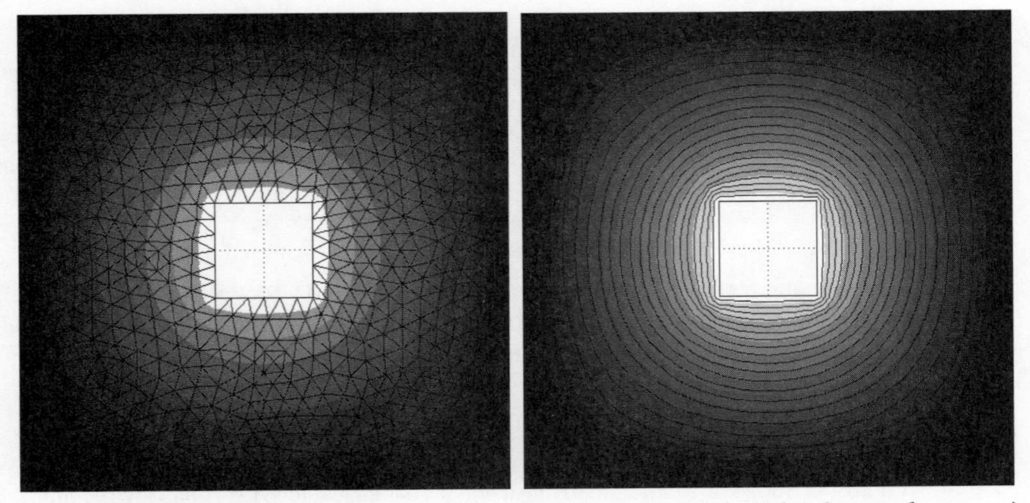

Figure 58: Refined FE mesh (left) and lines of constant potential (right) for an electrostatic computation of a coaxial cable with square conductors. Matlab PDE tool computation.

Another improvement may be obtained from an improved mesh generation. It is important to know that the shape of the mesh has a considerable influence on the accuracy of the results. To obtain good results, the elements should be as regular as possible. When triangular elements are used, this means that all angles of the triangles should be as close to 60 degrees as possible. Triangles with one angle that is much smaller than 60 degrees can cause high errors. Moreover, the size of the triangles should be adapted to the complexity of the field. Small triangles should be used in those areas where the field is rapidly changing, for example, near the corners of the inner conductor. Of course, the accuracy of the results may be improved by a refinement of the mesh as in Figure 58. As one can see, the mesh generator of the Matlab PDE tool [MATLAB,

1996] does not refine the mesh in these critical areas. Adaptive mesh generators are quite complicated and time-consuming and the implementation of such mesh generators is much more demanding than the implementation of simple FE codes.

Whitney elements

The extension of FE to electrodynamics is much more difficult than the extension of FD to dynamics because the variational formulation and the method of weighted residuals require good analytical knowledge. The same holds for the development of absorbing boundary conditions that are essential in electrodynamics. Moreover, simple FE extensions lead to various problems, such as spurious modes in the computation of guided waves and resonators. Spurious modes are unphysical solutions caused by the numerical model that must be filtered out. For these reasons, FE was rarely applied to electrodynamics for a long period of time whereas it was widely used in statics. The breakthrough came with Whitney elements or vector elements. In these elements one uses different vector basis functions to approximate the different fields. The E and H fields contained in the first two Maxwell equations in the curl operators are approximated by Whitney 1 forms that resemble a vector field rotating around a line, whereas the D and B fields contained in the third and fourth Maxwell equations in the div operators are approximated by Whitney 2 forms that resemble a vector field that is divergent from a point. The Whitney forms are well adapted to Maxwell's equations and lead to powerful FE codes for electrodynamics.

Although Whitney elements are very useful, it should be pointed out that these elements do not avoid the problems of finding absorbing boundary conditions and appropriate mesh generators. For more information see [Binns, 1963], [Binns, 1992], [Chari, 1980], [Silvester, 1983], [Whiteman, 1997], and

[http://www.engr.usask.ca/~macphed/finite/fe_resources/fe_resources.html].

Boundary elements

Remember that the method of weighted residuals used the first theorem of Green to obtain two integrals, one over the given domain and one over its boundary. The boundary integral was eliminated by an appropriate selection of the weighting function. Instead of these, one can also use the second theorem of Green (1.130) to derive a sum of a domain integral and of a boundary integral that allows one to replace the Laplace equation by a boundary integral equation. To solve this equation, only the boundary of the domain is then discretized by finite elements. As a consequence, the Boundary Element (BE) method has no problems with open domains and with absorbing boundary conditions. The BE method is a boundary method with 'subdomain' basis functions. Here, 'domain' means the domain of the integral, which is the boundary of the physical domain. For more detailed descriptions of the BE method see [Beer, 1992].

Transmission Line Matrix (TLM)

The idea of lumped finite elements for statics can be generalized to approximate the propagation of electromagnetic waves. Since a transmission line can be considered as a lumped element for wave propagation, one may replace space by a network of appropriate transmission lines. Each transmission line can then be replaced by some capacitors, inductors, and resistors (see the section on Transmission lines). Finally, the network may be analyzed with standard methods for network analysis.

For simplicity, we consider the propagation of electromagnetic waves in a loss-free, 2D domain. Since there are no losses, no resistors will be present and each transmission line is replaced by an inductance along the wires and a capacitor between the wires. There are two simple ways of implementing such transmission line grids: shunt networks and series connected networks. Figure 59 shows a shunt node in a 2D TLM network and the corresponding lumped element model, whereas Figure 60 shows the lumped element model of a series node.

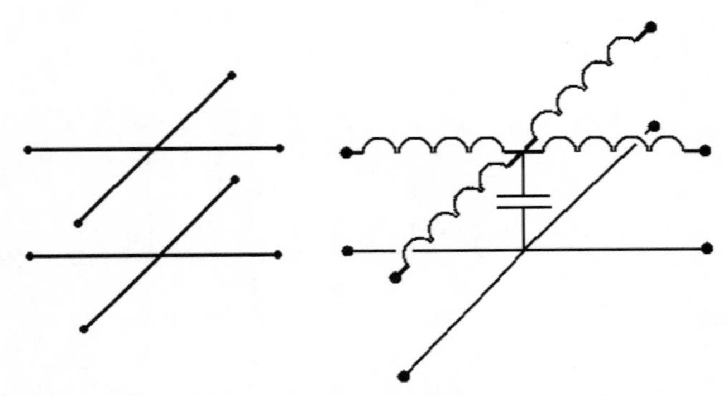

Figure 59: 2D TLM shunt node (left) and equivalent lumped element model (right).

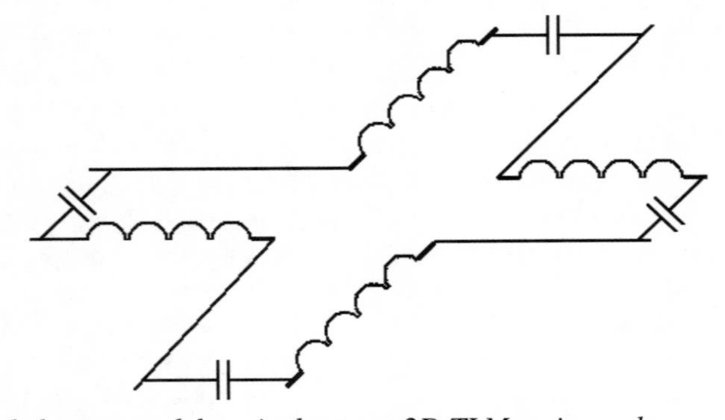

Figure 60: Lumped element model equivalent to a 2D TLM series node.

The shunt node has a voltage perpendicular to the *xy* plane. Therefore, the corresponding TLM model may be used for E waves propagating in the *xy* plane. The voltage of the series node is within the *xy* plane. This is appropriate for TE or H waves.

The TLM shows some similarities to the FE with lumped finite elements and therefore also to the FD methods. Both frequency domain and time domain TLM codes are available. For a more detailed description of the TLM with a computer code, see [Hoefer, 1991].

Method of Moments (MoM)

It is well known that an electromagnetic wave cannot penetrate through a wire grid when the mesh size of the grid is considerably shorter than a wavelength. Therefore, a wire grid may be used to replace a solid PEC plate, provided that one is not interested in an accurate computation of the field near the plate. Wire grids may therefore be used to approximate objects built of well conducting materials, for example, ships, airplanes, and cars, as well as many devices of technical interest. Obviously, one can consider the replacement of a solid plate by a wire grid as a traditional FE technique. In electromagnetics, such wire grids were widely used for the simulation of antenna and scattering problems with the Method of Moments (MoM). A typical example is the NEC code [http://www.dec.tis.net/~richesop/nec]. In mechanics, the MoM can be considered as a special case of FE that is obtained when a solid plate is replaced by lumped finite elements and the moments in these elements are expanded by the basis functions. The MoM for electromagnetics [Harrington, 1968] uses a similar procedure, but it expands the currents in the elements rather than the moments. Therefore, the term moment does not correctly describe the MoM for electromagnetics. Here, the MoM can be considered as a FE technique that expands the currents rather than the fields or the potentials.

The MoM procedure is advantageous for open domains because it removes the problems with absorbing boundary conditions when the currents are contained in a finite domain whereas the electromagnetic field extends to infinity. In the special case of PEC's, the MoM domain collapses to the boundaries of the field domains because one has only surface currents. Therefore, the MoM may simulate a boundary method although it is a domain method in general. When one compares FE, BE, and MoM, one must keep in mind that the domains in these methods may be different, depending on the material properties of the physical domains.

Also the MoM has been extended toward more sophisticated models with similar discretization as the FE methods. Moreover, the MoM theory was generalized in such a way that it also contains FE. Finally, the MoM is defined as a general domain method that expands a primary field by a series expansion and applies some projection technique to compute the unknown parameters as indicated in the chapter on 4 Generating matrix equations. 'For simplicity' the scalar product of the standard projection technique was replaced by a slightly different product [Harrington, 1968]. Moreover, integral formulations of the form (5.185) and Green's functions are favored in the MoM community. Finally, subdomain basis functions are typical for MoM codes.

Statics

To illustrate the MoM procedure, let us consider the 2D Dirichlet problem. A typical example is the coaxial cable in electrostatics. Here, one has no currents. The sources of the electric field are the surface charges on the conductors. Therefore, one can apply the Coulomb integral

$$\phi(\vec{r}_T) = \int_{\partial D} \frac{\varsigma(\vec{r}'_T)}{4\pi\varepsilon|\vec{r}_T - \vec{r}'_T|} \mathrm{d}\ell', \tag{6.18}$$

to obtain the scalar potential in the field domain D. Now, one introduces a series expansion for the charge density and inserts it in (6.18):

$$\phi^0(\vec{r}_T) = \int_{\partial D} \frac{\sum_{k=1}^{K} A_k \varsigma_k(\vec{r}\,'_T)}{4\pi\varepsilon|\vec{r}_T - \vec{r}\,'_T|} \mathrm{d}\ell' = \sum_{k=1}^{K} A_k \int_{\partial D} \frac{\varsigma_k(\vec{r}\,'_T)}{4\pi\varepsilon|\vec{r}_T - \vec{r}\,'_T|} \mathrm{d}\ell', \tag{6.19}$$

where ϕ^0 denotes the approximation of the scalar potential. This integral is required because there are no boundary conditions for the charge density that would allow one to compute the unknown parameters A_k. Therefore, one needs the boundary conditions either for the potential or for the electric field to compute these parameters. For the Dirichlet problem, the potential is known along the boundary ∂D. To fulfill the boundary conditions, one simply inserts (6.19) and projects this on a set of testing functions t_i defined on ∂D:

$$\phi^0(\vec{r}_T) = h \quad \Rightarrow \quad (\phi^0(\vec{r}_T), t_i) = \sum_{k=1}^{K} A_k \int_{\partial D} t_i(\vec{r}\,''_T) \cdot \int_{\partial D} \frac{\varsigma_k(\vec{r}\,'_T)}{4\pi\varepsilon|\vec{r}\,''_T - \vec{r}\,'_T|} \mathrm{d}\ell' \mathrm{d}\ell'' = (h(\vec{r}_T), t_i). \tag{6.20}$$

From this, one easily obtains a matrix equation. Note that all matrix elements are non-zero in general. This means that one has a dense matrix, which is typical for boundary methods, whereas sparse matrices are typical for the prominent domain methods such as FE, FD, and TLM. Each of the matrix elements contains two integrals. Since numerical integration may be expensive, one usually inserts simple testing and basis functions that allow one to evaluate the integrals analytically. For example, one uses basis functions that are constant in a finite element of the boundary and zero outside. Moreover, one often uses Dirac testing functions, i.e., collocation or point matching.

The resulting approximation of the potential fulfills the Laplace equation everywhere inside D. When 'subdomain' basis functions are used to model the charge density along the boundary, the resulting approximation of the charge density may be discontinuous. This can cause singularities of the field on the boundary. As a consequence, the results are most inaccurate near the boundary ∂D. Moreover, difficulties arise, for example, when matching points are set somewhere near these discontinuities. The Generalized Multipole Technique (GMT) presented in the following section provides an elegant way to overcome these difficulties.

Dynamics

In practice, the MoM is mainly applied in electrodynamics. Meanwhile, there is a huge variety of MoM implementations based on wire grids, current patches (rooftop functions), on Electric Field Integral Equations (EFIE, mainly equation (5.185)), on H Field Integral Equations (HFIE, mainly equation (5.186)), on Mixed Field Integral Equations (MFIE), on iterative procedures, etc. Several additional techniques have been developed to speed up MoM codes, for example Conjugate Gradients (CG), Fast Fourier Transforms (FFT), and Fast Multipole Methods (FMM) [Greengard, 1987].

Most of the MoM work has been in the frequency domain. Since the time domain approach became very successful for FD, TLM, and also for FE, there have been several attempts to implement time domain MoM codes [Kashyap, 1995]. Up to now, these implementations have shown severe stability problems that could not be solved in a general way. Note that time domain techniques are optimal when the matrices are sparse, especially for 3D domain methods, because such matrices are most easily handled with iterative matrix solvers. The trick of domain methods is that these work iteratively and update the time with each iteration step. Since boundary methods are typically represented by dense matrices, these methods are not well suited to the time domain approach. Although the MoM is formulated as a domain method, it often discretizes the boundaries of the physical domains and it also has rather dense matrices. Therefore, there is little hope that efficient and stable time-domain MoM codes will become available.

Generalized Multipole Techniques (GMT)

Generalized Multipole Techniques (GMT) have independently been developed by several groups under different names: Charge Simulation (CS) [Singer, 1974], [Steinbigler, 1979], Discrete Sources [Eremin, 1992], [Eremin, 1995], Fictitious Sources [Tayeb, 1991], Method of Auxiliary Sources (MAS) [Bogdanov, 1998], [Kupradze, 1967], Current Model Method (CMM) [Leviatan, 1991], Multi Origin Technique (MOT) [Akkermans, 1982], Multiple Multipole Program (MMP) [Hafner, 1980], [Hafner, 1983], SPherical EXpansions (SPEX) [Ludwig, 1989], Yasuura method [Okuno, 1998], [Yasuura, 1965], etc. The motivation and background of the different groups working on GMT was quite different. CS was restricted to electrostatics and was developed as a generalization of the mirror charges that were often used to obtain analytic solutions in special cases. MCM was derived from the MoM in order to remove problems with singularities of the field on the boundary. DS and MAS were influenced by the Vekua theory, whereas MMP was derived as a generalization of Mie theory and later also of Vekua theory.

However, GMT is a boundary method that expands the field in each domain with a set of basis functions that fulfill the Maxwell equations analytically. The most prominent basis functions are multipoles that have a local behavior which is numerically most useful. Local behavior means that the corresponding field is concentrated around the origin of a multipole and has a singularity at the origin. Some of the GMT versions use only zero order multipoles (monopoles), others use only first order multipoles (dipoles), and more general implementations use arbitrary integer orders. The multipole orders might also be real numbers in special situations, but real order multipoles were rarely used. Some implementations are restricted to multipole expansions only, whereas more general implementations include other analytic solutions of Maxwell's equations obtained either from the separation of variables (see section on Separation of variables) in special coordinates or from integration, especially of Green's functions (see section on Green's functions).

Statics

For simplicity, we consider the same 2D electrostatic problem as described in the section on the MoM. From (6.19) we see that the scalar potential is indirectly expanded by a series of basis functions that are Coulomb integrals over charge distributions. Therefore, we can simply write

$$\phi^0(\vec{r}_T) = \sum_{k=1}^{K} A_k \phi_k(\vec{r}'_T) \ , \tag{6.21}$$

instead of (6.19). Although we may obtain appropriate basis functions ϕ_k by use of the Coulomb integral

$$\phi_k(\vec{r}_T) = \int\limits_{\partial D} \frac{\varsigma_k(\vec{r}'_T)}{4\pi\varepsilon|\vec{r}_T - \vec{r}'_T|} \mathrm{d}\ell', \tag{6.22}$$

over the border of the domain *D*, we also may integrate over some path outside *D* that does not coincide with ∂D. Moving the path away from the border removes the singularity of the field on the border when the charge distribution ζ_k is discontinuous. This is the main idea of the CFM for electromagnetic scattering. Since the problem with the singularity is removed when we move the path away from the border of the domain *D*, we may integrate over arbitrarily 'ugly' charge distributions, including Dirac functions, which remove the integral and lead to the CS approach:

$$\phi_k(\vec{r}_T) = \frac{\lambda_k(\vec{r}'_T)}{4\pi\varepsilon|\vec{r}_T - \vec{r}'_T|}, \tag{6.23}$$

where λ_k is an infinite line charge perpendicular to the *xy* plane or a 2D monopole. Note that the CS approach was found directly without the MoM formulation, Coulomb integral, and Dirac type charge distribution. In the CS, integrals over given charge distributions of the form (6.22) were introduced later to improve the modeling of long sections of the boundary ∂D with a constant potential.

The main problem of the CS consists in setting a finite number of charges in such a way that the desired solution can be simulated as accurately as possible. Figure 61 illustrates the placement of charges for simulating the potential in a domain with a hole. Good results are obtained when one can associate a part of the boundary to each charge in such a way that the part of the boundary and the charge build more or less a triangle with equal sides. This construction is similar to constructing a first layer of triangular finite elements outside the domain *D*.

To compute the unknown parameters in (6.21), the simple point matching technique can be applied. To obtain accurate results, the placement of the matching points on the boundary is important. Usually, one associates one matching point to each charge, for example, in such a way that the matching point is the point on the boundary with the shortest distance from the charge.

Numerical problems may occur in the CS when several charges are located at almost the same position. First of all, the condition of the resulting matrix becomes very high. Moreover, one often observes very high values of the corresponding parameters. Remember that a dipole can be obtained from two multipoles of equal strength but opposite sign in the limit, where the distance between the monopoles goes to zero and the strength goes to infinity. Similarly, one can obtain higher order multipoles from clusters of several monopoles. Therefore, one can overcome some of the numerical problems of the CS by replacing several charges that are close together by an appropriate multipole expansion. Since several multipole expansions are used to model the field in a domain in general, this is called the Multiple Multipole Program (MMP) approach. The construction of a set of multipoles for a given domain D is similar to the construction of a set of charges for that domain (see Figure 61). Usually, fewer multipoles than charges are required for obtaining the same accuracy.

The MMP approach causes difficulties with the simple point matching technique. Since several unknown parameters are associated with each multipole expansion, one must also define several matching points near each multipole. In the simple point matching technique, one works with a square matrix. Therefore, one has to set as many matching points as unknowns when one boundary condition is imposed at each matching point. The correct position of the matching points is known to be essential for obtaining accurate results. Unfortunately, it turns out that it is difficult to find a set of appropriate matching points for MMP expansions in general. Therefore, the generalized point matching technique that works with overdetermined systems of equations, the error minimization technique, or the method of weighted residuals is often applied in GMT.

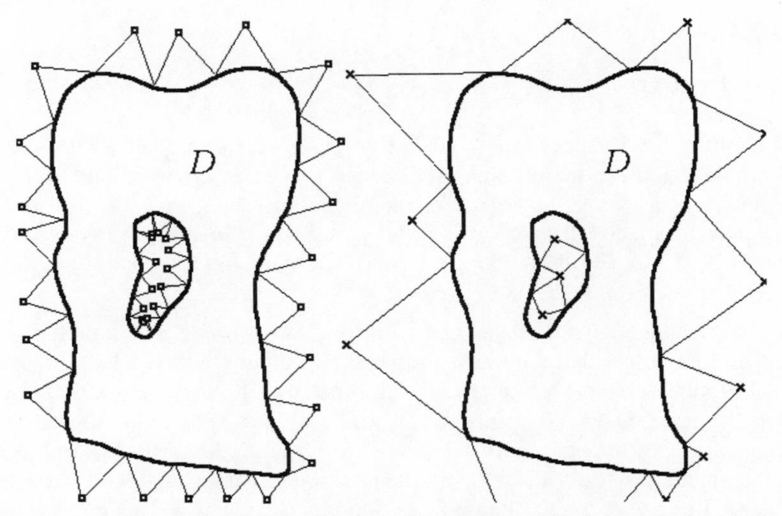

Figure 61: Position of the charges (monopoles) and multipoles used to expand the field inside the domain D. Squares indicate the positions of charges, crosses indicate the origins of multipoles. The thin lines indicate the range of the biggest influence of the corresponding basis function.

Dynamics

The extension of GMT to dynamics is simple and straightforward. Since GMT is a boundary method with dense matrices, it is best to work in the frequency domain. Then, the static Laplace equation is replaced by the Helmholtz equation and the static multipole expansions are replaced by dynamic multipole expansions (see section on Separation of variables). The placement of dynamic multipoles is essentially the same as for static multipoles. To obtain good results, the shortest distance of each multipole from the boundary should be less than half of the wavelength (in most cases).

Remember that the MoM requires the discretization of the boundaries only in the loss-free case, whereas domain discretization is required in the lossy case. Multipole expansions may simulate current and charge distributions on the surface of the given domain as well as any other charge distribution outside the given domain. Therefore, the GMT approach for lossy materials remains the same as for loss-free materials. In fact, GMT expands the electromagnetic field directly rather than any current and charge densities. Once the electromagnetic field is known, the current density inside a lossy body is obtained from Ohm's law from the electric field, and the surface current density on a PEC surface is obtained from the boundary condition (5.306) from the H field.

The GMT is considered in detail in the chapter on 8 MMP - A general boundary method implementation.

Special remarks

Accuracy and errors

The aim of all numerical methods is to provide accurate results with low errors within a short computation time. In most cases, the users of such codes would like to define the desired accuracy as a system parameter of the code, but only a few codes provide features that allow the user to specify the accuracy. Many codes do not even provide any information on the accuracy of the results or on the errors. Therefore, the validation of the results may be very difficult.

Assume that a code outputs an error number, for example, relative error 1.35%. What does this mean? Remember that the results of a code for computational electromagnetics are fields rather than numbers. To associate an error number to a field, an error definition is required. In the error definition, usually a norm is involved. Very often, the square norm is used (see the chapter on 2 Function analysis). Moreover, the relative error is the absolute error divided by some quantity that must also be defined. Assume that you know the approximation f^0 of a field f as well as the correct solution f. As soon as the norm is defined, you can define the relative error either as $\|f - f^0\| / \|f\|$ or as $\|(f - f^0) / f\|$. Note that these two definitions may lead to different results as well as the various definitions of the norm itself. The appropriate error definition depends on the problem to be solved. Sometimes, high maximum errors may be tolerated. Many codes for electromagnetics have regions where the local relative error of the field $(f - f^0)/f$ is 100%. Such errors may be acceptable when the field in the corresponding region is small or when the region itself is small. Therefore, it is often sufficient to consider the square norm of the error. In high voltage applications, the areas of highest electric field are most important. In these areas, the relative local error should be small. Here, big errors may be tolerated in areas of small fields. This means that the error distribution must not be uniform at all and that the local relative error is much more important than the global error defined with the norm. Therefore, numerical codes should not output only a single error number. In most cases, the user should know at least the following: 1) the maximum field strength, 2) the maximum absolute error, 3) the average field strength (averaged with the square norm), 4) the average absolute error.

From the field, one often derives secondary quantities that are of main interest for the user. Therefore, the user is interested in the accuracy of the secondary quantities rather than in the accuracy of the field. It is often difficult to estimate the error of a derived quantity from the error of the field. For example, in electrostatics, one might be interested in the capacity rather than in the electric field. When the numeric approximations of the E and D field are known, the capacity may be obtained from (5.119), i.e., from integration. The numeric evaluations of these integrals will cause additional errors. At the same time, some errors of the field computation may cancel each other during the integration. Therefore, it may happen that the error of the derived quantity is similar to, bigger than, or smaller than the error of the field.

Finally, the correct solution is rarely known. Therefore, the error output of a numerical code is usually an estimate that may be not reliable at all. Designers of such codes tend to define the error estimation routines in such a way that some small averaged errors are displayed rather than worst case estimates of maximum errors. Therefore, the users must perform several computations of one and the same problem to validate the results. Whenever possible, one should compare the results obtained with different codes or at least with different discretizations.

Convergence

When one compares the results of a code obtained with different discretizations, one expects that finer discretizations lead to more accurate results, or more precisely, that the error norm decreases with increasing discretization, i.e., with an increasing number of unknowns. In the limit of infinitely many unknowns, one expects that the error converges to zero. Note that this requires a reasonable error definition. Moreover, there are often many different discretizations of one and the same problem with the same number of unknowns, but with very different errors. Therefore, one cannot associate a unique error number e to a given number of unknowns n. Instead, one has an error range $e_{min}(n)...e_{max}(n)$. Poor discretizations often lead to 100% errors. Thus, the upper bound e_{max} is of no interest. Finding the lower bound e_{min} is very difficult (otherwise, the codes would be implemented in such a way that e_{min} is achieved). This requires the optimization of the discretization. For example, when FE is applied, all possible grids with a constant number of unknown parameters would have to be evaluated. Therefore, one wants to know a typical error range $e'_{min}(n)...e'_{max}(n)$ that is obtained with reasonable discretizations and one hopes that both $e'_{min}(n)$ and $e'_{max}(n)$ are functions that decay monotonically with increasing n. In fact, computational electromagnetics is not that simple. Many methods show a different behavior. The following behavior may often be observed: 1) the functions $e'_{min}(n)$ and $e'_{max}(n)$ grow at sufficiently small n and start shrinking when a sufficiently high n is reached, 2) the functions $e'_{min}(n)$ and $e'_{max}(n)$ 'oscillate' more or less randomly while decreasing, 3) the functions $e'_{min}(n)$ and $e'_{max}(n)$ show a staircase-ike behavior, 4) the functions $e'_{min}(n)$ and $e'_{max}(n)$ converge first and grow after a certain value of n (see Figure 62). Note that one can also have combinations of these types of convergence.

Designers of numerical methods often try to improve the convergence in such a way that a faster convergence is obtained. Although this is desirable for obtaining very accurate results, it should be pointed out that methods with a slow convergence may be very efficient when results with a moderate accuracy are desired. Since moderate or even low accuracy is sufficient for many engineering problems, one should be careful with techniques that improve the convergence. Such techniques might slow down the method for the most important applications. However, a good convergence simplifies the validation of the result, whereas the types of convergence illustrated in Figure 62 cause problems for the validation.

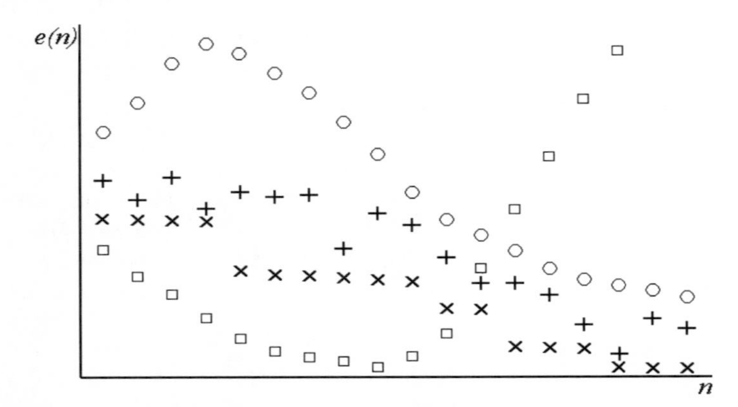

Figure 62: Various types of convergence. 1) o: convergence after an initial divergence, 2) +: convergence with random noise or oscillations, 3) ×: staircase convergence, 4) square markers: divergence after initial convergence (semi-convergent series).

Completeness

Completeness of the basis of series expansions (4.5) is certainly desirable, but also this term needs a detailed definition before it is to be used. Mathematical proofs of completeness may only be applied to simple situations where a very special set of basis functions is applied and where completeness means that any possible solution with some specifications and restrictions may be approximated with any desirable accuracy. A typical example was presented in the subsection on Completeness of multipole fields. In computational electromagnetics, one often searches for a very specific solution of a very specific problem. In this case, the basis must not be complete at all. It is sufficient that a finite set of basis functions allows one to obtain the desired results with the desired accuracy. When another solution of another problem is searched, one may replace the set of basis functions by another set.

Each set of basis functions may be used to approximate several solutions of several problems with a certain accuracy. The more specialized a set of basis functions is, the more efficient the corresponding code. Assume that you have FE mesh for solving a 2D Dirichlet problem for a given circular domain D. When this mesh consists of triangles of similar size, it can be used to approximate any solution for any boundary values that do not vary too much within one element. Assume now that the given boundary values are constant everywhere, except in a small area A of the boundary where the values change rapidly. One then knows that the potential will be almost constant on the entire mesh, except in the vicinity of A. Therefore, one should use small elements near A and big elements far away from A. Such an adapted mesh will allow one to solve the given problem more efficiently, but it is incomplete in the sense that it does not allow one to obtain solutions with a similar accuracy when the boundary values are different, for example, almost constant near A and rapidly varying at some other place. Similarly, one would put a couple of multipoles near the area A when the GMT is used. The GMT basis would then be incomplete in the same sense.

Since 'incomplete' sets of basis functions may always be better adapted to a specific problem, one has two classes of codes with different properties: 'complete basis' codes with a nonadapted discretization and 'incomplete basis' codes with an adapted or adaptive discretization. The former are often more user-friendly because the discretization is usually done automatically by the code. The latter may be more intelligent because they can be adapted or they can adapt themselves to a specific situation or to a specific class of problems. Adaptive discretizations can mimic the learning process.

Condition of matrices

When one searches for a set of basis functions that might approximate a specific solution of a problem very well, one often can find several basis functions that come close to the desired solution. A superposition of several 'almost correct' basis functions will cause numerical dependencies in the resulting matrix equation, i.e., the matrix will obtain a big condition number. Problems with ill-conditioned matrices may occur in any numerical method, especially when the discretization is adaptive. It has already been demonstrated in the section on Ill-conditioned matrices that one should be careful with techniques that reduce the condition number of the matrix equation to be solved. The statements made there can be verified in the context of many methods for computational electromagnetics.

Combination and generalization of methods

The aim of computational electromagnetics is the development of powerful codes that are accurate, reliable, efficient, and user-friendly, and have a wide range of applications. Today, it is quite clear that there is little hope that a single method allows one to implement a code that is optimal for all kinds of applications and requirements. Each method has its own advantages and also drawbacks. To select an appropriate method for a specific task one therefore needs to compare the different methods.

For code designers, the combination and the generalization of methods are the most promising procedures toward better codes. To obtain useful combinations and generalizations, a detailed analysis of the original methods is required.

Comparison of methods

Those who have implemented a new idea or even a new method compare the resulting code with results obtained from other codes mainly to demonstrate the power of their idea or method. These comparisons are often not really fair comparisons. Sometimes, 'quick and dirty' implementations of concurrent methods are used. Usually, only a few 'selected' concurrent methods are considered whereas really powerful concurrent codes are ignored. Although such comparisons may be helpful for advertising, they are useless or even misleading for those who want to improve codes and also for potential users.

To obtain more reliable comparisons, many different benchmark problems have been proposed. Potential code users may then select a benchmark problem that is close to the problem they want to solve and select the code that performed best on the benchmark problem. Unfortunately, there are many benchmark problems and those who design new codes are often not even aware of the benchmark collections or they have simply no time to work on all benchmark problems they could solve. Moreover, most of the codes can be tuned by the code designers or by experienced users in such a way that a relatively high performance is reached. Code designers who spend much time on solving benchmark problems may therefore obtain much better results than inexperienced users of the same code.

However, those who intend to improve codes need reliable and fair comparisons and a detailed analysis not only of the advantages, but also of the drawbacks of codes. In fact, removing drawbacks is the first step toward any improvement and this requires a correct analysis of the disadvantages. Since most publications focus on the demonstration of the advantages of a favored method, this analysis is quite demanding and should be based on experience rather than on published papers.

When one does not correctly detect the main problems of a method, all attempts to improve it will fail. For example, around 1970, the so-called Circular Harmonic Analysis (CHA) [Goell, 1980] was successfully used for several applications, but CHA codes written by other authors failed in other situations. These authors presented a wrong analysis of the failure of the CHA. They guessed that the CHA basis is incomplete. Therefore, the remedy would have been finding a complete CHA basis. In fact, Vekua theory [Vekua, 1967] had already proved that the standard CHA basis is a complete approximation basis. Therefore, the search for a complete CHA basis was vain and these attempts could not lead to any improvement of the CHA. The correct analysis of the CHA drawbacks [Hafner, 1980] showed 1) a bad convergence depending on the geometry of the domains where the CHA basis was used, 2) severe problems with the correct placement of the matching points used for computing the unknown parameters of the CHA basis, and 3) problems with ill-conditioned matrices. These problems could be removed by 1) replacement of the CHA basis by a MMP basis, 2) generalization of the point matching technique, and 3) direct solution of the overdetermined system of equations obtained from the generalized point matching instead of the solution of the corresponding normal equations that are also obtained from the error minimization technique or from the method of weighted residuals or from a projection technique with Galerkin choice of testing functions. For details, see the chapter on 8 MMP - A general boundary method implementation.

Hybrid methods

The analysis of the advantages and drawbacks of different methods sometimes shows that two methods are complementary in the sense that the first method has advantages where the second one has drawbacks and vice versa. In this case, the combination of the two methods, i.e., the development of a hybrid method, seems to be promising. Although the idea of hybrid methods is simple, the procedure for constructing useful hybrid methods is quite difficult. First of all, deep knowledge and a careful analysis of the methods to be combined is required. This is a big problem because most of the code designers are specialists on a single method. Furthermore, it is not guaranteed that the hybrid method removes the drawbacks and cumulates the advantages. Often, only the difficulties are cumulated. Therefore, the development of hybrid methods may be demanding or frustrating (see, for example, [Hafner, 1994/1], [Steinbigler, 1979]).

It is important to recognize that there are very different ways to combine two methods. Most importantly the coupling between the methods can be different. When the two methods result in a single, hybrid code, the coupling is strong. In this case, good results can only be expected when the two original methods are sufficiently close together. For example, one may combine

MoM and MMP, because both methods are based on similar concepts, work in the frequency domain, discretize the boundary when no losses are present, and so on. In fact, the strong coupling of two methods is usually a generalization of one of the methods in such a way that both original methods become special cases of the generalized method.

The weak coupling of methods is relatively simple. Instead of solving a problem with a single method, one splits the problem into several parts and solves each part with a different code. To simulate the coupling of the electromagnetic field between the parts, each of the codes must be designed in such a way that it is open to accept fields evaluated in another code. For example, one may combine a FD code with a MoM code. The former has problems with infinite domains, whereas the latter has problems with geometrically complicated structures. Therefore, one may subdivide the entire space in an inner domain D_i that contains a complicated structure to be solved by FD and an outer domain D_o that is open and is to be computed with the MoM code. The boundary ∂D between these two domains is then the interface between the two codes. The two codes must then be designed in such a way that they can communicate the field on this boundary. Assume that there is some radiating antenna in D_o that illuminates D_i. Then one can start with a MoM computation of the D_o without imposing any boundary conditions on ∂D. The MoM code then computes the field on ∂D and communicates this to the FD code. The FD code now computes the field inside D_i with a given incident field on ∂D. The resulting field will also radiate through ∂D. The FD code communicates this back to the MoM code. This code now computes the influence of the incident field from ∂D on the antenna. Obviously, the procedure is iterative. When the interaction between the antenna and D_i is weak, a few iterations will be sufficient.

Note that a weak coupling of domain methods with boundary methods is especially useful to compute the farfield when the domain method is used to compute the nearfield of a radiating structure. Here, no iterative procedure is required, because it is assumed that the radiated field is not reflected back into the domain where the domain method is applied. Therefore, one can simply start with the domain method, and transfer the field on the boundary to the boundary method that computes the desired far field. So-called near to farfield transformations, that are often used in FD codes, are nothing else than a coupling of FD with MoM.

Generalization

Generalization is one of the most powerful concepts in theoretical physics. Its main purpose is to unify theories, and unification of theories seems still to be the first goal of theoretical physics. Although the term generalization was explicitly used for the first time in Einstein's general theory of relativity, it was already playing an important role a long time ago. After Faraday's numerous and partially successful attempts at unifying forces, Maxwell's theory unified the theory of electricity and magnetism and the theory of light.

Although the design of numerical codes is somehow different from the design of physical theories, there are also many similarities. Finding a general method for solving any numerical problem corresponds to finding the great uniform theory in physics. Generalization is the key process for such activities.

Two steps are usually required in order to unify two distinct theories or methods: 1) Analysis of the two theories or methods, i.e., of their similarities, differences and the corresponding mathematical background; and 2) Generalization of one of the theories or methods in such a way that the other can be obtained or derived as a special case. The second step can require a generalization or even a completely new design of the corresponding mathematical formalism and often leads to a new interpretation of both theories or methods. Which of the theories or methods should be generalized is not obvious at all. It is more or less a matter of taste. However, this can lead to confusion. For example, one can generalize either the MoM or the method of Finite Elements (FE) in such a way that either the FE is a special case of the (generalized) MoM or the MoM is a special case of the FE.

The main effect of a generalization is that it leads to more freedom. This freedom allows one to consider "concurrent methods" as special cases, i.e., to unify different methods, which may be useful from a theoretical point of view. In practice, this plays a minor role. Here, one is interested in efficient codes and such codes are always a specific implementation of a method, rather than a general one. Since a generalized method provides more freedom, deriving a powerful code becomes more difficult. First of all, freedom leads to uncertainties. In order to overcome the problems caused by freedom, one has to derive appropriate rules. For example, when FD is generalized in such a way that arbitrary, irregular grids may be used, one needs rules that tell one how to discretize space, i.e., how to generate appropriate grids for a specific problem. As long as one admits regular grids only, the number of model parameters that define the grid is small. Therefore, the user has not much choice, but he can also not make many mistakes. When the code admits arbitrary grids, these grids are described by many parameters that must be set appropriately. Otherwise, the results obtained with irregular grids are worse than those obtained with regular ones.

When useful rules have been derived either from experience or from theoretical considerations, these rules may also be implemented in a code. Since experience may be gained from running such a code, the code may be implemented in such a way that it is able to improve its rules for defining the discretization parameters. This may result in 'intelligent' codes that learn have to solve problems more efficiently.

Specialization

General implementations of general methods typically lead to relatively complex codes that have a variety of system parameters. These parameters allow experienced users to tune the code for obtaining efficient solutions of a specific problem. Without an appropriate tuning, the code will be rather inefficient. Since tuning is often difficult, such codes are not user-friendly when the tuning needs to be done by the user. Therefore, specialized codes with a narrow range of applications are usually more user-friendly and efficient than general codes. To cover a wide range of applications, one can therefore work with several specialized codes rather than with a general one. Obviously, the handling of various codes based on different techniques is also not user-friendly. Therefore, a unique graphic platform may be designed as an interface to the user. The platform then needs some intelligent rules to decide which of the codes shall be applied to solve a problem specified by the user.

An interesting alternative to a platform that distributes work to a conglomerate of specialized codes is the design of a general code with an automatic tuning. To find appropriate tuning rules, one needs experience that may be obtained from a huge number of different computations only. This process may be considered as an optimization procedure. Therefore, one can link a general code with an external optimization procedure that optimizes the set of tuning parameters. Such an optimization procedure may be fed with a couple of testing examples that are typical for the area of interest of the user. The optimization procedure may run in the background while the computer has no other job. As soon as the optimization procedure has found better sets of tuning parameters, the user profits from the background optimization because the code will be able to solve the problems more efficiently – as long as the user does not change his area of interest, i.e., as long as the code works on similar problems. An example of such an optimization is shown in the section on the Eigenvalue Estimation Technique.

Prior knowledge

In order to achieve a maximum of user friendliness, some codes provide only a small set of tuning parameters and automatically perform the desired discretization. The user does not require any prior knowledge to run such codes. When there are many tuning parameters, prior knowledge or extended experience may be required to obtain accurate results or to reduce the computation time. Inexperienced users might hate such codes, but one should keep in mind that computational electromagnetics is a complex business. Usually, one is interested in simulations that are so demanding that none of the available codes is able to do these simulations quickly and easily. Therefore, tuning capabilities are very important and codes that allow the user to take advantage of his prior knowledge turn out to be most useful.

Since knowledge is obtained from running codes, the results obtained may be stored to provide some knowledge base for speeding up future computations.

How can a code be designed to benefit as much as possible from prior knowledge? Perturbation theory (see section on Perturbations) is one of the keys to such codes. When the solution of a problem is known that is close to the current problem, one can consider the current solution as a superposition of the known problem and a series expansion with a basis that only models the differences between the two problems. A useful technique is embedding the known solution in a connection (see the section on Generalized solutions - connections).

When iterative matrix solvers are applied, the number of iterations may be drastically reduced when one can start with a good initial guess of the parameter vector to be computed. Such a guess may be obtained from prior knowledge, i.e., from previous computations of similar problems. When several parameter vectors of previous computations are available, one can apply intelligent extrapolation routines to obtain an accurate guess of the parameter vector of the current computation. This is called the Parameter Estimation Technique (PET) [Hafner, 1994/2], [Fröhlich, 1997]. A detailed description of the PET for the MMP code is given in the section on the Eigenvalue Estimation Technique. Note that the standard time domain procedure may be considered as a zero order PET, because the iteration of each time step is started with the values obtained in the previous time step. Obviously, one may expect more accurate results from higher order PET, but this requires the storage of more parameter vectors, which may be large, especially when domain methods are applied. Incidentally, higher order extrapolations are much more demanding than one might believe. To achieve reliable higher order extrapolations, an appropriate extrapolation basis must be found. The standard power series and Fourier basis functions are usually not suited for extrapolations.

7 Generalizing Finite Differences

General remarks

The method of Finite Differences is one of the most prominent old methods for computational physics that has been applied to various problems including all sorts of problems in electromagnetics. The success of FD codes is based on the simplicity of this method. There are two main FD design problems that are closely related: 1) grid design, i.e., discretization of the given domain, and 2) design of appropriate operators that work on the grid and allow one to iteratively approximate the desired solution.

For simplicity, most of the prominent FD codes work on regular, Cartesian grids. Sometimes, slightly irregular grids are permitted and some codes provide tools for a local grid refinement. FE experts consider the restriction of FD codes to more or less regular grids as the biggest drawback of these codes. However, the FD method itself is not restricted to regular grids. Therefore, no generalization of the FD method is required when one intends to design more general FD codes for irregular grids. This means that one has to generalize FD codes rather than the FD method. When one works on irregular grids, one finds that one can easily find not only one FD operator, but also a huge variety of more or less complicated FD operators. Selecting the most appropriate operator turns out to be art rather than science, because much experience is required. In the following, we consider the design of a quite general FD implementation for electromagnetics that allows one to get experience with irregular grids and various FD operators.

The main tuning parameters of general FD codes are parameters that define the grid and the FD operators. These parameters may be optimized by an external optimization procedure. The optimization of operators is much more demanding than simple parameter optimization. Genetic Programming (GP) is a procedure that may be applied here. Unfortunately, the power of modern computers is not sufficient to design FD operators by GP, but it is expected that the generalization of GP will speed up the optimization in such a way that optimized FD operators for irregular grids will become available in the near future. In the following, the design of generalized FD codes that may be linked with external optimization routines is outlined.

All FD computations presented in this book were performed with the MaX-1 code [Hafner, 1998/S] that may also be used for more extended FD studies.

Iterations, convergence, and accuracy

In order to get more insight in the FD procedure, let us study the simple 2D electrostatic computation of a coaxial cable with circular wires (see Figure 52) and a homogeneous, isotropic dielectric between the wires. For this example, we can easily obtain an analytic solution. The potential in the domain between the wires has the form $a+b*\ln(r)$, where r is the radius, i.e. the distance of the field point from the center of the wire. The parameters a and b depend on the radii of the wires and on the potentials on the wires. Our test case is a not very realistic case with inner radius 0.25m and outer radius 0.95m, but this may easily be scaled. The inner wire is on potential one, the outer one is on potential zero.

For simplicity, we use a regular grid with the same number of grid lines in the x and y directions, i.e., $n_x=n_y=n$. The origin of the Cartesian coordinate system is set in the center of the wire and the square area x=-1m...+1m, y=-1m...+1m is modeled with the FD grid. Now, we compute the field in all grid points iteratively by applying the 5-point-star operator, as indicated in the section on Finite Differences (FD) and Finite Integrals (FI). Since the locations of the grid points vary for different grids, we use bilinear interpolation for computing the field at specified points. To reduce the amount of data, we only observe two points, (0.707,0) and (0.5,0.5). The first point is on the x axis, whereas the second one is on the diagonal between the x and y axes. Both points are at the same distance from the center of the wire. Therefore, the potential in both points should be identical and equal to approximately 0.221 for the test case. Selecting these points is reasonable because the regular FD grid causes some structure in the 2D space that breaks the rotational symmetry of the cable. On the grid, points on the x and y axes and points on the diagonals are most different.

Figure 63 shows the values at the two different points (0.707,0) and (0.5,0.5) obtained with 400 iterations on different grids. As one can see, the grid causes different values at the two points, because it does not fit well to the rotational symmetry. Note that the value at the first point is sometimes higher and sometimes lower than the value at the second point. Obviously, the convergence is very 'noisy'. Most of the numerical values are below the correct value obtained from the analytic solution. With increasing numbers of grid lines, the differences between the two points vanish, but they also decay and do not converge toward the correct value. The reason for this almost linear decay is that an increasing number of iterations is required for an increasing number of grid lines; 400 iterations are not sufficient for grids with 40 or more grid lines in the x and y directions.

When the number of iterations is increased from 400 to 1000, the zone of incorrect results caused by an insufficient number of iterations is shifted from about 42 to 75 (see Figure 64). From this, we can roughly estimate how many iterations should be performed for a given number of grid lines n in both the x and y directions. It seems that the number of iterations is approximately proportional to $n*\log(n)$, at least for the simple 2D static case that is considered here.

Since the computation time for each iteration is proportional to the number of grid points, i.e., n^2, the computation time is approximately proportional to $n^3*\log(n)$.

From the 'noise' we can roughly estimate the accuracy of the FD results. In Figure 64 we see that more than 10% errors of the correct value are observed for either $n<60$ or $n>73$. The latter is explained by the insufficient number of iterations.

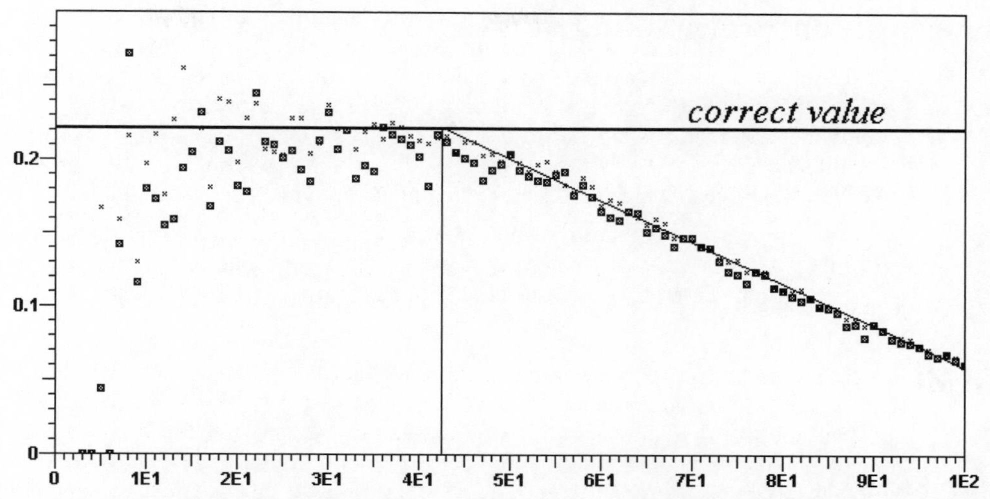

Figure 63: Electrostatic FD computation of the potential at the points (0.707,0) and (0.5,0.5) of a coaxial cable with circular wires, inner radius 0.25, outer radius 0.95. Results after 400 iterations with a 5-point-star approximation of the Laplacian. Regular grid with a variable number of grid lines in the x and y directions (n_x=n_y=n=3...100). Potential of the inner wire 1, outer wire 0. The correct value at the points is 0.221. Note that much 'noise' is generated because of the approximation of the circular boundaries. This noise is reduced with increasing n. It may be used to estimate the accuracy of the results. For more than 42 grid lines, severe inaccuracies are caused because 400 iterations are not sufficient. This causes an almost linear decay of the results for increasing n. Square plus × markers indicate the values at the point (0.707,0), × markers without squares indicate the values at the point (0.5,0.5).

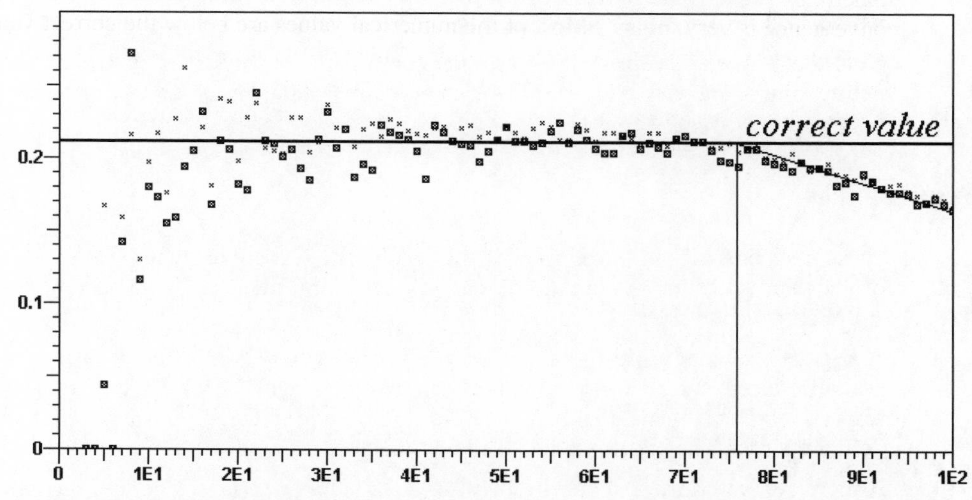

Figure 64: Same computation as in Figure 63 with 1000 iterations instead of 400 iterations. As one can see, the linear decay is shifted from n=42 to n=75.

We also see that 20% error is observed for n=41 and n<36. The former is a quite exceptional point. Lastly, 40% error is obtained for n<18. From this, we see that doubling the number of grid lines in both directions should reduce the error by a factor of approximately two. This means that the convergence is more or less linear with respect to n, which is quite poor. For obtaining 1% error, we can estimate that n should be about 600 and that approximately 20,000 iterations are required. Therefore, the 5-point-star operator must be evaluated approximately 7200 million times, which causes already long computation times on a PC.

Advanced techniques that allow one to obtain accurate results with fewer iterations and fewer grid lines are certainly desirable. For achieving this goal, one can search for either better FD grids or better FD operators. Let us start with FD grids.

Grids

Regular grids

Usually, a regular grid is defined as a grid with lines that are parallel to the axes of a Cartesian coordinate system. Moreover, the distances between neighbor grid lines are constant. Therefore, any grid point can be characterized by integer numbers rather than by real coordinates. In 2D space, the Cartesian coordinates x,y may be replaced by the integer numbers i,j and one has $x=i*d$, $y=j*d$, where d is the distance between neighbor grid lines. Such a regular grid consists of squares with side length d. It may be completely defined by 1) the origin and the orientation of the Cartesian coordinates, 2) the distance d between neighbor grid lines, 3) the numbers of grid lines n_x, n_y in the x and y directions.

The most important drawback of regular grids is that boundaries of a given domain are only approximated very roughly as one can see, for example, in the computation of a coaxial cable with circular wires (see Figure 52 and the previous section).

Note that one could replace the Cartesian coordinates by any other system of coordinates and define some 'regular' grid on such a coordinate system, but we will consider such grids as irregular ones.

Irregular grids

As soon as one tries to introduce irregular grids, one recognizes that there are different classes of irregular grids that require a different amount of information for an accurate definition. For simplicity, we consider 2D space only.

A slightly irregular grid is obtained when d is replaced by two different distances d_x, d_y which are introduced for the distances between neighbor grid lines in the x and y directions respectively. This 'generalization' requires a very small increase of the information required to define the grid. The corresponding grid lines become rectangles.

A more general, Cartesian grid is obtained when variable distances between neighbor grid lines are admitted. In general, one can define the locations x_i of the grid lines parallel to the y axis and the locations y_j of the grid lines parallel to the x axis (see Figure 65). For this, one must store n_x+n_y real numbers. Sometimes, one has simple algorithms to compute the real numbers x_i and y_j from a formula that requires less storage than n_x+n_y real numbers. This means that the algorithmic information content of such grids may be less than the information content of n_x+n_y real numbers. Whether the grid lines are defined by algorithms or by arrays of real numbers depends on the preference of the code designer.

Cartesian grids with variable distances between neighbor grid lines permit a better approximation of the boundaries, but such grids cause also fine discretizations in areas where a rough discretization would be sufficient. This causes an increase of memory and computation time. To avoid this, one can work with locally refined grids as illustrated in Figure 65. Obviously, additional storage is required for definition of the location of the grid refinement. Moreover, appropriate routines for handling the refinement are needed.

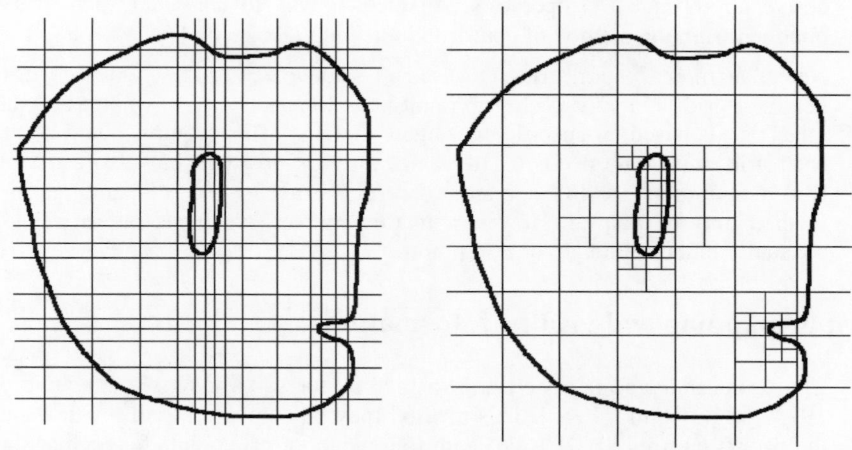

Figure 65: Irregular Cartesian grid with variable distance between grid lines (left) and with locally refined grid (right).

Although one can approximate a given boundary quite well with irregular and locally refined Cartesian grids, the shape of the boundary approximation remains essentially similar to a staircase, which is far from a smooth boundary approximation. Such approximations may be obtained by several techniques. First of all, the Cartesian coordinates *x,y* can be replaced by arbitrary coordinates *u,v*. The definition of such a coordinate system requires only the metric coefficients. For orthogonal, 2D coordinates one has only two metric coefficients g_u, g_v. Note that these coefficients are only constant for Cartesian coordinates and for similar coordinates with straight coordinate lines. For more general coordinates, at least one metric coefficient is not constant. Therefore, the metric coefficients are defined by formulae rather than by real numbers. As soon as the coordinate system has been defined, one can proceed as for Cartesian coordinates.

Instead of defining irregular grids with a non-Cartesian coordinate system, one can define directly the Cartesian coordinates x_{ij}, y_{jj} of the resulting grid points. Since one has n_x*n_y grid points, this requires the storage of 2*n_x*n_y real numbers, which is already of the same order of magnitude as the storage required for the electromagnetic field at each grid point. Depending on the definition of the FD operators for such a grid, additional information, for example, the metric coefficients at each grid point, might be required. From this, one can recognize that irregular grids can cause a drastic increase of the memory requirement.

To reduce the memory required for storing grid data, one may apply algorithms that evaluate these data, because the algorithmic information content of irregular grids is often much less than the information content of $2*n_x*n_y$ or even more real numbers. The evaluation of such algorithms may slow down the performance of the resulting FD codes. Therefore, there is a tradeoff between memory and speed.

Grids that may be defined from a specific coordinate system are most useful for the development of appropriate FD operators because one may write down the Maxwell equations and derived equations, for example, Helmholtz and Laplace equations, in these coordinates (see the chapter on 1 Geometry, differential and integral forms). From these notations, one can derive appropriate FD operators. An elegant way to construct such grids for 2D space is obtained from the method of conformal mapping (see below).

Note that more general grids are used in standard FE codes. On such grids, the number of neighbor grid points for each grid point is no longer constant, whereas it was equal to four for all 2D grids based on specific coordinate systems. This requires again additional storage. For each grid point, one needs to know not only its position, but also a list of neighbor points. Deriving appropriate FD operators for such grids is very demanding. In this case, the FI method may be helpful. However, in this chapter, we focus on irregular grids that have a constant number of neighbor grid points.

Conformal mapping and grid transformations

Conformal mapping is a well known method for solving simple 2D static problems [Henrici, 1974-1986], [Ramo, 1965]. This method may be combined easily with a generalized FD code that works on irregular 2D grids with four neighbor grid points for each grid point.

Conformal mapping works in the complex plane $z=x+i*y$, where x, y are real numbers that may be interpreted as the Cartesian coordinates in 2D space. Now, one may define a transformation $w=f(z)$ that maps each point of the z plane on a point in the complex w plane. The transformation f is conformal when the angles between two lines in the z plane are equal to the angles between the transformed curves in the w plane. Conformal transformations therefore transform the orthogonal Cartesian coordinate lines x=const. and y=const. into orthogonal lines in the w plane. This is very useful, for example in 2D electrostatics, where the electric field lines are orthogonal to the lines of constant potential. When the field in a capacitor with homogeneous material between two electrodes of constant potential is searched, one may search a conformal transformation that maps two lines, for example, $x=a$ and $x=b$ on the two boundary lines of the electrodes. This means that the lines $w_a=f(a+iy)$ and $w_b=f(b+iy)$ in the complex w plane coincide with the electrodes and therefore are lines of constant potential. From this, it follows that all lines $w_c=f(c+iy)$ with a constant value c are also lines of constant potential and that all lines $w_d=f(x+id)$ with a constant value d are field lines.

A typical example is the conformal transformation $w=\exp(z)$ that maps lines parallel to the y axis on concentric circles. Thus, two parallel plate electrodes in the z plane can be mapped on two concentric, circular electrodes. Note that exp is periodic with respect to the imaginary part of the argument, i.e., in the y direction. The period is 2π. To obtain full circles, one may map for example the rectangular area $x=-1...+1$, $y=-\pi...+\pi$ on the w plane as illustrated in Figure

66. This allows one to obtain the analytic solution of the electrostatic field in a coaxial cable with circular conductors. Note that the inverse transformation of $w=\exp(z)$ is $z=\mathrm{Log}(w)$, where Log is the complex logarithm. From this, it follows that the radial dependence of the potential in the coaxial cable is logarithmic.

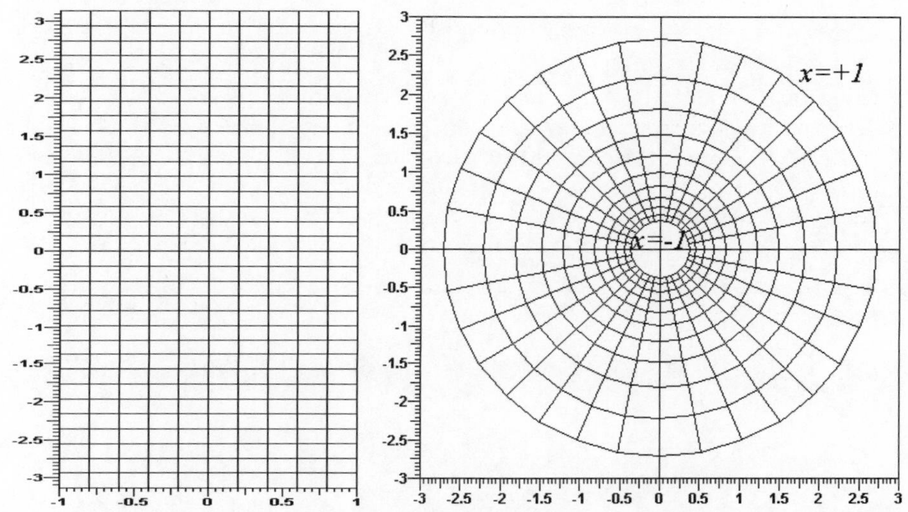

Figure 66: Conformal mapping of the rectangular area $x=-1...+1$, $y=-\pi...+\pi$ in the z plane (left) on an area between two concentric circles in the w plane (right) with the conformal transformation $w=\exp(z)$.

Conformal transformations can be nested or combined in such a way that more complicated grids are obtained. For example, the Moebius transform $w'=(aw+b)/(cw+d)$ may be applied to the w plane to obtain a non-coaxial cable with circular conductors, or the exp transform may be applied once more to obtain the grid illustrated in Figure 67.

Obviously, conformal mapping can be easily combined with 2D FD codes. This combination allows one to generate irregular FD grids that may be well adapted to the given geometry. Figure 68 shows an example where the grid lines do not exactly coincide with the given boundaries of the field domain.

Grids obtained from conformal mapping may also be useful when conformal mapping alone cannot be used to obtain an analytic solution, for example, when the geometry is too complicated, when different materials are present, for non-static problems, and so on.

When conformal transformations are exclusively used to generate appropriate, irregular FD grids, the conformal property is not really necessary. Therefore, one can also use non-conformal transformations. This is especially important in 3D space, because conformal mapping is restricted to the complex plane, i.e., to 2D space. However, general transformations in 3D space are much more demanding than conformal mapping and much less work has been done on such transformations. Even the visualization of the corresponding 3D grids is hard.

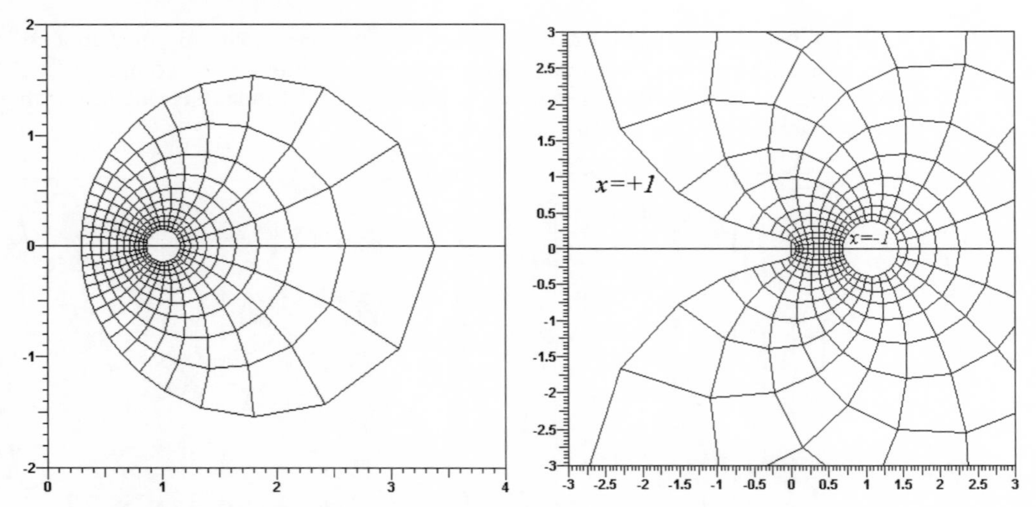

*Figure 67: Nested conformal mapping w'=(aw+b)/(cw+d)=(a*exp(z)+b)/(c*exp(z)+d) with a=c=1, b=-0.2, d=0.2 (left) and w'=exp(w)=exp(exp(z)) (right).*

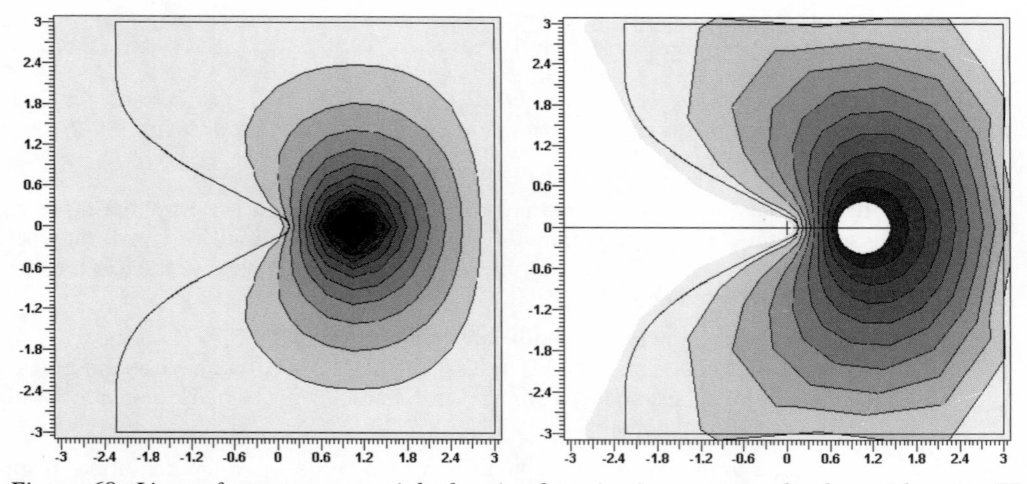

*Figure 68: Lines of constant potential of a circular wire in a rectangular box with a tip. FD computation with a regular grid, 19*19 grid lines (left), FD computation with an irregular grid obtained from the conformal transformation exp(exp(z)), 11*31 grid lines (right). Note that the computation with the irregular grid is much more accurate, and that almost the same number of grid points are used.*

Operators

As soon as one introduces irregular grids, one has also to find appropriate FD operators that are usually more complicated and time-consuming than the well-known FD operators for regular grids. To illustrate the procedure, let us start with a one-dimensional, second-order derivative on an irregular grid. A simple way to obtain an approximation of the second-order derivative of a function $f(x)$ that is discretized in the grid points x_i starts with a second-order power series expansion of the function

$$f(x) \approx a + bx + cx^2. \tag{7.1}$$

The parameters a,b,c can be evaluated from the values of f in three neighbor points:

$$\begin{aligned} a + bx_i + cx_i^2 &= f(x_i) = f_i \\ a + bx_{i-1} + cx_{i-1}^2 &= f(x_{i-1}) = f_{i-1} \\ a + bx_{i+1} + cx_{i+1}^2 &= f(x_{i+1}) = f_{i+1} \end{aligned} \tag{7.2}$$

which is a linear system of equations for the three unknowns a,b,c. Now, the second order derivative is

$$f''(x) \approx \frac{\partial^2}{\partial x^2}\left(a + bx + cx^2\right) = 2c = 2\frac{\dfrac{f_{i+1} - f_i}{x_{i+1} - x_i} - \dfrac{f_i - f_{i-1}}{x_i - x_{i-1}}}{x_{i+1} - x_{i-1}}. \tag{7.3}$$

Note that this becomes equal to the second-order derivative (6.3) obtained with central finite differences, when a regular grid with $x_{i+1}-x_i=x_i-x_{i-1}=d$ is given. Of course, we can use the same formula for a second order derivative in the y direction when we would like to obtain an FD approximation of the 2D Laplacian:

$$\begin{aligned} \Delta_T f(x,y) \approx\ & 2\frac{\dfrac{f(x_{i+1},y_i) - f(x_i,y_i)}{x_{i+1} - x_i} - \dfrac{f(x_i,y_i) - f(x_{i-1},y_i)}{x_i - x_{i-1}}}{x_{i+1} - x_{i-1}} \\ & + 2\frac{\dfrac{f(x_i,y_{i+1}) - f(x_i,y_i)}{y_{i+1} - y_i} - \dfrac{f(x_i,y_i) - f(x_i,y_{i-1})}{y_i - y_{i-1}}}{y_{i+1} - y_{i-1}} \end{aligned} \tag{7.4}$$

which is already much more complicated than (6.4). Note that this FD approximation may be used for slightly irregular grids with grid lines that are parallel to the x and y axes.

Deriving an appropriate FD approximation for irregular grids is considerably more difficult. When the grid is based on an orthogonal, curved system of coordinates, one may use (1.98) and (1.99) to obtain:

$$\Delta_T f(u,v) = \operatorname{div}_T \operatorname{grad}_T f = \frac{\left(g_v \dfrac{f_{,u}}{g_u}\right)_{,u} + \left(g_u \dfrac{f_{,v}}{g_v}\right)_{,v}}{g_u g_v} = \frac{f_{,uu}}{g_u^2} + \frac{f_{,vv}}{g_v^2} + \frac{f_{,u}\left(\dfrac{g_v}{g_u}\right)_{,u} + f_{,v}\left(\dfrac{g_u}{g_v}\right)_{,v}}{g_u g_v} \tag{7.5}$$

for an orthogonal irregular grid. In (7.5) one then can approximate all derivatives by finite differences. The problem with this formulation is that the metric coefficients and their derivatives must be known. This can be quite tedious for complicated grids. Moreover, the metric coefficients are functions of the 2D plane and must be either stored in the grid points or defined as functions that can easily be computed. Finally, such grids are still not completely irregular.

When we do not want to store metric coefficients, we may generalize the procedure used for the one-dimensional case, i.e., 1) we expand the field $f(x,y)$ by a series with unknown parameters, 2) we compute the parameters from the values of f at a grid point (x_{ij},y_{ij}) and at several neighbor grid points, 3) we approximate the second order derivatives with respect to x and y by the corresponding derivatives of the series expansion. From this, we finally obtain a FD approximation of the Laplacian. Since the series expansion used to approximate f is not unique, this procedure may lead to various FD operators. For example, we can start with the series expansion:

$$f(x,y) \approx a + bx + cx^2 + dy + ey^2. \tag{7.6}$$

Since we have five parameters a,b,c,d,e, we need at least four grid points that should be close to a given grid point (x_{ij},y_{ij}), exactly as in the 5-point star operator. Note that the term xy is missing in (7.6). When we would like to add this term for obtaining a better approximation of f, we would need an additional grid point. However, we can compute the five parameters from five equations. When we have an irregular grid, where each grid point can be characterized by two integer numbers i and j, we assume that the points (x_{i+1j},y_{i+1j}), (x_{ij+1},y_{ij+1}), (x_{i-1j},y_{i-1j}), (x_{ij-1},y_{ij-1}) are the four closest neighbors of (x_{ij},y_{ij}) and we obtain

$$\begin{aligned}
a + bx_{ij} + cx_{ij}^2 + dy_{ij} + ey_{ij}^2 &= f(x_{ij}, y_{ij}) = f_{ij} \\
a + bx_{i+1j} + cx_{i+1j}^2 + dy_{i+1j} + ey_{i+1j}^2 &= f(x_{i+1j}, y_{i+1j}) = f_{i+1j} \\
a + bx_{i-1j} + cx_{i-1j}^2 + dy_{i-1j} + ey_{i-1j}^2 &= f(x_{i-1j}, y_{i-1j}) = f_{i-1j} \\
a + bx_{ij+1} + cx_{ij+1}^2 + dy_{ij+1} + ey_{ij+1}^2 &= f(x_{ij+1}, y_{ij+1}) = f_{ij+1} \\
a + bx_{ij-1} + cx_{ij-1}^2 + dy_{ij-1} + ey_{ij-1}^2 &= f(x_{ij-1}, y_{ij-1}) = f_{ij-1}
\end{aligned} \tag{7.7}$$

which can be solved for a,b,c,d,e. From (7.6) we see that the 2D Laplacian can be approximated by

$$\Delta_{\mathrm{T}} f(x,y) \approx 2c + 2e. \tag{7.8}$$

Although this formula is simple, the evaluation of the parameters c and d from (7.7) causes quite long expressions. When we generalize the series expansion (7.6), by introducing K arbitrary basis functions b_k:

$$f(x,y) \approx \sum_{k=1}^{K} a_k b_k(x,y), \tag{7.9}$$

we can proceed as before, i.e., we write down a system of K equations at the grid point (x_{ij},y_{ij}) and K-1 neighbor points and we obtain an FD approximation of the 2D Laplacian of the form:

$$\Delta_{\mathrm{T}} f(x,y) \approx \sum_{k=1}^{K} a_k \Delta_{\mathrm{T}} b_k(x,y) = \sum_{k=1}^{K} a_k \left(\frac{\partial^2}{\partial x^2} b_k(x,y) + \frac{\partial^2}{\partial y^2} b_k(x,y) \right). \tag{7.10}$$

Note that we can also write down a system of more than K equations by adding more neighbor points and solve this overdetermined system in the least squares sense. The basis functions can be selected quite arbitrarily, but the second order derivatives of the basis functions with respect to x and y should be known. Moreover, at least one of the second order derivatives of the basis functions should be different from zero.

The quality of a FD procedure on an irregular grid obviously depends on the selection of the grid and of the corresponding FD operator. An irregular grid that is well adapted to a given problem is useless without an appropriate operator. To obtain FD results (see Figure 68) on a 'double exponential' conformal grid (see Figure 67), the FD operator (7.4) was applied. This operator is not well adapted to the 'double exponential' conformal grid. Despite this, the results seem to be much better than the results obtained with a regular grid and the 5-point-star operator that is well adapted to regular grids.

Generalized FD operators

So far, we have considered methods to adapt FD operators to a given grid. Such operators do not depend on the problem to be solved or they depend only indirectly on the given problem, when the grid is adapted to the problem. The question now is whether we can adapt a FD operator to a given problem in such a way that the problem can be solved more quickly (whereas other problems would be solved less quickly with this FD operator) than with a standard FD operator.

Remember the over/under-relaxation technique (6.7). With an appropriate choice of the relaxation factor, this technique can speed up the FD convergence. Obviously, we can easily generalize this formula by introducing five weighting factors to each term rather than a single relaxation factor:

$$f_{ij}^{n} = \frac{1}{4d^2}\left(\alpha_{i+1j} f_{i+1j}^{n-1} + \alpha_{i-1j} f_{i-1j}^{n-1} + \alpha_{ij+1} f_{ij+1}^{n-1} + \alpha_{ij-1} f_{ij-1}^{n-1} + \alpha_{ij} f_{ij}^{n-1}\right). \tag{7.11}$$

We can now try to adapt these factors to a given problem in such a way that the FD iterations converge more rapidly. Note that FD approximations of the Laplacian on irregular grids may have the same form as (7.11) with weighting factors that may depend on the location of the grid point. Therefore, we should call them weighting functions.

For example, let us consider a polar grid as illustrated in Figure 66. For the Laplacian in polar coordinates (r, φ) we have

$$\Delta_T f(r,\varphi) = \frac{\partial^2 f}{\partial r^2} + \frac{1}{r}\frac{\partial f}{\partial r} + \frac{1}{r^2}\frac{\partial^2 f}{\partial \varphi^2}. \tag{7.12}$$

Assume that the variable grid steps d_r=d*r and $d\varphi$=d*r are used in the two directions, which causes a finer grid for small r. Let the index i denote the radial direction and j the angular direction. When we approximate the derivatives with respect to r and φ by finite differences, we obtain

$$f_{ij}^{n} = \frac{1}{d^2 r^2}\left((1+d/2) f_{i+1j}^{n-1} + (1-d/2) f_{i-1j}^{n-1} + (1/r^2) f_{ij+1}^{n-1} + (1/r^2) f_{ij-1}^{n-1} - (2+2/r^2) f_{ij}^{n-1}\right). \tag{7.13}$$

This formulation is not unique, but its form is the same as (7.11), provided that the weighting functions are properly selected. Depending on the FD approximation, the weighting functions may turn out differently. Therefore, optimizing the weighting functions is the same as optimizing the FD approximation.

Since only five weighting functions must be optimized, this seems to be not very difficult. Here, it is important to recognize that these factors are functions of the 2D plane rather than constants. In (7.13), these functions depend on r only, because of the symmetry of the grid.

When we consider the special case of a coaxial cable with circular conductors, we know in advance that the field is rotationally symmetric. When we use a polar grid, it follows that the field is constant in the angular direction. Consequently, we can set the weighting functions α_{ij+1} and α_{ij-1} to zero without affecting the results, where the index j numbers the grid lines in the angular direction. This reduces the 5-point FD operator to a 3-point operator that can be evaluated more quickly. The 3-point operator is specialized to rotationally symmetric fields, i.e., to a special class of problems that is less general than the class of problems that can be handled with the 5-point operator (7.13).

When we work with the generalized 5-point operator (7.11), finding appropriate weighting functions implies either a tricky analysis or solving a demanding optimization problem. The latter might be done with the GGP concept presented in the section on Generalized Genetic Programming. To design appropriate weighting functions, GGP must be fed with several test cases, i.e., several problems with known solutions. The weighting functions proposed by GGP will then be inserted in the generalized 5-point operator. Afterwards, the resulting operator is used for solving all of the test cases. Then, the differences of the solutions obtained with the weighted operators and the known solutions are used (together with the computation time) to evaluate the fitness value of the corresponding weighting functions.

MEI technique

Since each FD operator acts locally, one may try to use 'local' test cases, so-called metrons, for optimizing the weights of generalized FD operators. This is the core of the Measured Equation of Invariance (MEI) technique proposed by Mei [Mei, 1992]. The procedure may be outlined as follows: 1) Generalize the FD operators by introducing N weighting factors. 2) Select N metrons that are appropriate for the problem to be solved. 3) Compute the weighting factors from the metrons. 4) Solve the problem by use of the weighted FD operators.

In general, one may use one set of metrons for each grid point and for the corresponding FD weights. This may require great computational effort when the number of grid points is high. This effort is reduced when the same metrons are used for a group of grid points.

To illustrate the procedure, let us consider the scattering of a plane wave incident on a PEC cylinder with an arbitrary cross section. Assume that the plane wave has an E field parallel to the axis of the cylinder and that the field propagates in the transverse plane, i.e., one has a 2D problem with an E wave propagation described by a scalar Helmholtz equation for the z component of the electric field outside the PEC:

$$(\Delta_T + \kappa^2)\begin{Bmatrix} E_z^{inc} \\ E_z^{scat} \\ E_z^{tot} \end{Bmatrix} = 0, \tag{7.14}$$

where the superscripts *inc*, *scat*, and *tot* indicate the incident field, the scattered field, and the total field. The transverse propagation constant κ is equal to the wave number k because there is no propagation in the z direction. On the PEC surface, the tangential electric field should vanish. Therefore, we have

$$E_z^{tot} = 0 \;\Rightarrow\; E_z^{scat} = -E_z^{inc}. \tag{7.15}$$

Since the incident field is given, the scattered field is known on the PEC boundary and fulfills the Helmholtz equation outside. Therefore, we may essentially apply generalized FD operators of the form (7.11) to compute the scattered field.

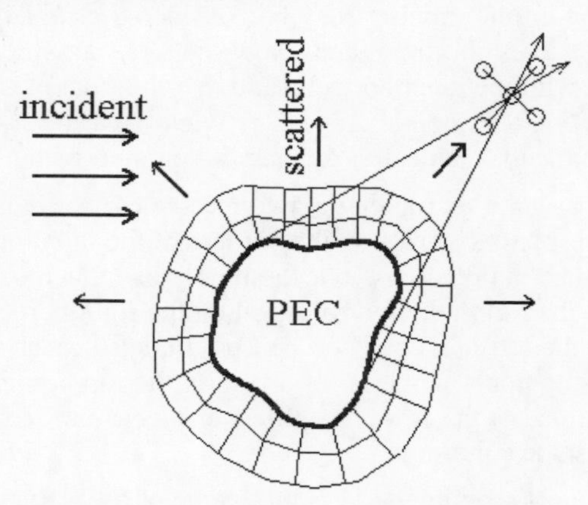

Figure 69: Plane wave incident on a PEC cylinder, adapted FD grid near the cylinder and a grid point further away. The scattered field at this point can be approximated by a superposition of plane waves that propagate within a relatively small angle (indicated by arrows).

When we observe a grid point at some distance from the PEC surface, we can assume that the scattered field may be approximated by a superposition of plane waves propagating in different directions. The standard 5-point-star operator is symmetric, i.e., it is appropriate for plane waves propagating in all directions. As one can see from Figure 69, the plane waves that might be observed at our grid point propagate more or less in the same direction. Therefore, we can use plane waves propagating in the expected directions as metrons to compute the weights of the FD operator. Such an adapted FD operator can easily handle waves propagating as expected, but it more or less suppresses waves propagating in a completely different direction. This has a desirable side effect: reflections caused by imperfect ABC's are traveling in the 'wrong' directions. Therefore, the adapted FD operator tends to suppress them. The MEI method therefore allows one to introduce ABC's that are far from being perfect a few grid lines away from the PEC surface. As a consequence, the domain to be modeled may be a quite thin strip around the PEC.

Although there was much interest in the MEI method for a short period of time, this method was almost completely abandoned by the FD community. There may be two reasons for that. First of all, the first papers exaggerated the method's usefulness and caused too high expectations and therefore frustrations. Although this method is promising, it requires much more experience than standard FD implementations and this experience must be gained from time-consuming work. Maybe, increased computer power can provide the desired experience and might cause a MEI revival.

Cellular Automata (CA)

Cellular Automata (CA) were widely used by computer freaks for generating nice pictures and by scientists who tried to find alternative formulations of complex processes [Gleick, 1991], [Rucker, 1989], [Toffoli, 1987], [Wolfram, 1994]. First, CA worked on the pixels of monochrome computer monitors. Each pixel is then represented by a binary number. Therefore, everything that is visible on a monitor may be described by a binary array B(i,j). Any animation on the monitor may be considered as a sequence of such arrays $B_k(i,j)$, where the index k denotes the time step. Now, one can assume that the pixels represent cells that communicate with the neighbor cells and that the state of a cell at time step k is defined by the state of the cell and its neighbors at k-1. When only the N closest neighbors have an influence on the state of a cell, each cell is a binary automaton with N+1 inputs and a single output.

Standard CA assume that a single automaton describes all cells. This leads to problems with cells that correspond to pixels on the borders of the monitor. When one does not want to invent special boundary automatons, one assumes that the monitor has hypercube topology. This means that 1) the right neighbor of a cell on the right border is the cell on the left border in the same row, 2) the left neighbor of a cell on the left border is the cell on the right border in the same row, 3) the upper neighbor of a cell on the top border is the cell on the bottom border in the same column, 4) the lower neighbor of a cell on the bottom border is the cell on the top border in the same column.

Probably the most prominent CA is the 'game of life' [Rucker, 1989]. This is a binary CA with N=8, i.e., the neighbors of the cell (i,j) are (i+1,j), (i-1,j), (i,j+1), (i,j-1), (i+1,j+1), (i+1,j-1), (i-1,j+1), (i-1,j-1). A cell is either dead (binary 0) or alive (binary 1). Now, the automaton counts the number of inputs that indicate cells that are alive. From this sum and from the actual cell state, it decides whether the cell will be dead or alive in the next time step:

```
Cell alive and (input sum < n1 or input sum > n2)
      → cell dies, otherwise cell remains alive
Cell dead and (input sum < n3 or input sum > n4)
  → cell remains dead, otherwise cell becomes alive
```

The life parameters n1, n2, n3, n4 determine the behavior of this game. Interesting behavior is observed for n1=2, n2=3, n3=3, n4=3, which is also called 2D live 2333. When one randomly initializes the cells and runs the 2D live 2333 rule several times, it seems as if structures of cells are generated that behave like small individuals in a 2D world. Some of these structures are stable and remain unchanged when they are not disturbed by another structure. Other structures change between two or more states and remain at the same location or move over the screen. The latter are called gliders. Now, one can design special life structures consisting of several cells that must be properly initialized. An interesting structure is the glider gun. This structure generates a sequence of gliders. Since the gliders can be considered as pulses, one can invent logical structures with gliders as input. If a sufficiently large computer is available, the game of life might even simulate a smaller computer.

Meanwhile, CA have also been developed to simulate chemical reactions and processes of interest for engineers. To obtain higher flexibility, the CA concept can be generalized. Obviously, FD schemes can be considered as special, non-binary CA. This allows one to obtain a new insight into FD and to design FD in a very general way: a FD operator can be considered as an automaton with a given number of inputs and outputs.

User definable operators

As soon as a FD operator is considered as a cellular automaton, one arrives at a FD definition that is so general that it includes all known FD operators as well as all FD operators that will be invented in the future. This allows one theoretically to write generalized FD codes in such a way that any problem of computational electromagnetics can be solved with one and the same code. Such codes play the role of general theories in physics.

The main problem of generalized FD codes is that the user must define the FD operators in form of formulae that describe the automaton. For doing that, one has several alternatives:

- Compiled standard computer codes: The user must write, compile, and link appropriate subroutines. This is tedious for the user, but it is general and results in fast performance.
- Using some symbolic mathematics platform such as Mathematica [Wolfram, 1991] or Maple [Char, 1991]: This is more user-friendly and also general, but the performance is much slower.
- Compiled code that contains a formula interpreter: This allows the user to define operators without the need of recompilation. The designer of such codes must write a formula interpreter. A sophisticated formula interpreter causes generates a lot of code with a slow performance. Therefore, it is reasonable to implement a simple formula interpreter that is well adapted to the need of FD users. This restricts the definition of the FD operators and results in a performance somewhere between compiled and symbolic math codes. MaX-1 is based on this concept.

Generalization of FD operators provides much freedom in the definition of the operators. Although this provides the chance of finding operators that are much better than the traditional ones, it also includes the opportunity of finding a huge number of bad operators. Finding a good or even optimal operator is therefore a very demanding search or optimization process.

Outlook: Automatic FD design

Users working with generalized FD codes may obtain much experience with irregular FD grids and with more or less well adapted FD operators. At the same time, the design of such operators is much more cumbersome and time-consuming than one might expect. A single user cannot design and test many different operators for optimally solving a given class of problems. Therefore, it is reasonable to let a computer do this work. Evolutionary strategies, such as GA, GP, and GGP (see section on Evolutionary Algorithms (EA)) are most promising for such complex optimizations. Linking such strategies with compiled standard codes is impossible, whereas linking them with symbolic math codes is difficult and results in performance that is much too slow. Therefore it is best to link evolutionary strategies with a compiled FD code that contains a formula interpreter for user-definable FD operators (see previous section). How can this be done?

First of all, the user should define a set of test cases within his area of interest. These test cases define the range of applications for which the FD operators are to be adapted. For each test case, a sufficiently accurate solution must be known. Such a solution is computed either with some other code or with a standard FD code with a sufficiently fine grid. The definition of the test case also includes the definition of the FD grid to be used. For FD codes with automatic grid definition, this might be omitted.

The generalized FD code must be designed in such a way that it is able to compute all test cases with an FD operator that is defined externally by the evolutionary strategy. Therefore, the FD code must be able to read in both operators and problems to be solved. It must automatically compute all test cases, compare the results with the known solutions and associate a fitness number to the FD operator. This means that the FD code must be fully automated. Since the evolutionary strategy may feed the FD code with crazy operators, the FD code must be completely robust, i.e., it should never crash. Therefore, the design of a generalized FD code is much more demanding than the design of standard FD codes.

The evolutionary strategy 'invents' new FD operators and forwards them to the FD code that returns the fitness values to the evolutionary code. Since the fitness evaluation is very time consuming, the evolutionary strategy must be able to find useful operators with a relatively low number of trials. When a population based strategy is used, this means that the population size must be as small as possible, typically in the order of 100. Here, the GGP concept seems to be most promising.

To illustrate the procedure, assume that you want to optimize a FD operator for 2D electrostatics. The 5-point-star operator is the most prominent operator for such applications. This is an automaton with five scalar inputs and a single scalar output that can be described by a function with five variables. Since the desired FD operator may depend on the location on the grid or on additional grid data, such as the metric coefficients, the function may have two or more parameters. These parameters may be considered as additional variables. However, optimization codes based on the GGP concept may design such functions.

The search space for functions with seven or even more variables is huge. Moreover, the fitness evaluation per function is high even in the simple 2D electrostatic case. Therefore, we must either wait for more powerful computers or find ways to speed up the optimization process. This can be achieved by several means that reduce the search space or that support the search process. Instead of the completely general design of a function with at least seven variables, one can optimize the weights of less general FD operators of the form (7.11). Since the five weights may be functions of space or of the location on the grid, GGP has to design five basis functions with two or more variables that will be interpreted as weighting functions. Note that GGP is designed for a simultaneous evolution of a set of arbitrarily many basis functions, whereas the standard GP handles a single function at a time. Since the 5-point-star operator is an operator of this type with most simple, constant weights, GGP may find this operator with almost correct weights when the testing examples cover a wide range of applications. According to a private communication with Jürg Fröhlich [Fröhlich, 1997], more efficient FD operators with a restricted range of applications could be found, but these results have not yet been published.

Because of the rapid growth of computer speed and memory, evolutionary strategies will certainly be able to optimize various FD operators for arbitrary applications in the near future. For current personal computers, this task is still too demanding, but developing such codes and testing them in simple situations is already possible. Since such codes are based neither on Maxwell theory nor on any other theory, one can call them 'Theory Free Codes' (TFC's). TFC's might be the basis of future computers that would be able to analyze experimental data within any area of interest. These machines might find an appropriate algorithmic description, i.e., an internal theory, for the measured data without any human support. CA's or generalized FD schemes linked with evolutionary strategies open the door to such TFC machines.

8 MMP - A general boundary method implementation

General remarks

The Multiple Multipole Program (MMP) was proposed by the author in 1980 [Hafner, 1980] as an extension of the Circular Harmonic Analysis (CHA) [Goell, 1980], Mie theory for wave propagation on two wires [Mie, 1900], and Point Matching. The first goal was the design of a computer code that would give accurate and reliable solutions of problems that cannot be solved analytically. This code should allow one to estimate the influence of additional assumptions necessary for obtaining analytical results. For example, when the propagation constant of a mode on a transmission line is computed, several assumptions are made that lead to theoretical inconsistencies. For example, the displacement current is ignored, though there are areas along the transmission line where the displacement current is dominant. The corresponding electrostatic and magnetostatic solutions lead to an electromagnetic field that does not fulfill Maxwell's equations exactly and violates Poynting's version of the energy conservation law.

As soon as some numerical problems had been removed, MMP turned out to be also efficient and useful for simulations of practical interest. After 1980, several MMP codes with a growing range of applications were developed by the author and his colleagues at the ETH [Bomholt, 1990], [Fröhlich, 1997], [Gnos, 1997], [Klaus, 1985], [Kuster, 1992], [Leuchtmann, 1987], [Novotny, 1996], [Regli, 1992]. From 1980 until 1987, the following Fortran codes were developed on CDC mainframes: 2D MMP for guide waves (Hafner), 2D MMP for EM scattering (Kley, Hafner), 2D MMP for electrostatics (Hafner), 2D MMP for magnetostatics (Ballisti, Hafner), and 3D MMP for EM scattering and statics (Klaus). From 1982 until 1988 a Pascal version for statics on a HP desk computer was developed by Leuchtmann. After 1987, 2D and 3D MMP versions for personal computers were designed and published as book-software packages [Hafner, 1990/S], [Hafner, 1993/S], [Hafner, 1998/S]. Later on, add-ons for 3D MMP were developed by Gnos, Zheng, Fröhlich, Leuchtmann, and others. Finally, an MMP code for acoustics and elastic waves was implemented at the MIT [Imhof, 1996]. Today, MMP is the most sophisticated and extended implementation of the GMP (see section on Generalized Multipole Techniques (GMT)).

In the following, the most important MMP features are presented to illustrate how numerical methods my be improved by generalizations.

Series expansions

MMP is a pure boundary method implementation. Therefore, the field inside each domain is approximated by a series expansion of the form (4.5). The basis functions must fulfill Maxwell's equations or equations derived from Maxwell's equations within the corresponding domain. Some procedures to find such functions were outlined in the section on Analytic solutions. Among these solutions, MMP selects the numerically most promising and useful ones. Note that MMP can also work with multi-domain basis functions that have a non-zero field within more than a single domain, but most of the common MMP expansions are non-zero in a single domain, i.e., single-domain basis functions.

MMP is a frequency domain code. This implies that the fields are complex and that the time dependence is characterized by the angular frequency ω. For simplicity, we focus on 2D basis functions, i.e., cylindrical basis functions that may entirely be described by the electromagnetic field in the transverse plane and by the propagation constant γ. In the static limit, one has $\omega = 0$ and $\gamma = 0$, but we start with the more general dynamic case.

For simplicity, we assume that all domains are linear, homogeneous, and isotropic, and that the material properties are not time-dependent. In [Piller, 1998] a method has been presented to obtain basis functions for anisotropic domains.

Plane waves

With Cartesian coordinates (x,y,z), one easily obtains plane waves as solutions of Maxwell's equations. For more information, see the subsection on Cartesian coordinates - plane wave solutions. For 2D applications, the wave vector is simply decomposed into a transversal and a longitudinal part, according to

$$\vec{k} = k_x\vec{e}_x + k_y\vec{e}_y + k_z\vec{e}_z = \vec{\kappa} + \gamma\,\vec{e}_z\ , \tag{8.1}$$

where $\vec{\kappa}$ is the transverse wave vector and γ the propagation constant along the z axis. Note that one can distinguish two types of such plane waves, E or TM waves and H or TE waves. The z component of the magnetic field of TM waves is zero. These waves may be entirely described by the z component of the electric field E_z. The transverse field components may be derived from (5.95) and (5.96). Similarly, the z component of the magnetic field of TM waves is zero. A general 2D plane wave is a superposition of a TM and a TE wave. Note that there is an exception when the plane wave propagates in the z direction, i.e., when $\gamma = k$. This is a TEM wave. A TEM wave may be also decomposed in two parts with different polarization. The special case $\gamma = k$ is only obtained on loss-free cylindrical structures with a single dielectric.

A superposition of plane waves that propagate in arbitrary directions may be used to approximate any solution in a given domain. Obviously, this may be considered as a spatial Fourier transform of the problem, which is the basis of the Spectral Domain Analysis (SDA). For example, the superposition of two identical plane waves propagating in opposite directions results in a standing wave as illustrated in Figure 70. As one can see, this superposition may be used to obtain propagating modes in parallel plate waveguides. Similarly, one can expand modes in rectangular PEC waveguides by a superposition of four plane waves when the directions of the four wave vectors are correctly selected.

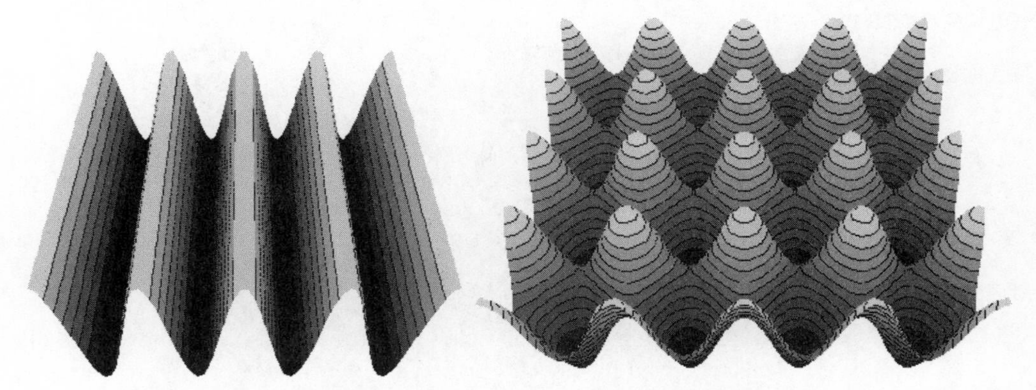

Figure 70: Superposition of two plane waves propagating in x and –x directions (left) and of four plane waves propagating in x, -x, y, -y directions (right). The superposition of two plane waves is appropriate for modeling waves in a parallel plate waveguide consisting of two parallel PEC plates, whereas waves in a rectangular waveguide with four orthogonal PEC plates can be modeled with four plane waves.

For usual plane waves, all Cartesian components of the wave vector are real valued. Since time-harmonic fields are described by complex numbers, one can easily generalize the plane wave definition by admitting complex components of the wave vector. The only constraint is the square norm of the wave vector that must be equal to $\omega^2\mu\varepsilon$. The amplitude of generalized plane waves is no longer constant over the entire domain, because the imaginary parts of the components of the wave vectors cause an exponential behavior in the corresponding direction. Figure 71 shows such a wave with real k_x and imaginary k_y. Plane wave expansions with complex components of the wave vector are useful for modeling the field in the exterior domains of dielectric waveguides.

Harmonic expansions

The dependence of the field of plane waves in the x,y,z directions is given by exponential functions. The superposition of two plane waves with equal amplitudes propagating in the x and $-x$ directions leads to a standing wave that is characterized by a sin or cos dependence of the field in the x direction. This can be considered as a new type of expansion. In general, one may have a field with sin, cos, or exp dependence in the x,y,z directions. We call this a harmonic expansion. For a given frequency, a harmonic expansion in a given domain is completely characterized by 1) E or H wave type, 2) the function types (sin, cos, exp+, exp-) in the three Cartesian directions, and 3) the components k_x and $k_z=\gamma$ of the wave vector. Note that the y component of the wave vector can be evaluated as

$$k_y = \sqrt{\omega^2\mu\varepsilon - k_x^2 - \gamma^2}\,. \tag{8.2}$$

A single harmonic expansion describes plane waves, parallel plate modes, and rectangular waveguide modes. Therefore, harmonic expansions can be considered as a simple generalization of plane waves.

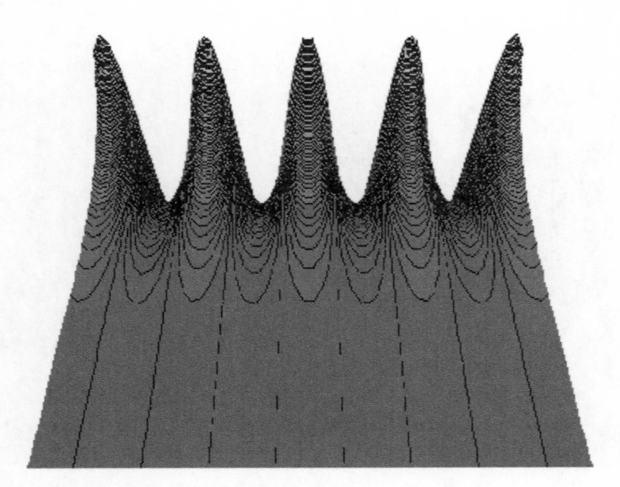

Figure 71: Plane wave with imaginary value of k_y.

Harmonic series expansions may cause numerical problems. First of all, the components k_x and $k_z = \gamma$ of the corresponding wave vectors must be selected properly, which may be difficult for complex geometry of the domain. Moreover, severe cancellations can occur in areas with relatively small fields, for example, in the shadow region of a scatterer.

To observe what happens with the superposition of plane waves or of harmonic expansions, let us construct a solution with rotational symmetry. Such a solution would be obtained by a superposition of infinitely many plane waves with the same amplitude, propagating in all directions of the 2D plane, which requires an integration of the angle of the propagation direction from 0 up to 2π. When we replace this integral by a finite sum of n plane waves with propagation directions $k/2\pi n$, k=0,1,...,n-1, we obtain approximations as illustrated in Figure 72.

Figure 72: Superposition of eight plane waves (left) and 16 plane waves (right) with constant amplitudes, propagating in the xy plane in the directions characterized by the angles $k/2\pi n$, k=0,1,...,n-1.

As one can see, this superposition leads to almost rotational symmetric solutions near the origin $x=y=0$, but further away, the symmetry disappears completely. With increasing n, the range of a good approximation of the rotational symmetry is increased. Obviously, one may use such superpositions for computing modes in circular waveguides, but the approximation of high order modes becomes very inefficient.

Much more severe problems are observed when one tries to approximate the field in geometrically more complicated domains. Therefore, we look for other types of solutions of Maxwell's equations. Note that the set of harmonic expansions is complete in the sense that any regular solution in a domain with sufficiently smooth boundaries might be approximated. Therefore, the 'new' solutions are superpositions of harmonic expansions, i.e., not completely new. What we are looking for are solutions that can be computed within short computation times and that simplify the modeling of fields in geometrically complicated domains.

Bessel expansions

When we continue with the superposition of plane waves of the previous section, we find in the limit of infinite n, a rotationally symmetric solution, which is nothing else than the Bessel expansion or normal expansion of zero order. This expansion was found with polar coordinates in the section on the Separation of variables. In addition to zero order Bessel functions, we also found higher order Bessel expansions that had no rotational symmetry (see Figure 73). Note that these expansions describe asymmetric modes in circular waveguides. Moreover, a superposition of Bessel expansions may be used to expand, for example, modes in rectangular waveguides. Theoretically, a series of Bessel expansions of all integer orders from 0 up to a maximum order n, allows one to approximate any regular solution of Maxwell's equations in a simply connected domain with a sufficiently smooth outer boundary. 'Sufficiently smooth' means that the boundary is continuous and that the first order derivative is also continuous, except at a finite number of points.

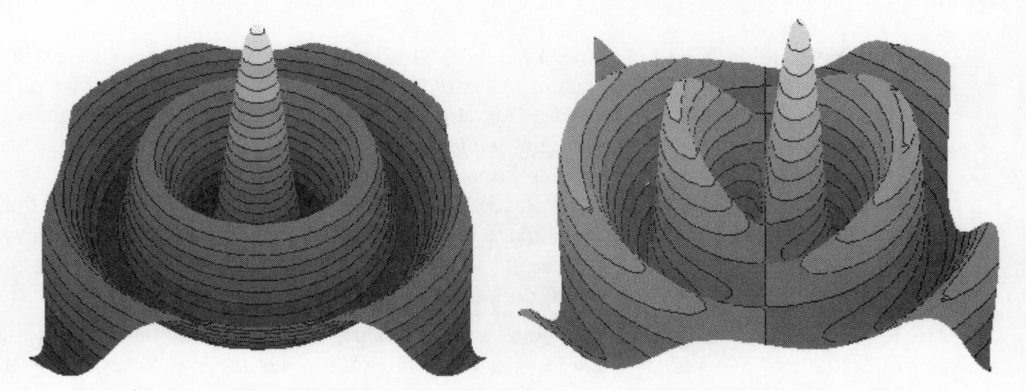

Figure 73: Zero order (left) and first order (right) Bessel expansions in the xy plane.

As for harmonic waves and plane waves, the completeness does not say anything about the number of basis functions required for obtaining a good approximation of a specific problem. It turns out that very high values of n are required when the boundary of the domain is not close to a circular boundary or when the field on the boundary is not very smooth. Therefore, pure Bessel series are useful only for a very limited class of problems.

Note that harmonic expansions are characterized by the wave type (E or H wave), the function types (sin, cos, exp+, exp-) in the Cartesian directions, and two complex numbers k_x and γ. The latter describes the propagation along the z axis. In 2D problems, i.e., cylindrical problems, γ is one and the same constant for all expansions. Therefore, harmonic expansions are characterized by the wave type, the function type, and a complex constant k_x. Bessel expansions are characterized by 1) the wave type (E or H), 2) the angular function type (sin, cos, exp+, exp-), 3) the order, and 4) the origin r=0 of the polar coordinate system. When the origin of the Bessel expansion is inside the given domain, the order must be an integer number. Thus, series of Bessel expansions are much more natural than series of harmonic expansions characterized by complex numbers. Moreover, the selection of appropriate integer orders is much simpler than the selection of appropriate complex k_x for modeling a given problem.

To overcome the convergence problems of Bessel expansions, we can generalize the Bessel expansion definition. We can use superpositions of Bessel expansions with different origins. When we move the origin outside the given domain, we can even admit real orders. This is similar to replacing orthogonal Fourier series by non-orthogonal ones. For a circular domain, an integer order Bessel series with origin in the center of the circle is nothing else than a Fourier series for the boundary values on the circle. With displaced, non-integer order Bessel expansions, the boundary values are expanded by a non-orthogonal basis of non-harmonic functions. Despite this, Bessel functions contain an intrinsic problem that is not removed by such generalizations. Essentially, these functions represent standing waves and cannot be used to approximate propagating waves, for example, the scattered waves. This is the reason why the given domain must be simply connected when Bessel expansions are applied. Multiply connected domains contain holes and the field in such domains is scattered at the borders of the holes.

Multipoles

Multipole fields have the same form as Bessel expansions, but these solutions of Maxwell's equations are singular at the origin. As one can see in Figure 74, the zero order multipole field may look similar to the zero order Bessel field, but the multipole field propagates either away from the origin or toward the origin, whereas the Bessel expansions are standing waves. In the time average, the multipole field is singular, whereas the Bessel field is regular. Because of the singularity, multipoles must be placed outside the given domain – provided that a regular solution of Maxwell's equations inside the domain shall be approximated. From Vekua theory [Vekua, 1967], we know that one multipole expansion with outgoing waves must be placed in each hole of a multiply connected domain.

Remember that harmonic expansions are characterized by the function types in the Cartesian directions x,y,z, whereas Bessel expansions and multipoles are characterized by the function type in the angular direction only. For cylindrical problems, the z direction plays a special role. It is separated in general. Therefore, all expansions (harmonic, Bessel, multipole) have one and the same z dependence. In most cases, this is exp($+\gamma z$). When exp($+\gamma z$) is replaced by exp($-\gamma z$), this leads to essentially the same solution with waves propagating in the $-z$ rather than the $+z$ direction. Finally, the superposition of exp($+\gamma z$) and exp($-\gamma z$) leads to standing waves along the z axis with either cos or sin dependence. The transverse field for all wave types in the z direction is one and the same. Therefore, one mainly has two function types for harmonic expansions and only one for the Bessel and multipole expansions. In fact, the radial function type of these functions is different because Bessel expansions are characterized by the radial Bessel function J_n and multipole expansions by Hankel functions H_n. One can therefore unify Bessel and multipole expansions and characterize them by the radial function type. Note that one has two kinds of Hankel functions. In the MMP code, the first kind represents outgoing waves and the second kind represent incoming waves. Note that opposite notation may be found in other codes. The Hankel functions correspond to the exp functions that also describe propagating waves. From the superposition of these functions, one obtains standing waves that are characterized by sin and cos (harmonic expansions and angular dependence of Bessel and multipole expansions) or by J_n and Neumann functions Y_n. The latter are singular in the origin.

Figure 74: Zero order multipole field at a certain time (left) and time average of the field (right).

In the MMP codes, Bessel and Hankel functions of the first kind are most frequently used. According to Vekua theory, these expansions form a complete approximation basis for multiply connected domains. It turns out that high orders of multipoles are used when the geometric shape of the holes in the given domain is not simple or when the field on the corresponding boundaries is complicated. Therefore, a further generalization is desirable. Note that the Vekua theory specifies that a multipole must be placed in each hole, but it does not specify exactly where to place the multipole. Because of the singularity, each multipole has a more or less local behavior, i.e., its field is mainly concentrated around its origin. When a multipole is placed close to the boundary, it can model the field in a close vicinity very well, but not far away. When one considers a wave that hits a boundary of an object, one can assume that a circular wave is excited at each boundary point and that this wave is scattered back into the given domain. This means that the scattered field is a superposition of monopole fields distributed along the boundary, i.e., the scattered field is an integral of monopole fields over the boundary. When this integral is approximated by a sum, the resulting superposition will have singularities on the boundary, but further away, the field is smoother. Figure 75 shows that the field of eight monopoles that are uniformly distributed on a circle can approximate the field of a zero order Bessel expansion inside the circle.

To remove the problems with the undesired singularities on the boundary, one may move the origins of the monopoles away from the boundary. This is the core of several numerical methods, namely the Method of Auxiliary Sources (MAS) [Bogdanov, 1998] and the Charge Simulation (CS) method [Singer, 1974] of electrostatics. By doing this, one obtains a much smoother field along the boundary, which is desirable in most applications. Note that the distance of the 'sources', i.e., monopoles is not predefined. To obtain good results, some correlation between the distance of a monopole from the boundary, from its neighbors, and the shape of the boundary value function should be established. First of all, the distance between neighbor monopoles and the distance from the boundary should be similar. Moreover, monopoles should be placed close to the boundary at those locations where the field is complicated, i.e., where the boundary values change rapidly.

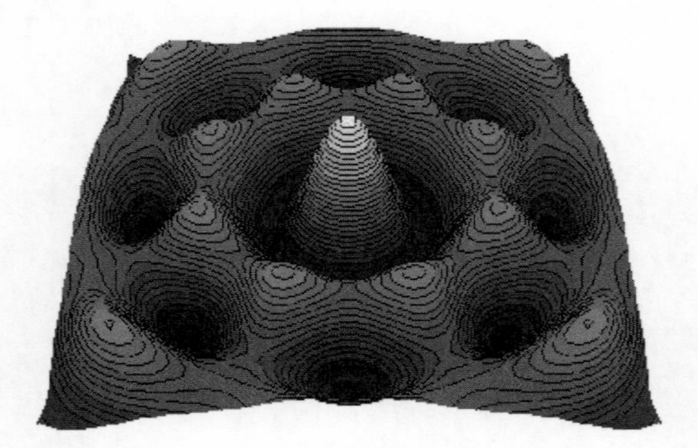

Figure 75: Field of eight zero order multipoles with equal amplitudes, uniformly placed on a circle. Inside the circle, a Bessel expansion of zero order is approximated.

When one replaces each monopole of the MAS or CS approach by a multipole expansion, one obtains the original Multiple MultiPole (MMP) expansion. This allows one to increase the accuracy or to reduce the number of origins required for obtaining the desired accuracy. Note that the MMP expansions contain monopole expansions as well as the Vekua expansions as special cases. Since both monopole expansions and Vekua expansions are complete in the numerical sense, MMP expansions are certainly also complete. Since MMP uses even more than a complete set of basis functions, one might expect that this is too much. In fact, MMP and all other numerical methods work with a finite number of basis functions and this is never too much. Inexperienced MMP users may construct bad multipole sets that cause numerical problems or sets that do not allow them to model the solution with reasonable accuracy, but the placement of multipoles is rather simple. In the original MMP codes (Hafner and Klaus), simple geometric rules for placing the multipoles were used. Essentially, one draws circles around each multipole with a radius that is 1.1 up to 1.6 times bigger than the distance of the multipole from the boundary and one places the neighbor multipoles slightly outside these circles. More sophisticated optimization techniques for the multipole placement were tested by Leuchtmann [Leuchtmann, 1987], but these attempts turned out to be time consuming rather than efficient. In fact, the convergence of MMP expansions is very fast and the placement becomes robust as soon as the number of multipoles is high enough. The multipole placement is only critical when one tries to push the matrix size to a minimum. We will consider these problems later.

Multipoles placed outside the exterior boundary of a given domain may have a real order. This is a generalization of the standard MMP approach that allows one to exactly model the singularities of the field that may occur at sharp corners, i.e. at discontinuities of the first derivative of the boundary. Although this generalization has some useful applications, it has not been implemented in the standard MMP codes because singularities cause numerical problems anyway and because working with real orders is numerically more time-consuming and also more difficult for the user.

Another generalization of the MMP expansion is obtained when one puts the origins of the multipoles in complex rather than in real space. Complex origins have interesting benefits and were successfully used by several authors for different applications [Akkermans, 1982], [Boag, 1994], [Eremin, 1995]. Both MAS and Discrete Sources (DS) codes work with complex origins, especially for the modeling of 3D objects with rotational symmetry. Obviously, the correct placement of complex origins is much more demanding than the placement of real origins. As all reasonable generalizations, complex origins provide more freedom, a higher chance of finding the desired solution with fewer unknowns, but also a reduced user-friendliness and a higher complexity of the code. The main effect of complex origins is that the singularity of the multipole is somehow distributed in the real space along a cut. The proper definition of the cut may be difficult for domains with a complicated shape. Therefore, applying complex origin multipoles is considerably more difficult than applying standard multipoles.

Standard multipoles radiate electromagnetic energy in all directions. From superpositions of different orders of a multipole one can obtain a beam-like radiation pattern (see Figure 77). Beam multipole techniques have been used [Boag, 1994] for reducing the condition number of the MMP matrix, which may be necessary for huge problems. Although a reduction of the condition number is desirable for the matrix solver, such techniques also cause undesired errors (see the section on Ill-conditioned matrices).

Instead of superpositions of different orders of one and the same multipole, one can also consider the superpositions of identical orders of multipoles located at different places. Superpositions of multipoles located along a line are called line multipoles [Gnos, 1994]. Similarly, one can also distribute multipoles on a surface and generate surface multipole expansions. Such distributed multipoles require not only the geometric definition of the line or surface where the multipoles are placed, but also the definition of some weighting functions along the line or on the surface. The proper use of distributed multipoles is therefore more difficult than one might expect. Therefore, the line multipole implementations in the 3D MMP code [Hafner, 1993/S] were rarely used. Since one can easily use so-called connections (see below) to approximate distributed multipoles, no such multipoles were implemented in the MaX-1 code.

Dependent expansions - Rayleigh expansions

The expansions we have considered up to now are completely independent from each other. This means that each expansion is uniquely defined by some parameters that do not depend on the parameters of any other expansion. For example, when we consider a simple, planar problem of a plane wave incident on a planar film on a substrate (see Figure 76), we have three domains: 1) the top layer, where the incident plane wave is defined and where we expect a reflected plane wave, 2) the medium layer, where we have a superposition of two plane waves or two harmonic expansions, 3) the bottom layer or substrate, where we have a transmitted plane wave.

To obtain the analytic solution, the angles of the unknown refracted and reflected plane waves in the three layers are computed from Snell's law and the amplitudes are computed from the boundary conditions between the layers. When a code based on a boundary method is applied, the code can compute the amplitudes from the boundary conditions, but the user must provide the basis functions, i.e., the angles of the plane waves to be used in the series expansion. Although Snell's law [Born, 1980] is simple, such a procedure is not user-friendly. Therefore, it is desirable to have plane wave and harmonic expansions that depend on the specific data of the incident plane wave, especially on the angle of incidence.

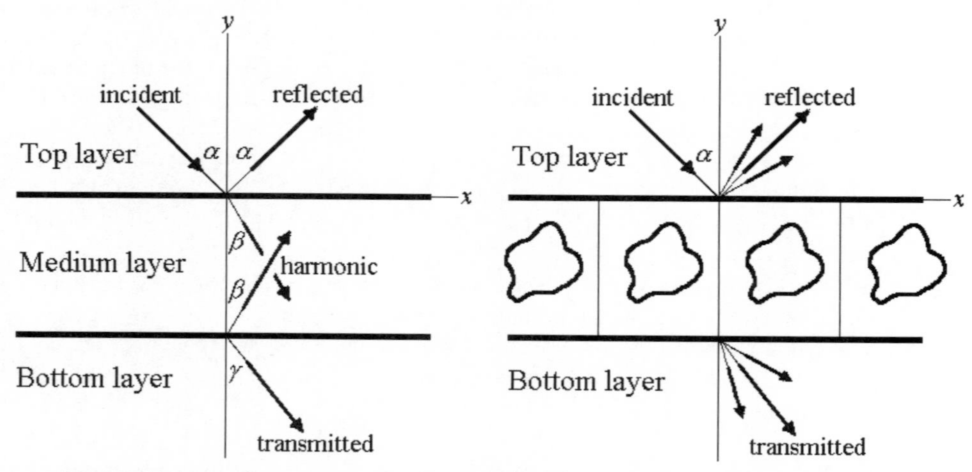

Figure 76: Multilayer structure with three homogeneous layers (left) and a grating in the medium layer (right). The incident plane wave in the top layer and the other plane waves of the analytic solution are indicated by arrows. The two plane waves in the medium layer may be replaced by harmonic expansions when the layer is homogeneous. For a grating, one has several reflected plane waves in the top layer and several transmitted plane waves in the bottom layer and a field in the medium layer that must be approximated by series expansions depending on the grating structure.

A similar, but more difficult situation is observed when gratings are analyzed. Here, one has infinitely many reflected plane waves in the top layer instead of a single one. A finite number of these plane waves are propagating with a real wave number in the y direction; all other plane waves are evanescent, i.e., have an imaginary component of the wave vector in the y direction. The same holds for the transmitted waves in the bottom layers. The data for all reflected and transmitted waves depend on the incident wave and on the periodicity of the grating. The corresponding 'dependent' expansions are called Rayleigh expansions. Note that Rayleigh expansions of zero order are the plane waves obtained when the grating is replaced by a homogeneous layer. Therefore, zero order Rayleigh expansions are useful for solving homogeneous multilayer problems.

Note that the concept of 'dependent' expansions may be generalized and applied to more complicated problems. This is not detailed here.

Multi-domain expansions

Rayleigh expansions may be used to easily expand the field in multilayer structures, but these expansions do not automatically fulfill the boundary conditions. Since the boundaries of such structures are infinite, only a part of the boundaries may be discretized. This may cause numerical problems. When the incident field is a plane wave and the scattered field is expanded by zero order Rayleigh expansions, the amplitudes of the Rayleigh terms may be computed from the boundary conditions at one point on each boundary. The resulting solution will automatically fulfill the boundary conditions everywhere. This can also be done analytically. As soon as the analytic solution is known, it can also be used as a new basis function for MMP codes. Such basis functions are called multi-domain expansions because the corresponding field is non-zero in more than a single domain.

Multidomain expansions are especially useful for solving problems with infinite boundaries that cannot be discretized. The implementation of multidomain expansions may be very demanding, because such expansions depend on the geometry and material properties of several domains. A multidomain expansion for a plane wave incident on a multilayer structure is relatively simple, whereas a multidomain expansion for a dipole on a multipole structure is already very demanding, especially when some layers are lossy. This leads to complicated integrals that must be solved numerically [Novotny, 1996].

Note that multidomain expansions are usually combined with other expansions, i.e., multidomain expansions usually solve only a part of the entire problem. A typical example is a particle on a multilayer structure.

Connections

Connections are superpositions of basis functions with predefined amplitudes. Typically, connections are obtained from solutions of simplified problems or from solutions of a part of a given problem. For example, when one considers the scattering of the field of an antenna at some object, one may first compute the antenna without the scattering object. Then, one can collect the resulting basis functions and the corresponding amplitudes in a connection. This connection is a basis function that models the antenna field. Now, one may use the connection as the excitation of a scattering problem with the given object. The resulting scattered field has no influence on the antenna, because the antenna is not modeled for the scattering problem. Therefore, the interaction between antenna and scatterer is suppressed. This is acceptable when the distance between the antenna and the scatterer is sufficiently big. Otherwise, the scattered field may be packed in a second connection that is now used to illuminate the given antenna, which produces a field scattered at the antenna. This field can again be packed in a connection and the procedure is iteratively repeated until a steady solution is obtained. Although such iterative procedures may be applied to solving problems that are too big to be solved in a single step, they were rarely applied in the past because they are very demanding for the user. Therefore, connections are most frequently used to split parts of a problem with weak interactions.

Note that a connection that models an antenna field can also be used to model finite sets of identical antennae with weak interactions. In this case, the connection must be computed only once.

Connections may model the field in a single domain, i.e., connections may be single domain expansions. In general, connections model the field in more than one domain, i.e., connections are multi-domain expansions. For multi-domain expansions, the boundaries of the corresponding domains should be known. In the 3D MMP code [Hafner, 1993/S], expansions and boundaries are independent objects. This makes the handling of multi-domain connections quite tricky. The MMP implementation in the MaX-1 code [Hafner, 1998/S] is different. Here, a connection may not only contain expansions and the corresponding amplitudes, but also definitions of boundaries.

Connections may also be nested, i.e., a connection may contain another connection. This allows one to model complex structures in an extremely flexible way, step by step. Moreover, one can take advantage of prior knowledge, i.e., prior computations of models similar to the current one. Embedding of known solutions in connections allows one to build some knowledge base.

Connections may also be helpful for testing new ideas of basis functions instead of implementing them explicitly. For example, one can generate multipole beams either from a multipole with appropriately defined amplitudes of all orders or from line multipoles. Line-multipoles that are obtained from a uniform multipole distribution along a given line may be approximated by a finite set of multipoles on this line and these multipoles can be included in a connection that simulates the desired line multipole. Figure 77 shows two beams approximated by connections.

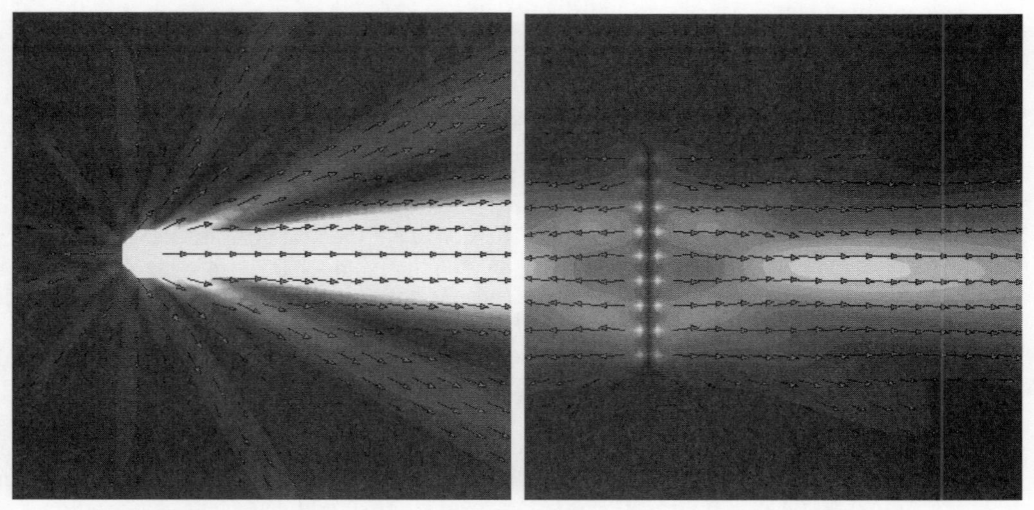

Figure 77: Approximation of a beam expansion with a single multipole with ten orders (left) and with an array of nine dipoles. The former is only useful sufficiently far away (several wavelengths) from the origin. The corresponding picture shows an area that is large compared with the wavelength (side length is 40λ). The dipole array generates two main beams. Both pictures show the time-average of the Poynting vector.

Boundary conditions

Since MMP is based on a boundary method, the boundaries are discretized and the boundary conditions (see section on Boundary and continuity equations) are used for computing the linear parameters in the series expansion of the form (4.5). Remember that there are many different boundary conditions depending on the problem to be solved. Usually there are 'minimal complete' sets of boundary conditions to be solved for obtaining analytic solutions, but these sets are not unique. For example, on a boundary between two dielectric domains, the tangential components of the E field and of the H field must be continuous. Since there are two orthogonal tangential directions, this is a set of four independent scalar components. When all four continuity conditions are met everywhere analytically and when the field inside the domain is expanded by analytic solutions of Maxwell's equations, all other boundary conditions are automatically met as well. Therefore, such a set of boundary conditions is complete. Moreover, it is minimal because wrong solutions would be obtained when one of the boundary conditions would be omitted. This set is not unique, because one of the tangential continuity conditions for E and H may be replaced by the normal continuity conditions for the D and B fields. When one introduces potentials, one can also derive continuity conditions for the potentials and so on. Moreover, one can construct linear combinations of some boundary conditions that are again boundary conditions. From the huge variety of boundary conditions, one can select many different 'minimal complete' sets of four scalar conditions. When such sets are used for obtaining numerical solutions, the results may turn out to be quite different because a numeric error minimization of four selected boundary conditions does not also minimize the error of other boundary conditions. Therefore, a proper selection of a 'minimal complete' set of boundary conditions may be important.

To avoid problems with undesired inaccuracies in the boundary conditions that are not explicitly used in the 'minimal complete' set, one may work with 'non-minimal' sets of boundary conditions. This is usually done in the MMP code because it leads to more robust solutions and it avoids the problem of a proper selection of a 'minimal complete' set. The latter would either reduce the user-friendliness of the code or increase the complexity of the code. The MMP code uses the following sets of boundary conditions for electrodynamics:

- Standard set between two linear, homogeneous, isotropic domains: six continuity conditions (tangential E and H, normal D and B components). This set is also used for fictitious boundaries and for periodic boundaries. In this case, 'minimal complete' sets have only four conditions. As one can see from Figure 78 and Figure 79, the 'minimal complete' sets lead to slightly higher errors and to less balanced error distributions. These effects may be more pronounced in more complicated problems.

- PEC set on a PEC surface: 3 continuity conditions (tangential E, normal B components). In this case, 'minimal complete' sets have only 2 conditions.

- PMC set on a PMC surface: 3 continuity conditions (tangential H, normal D components). In this case, 'minimal complete' sets have only 2 conditions.

- SIBC set on the surface of a good, but imperfect conductor: 2 continuity conditions (tangential E component). This is a 'minimal complete' set.

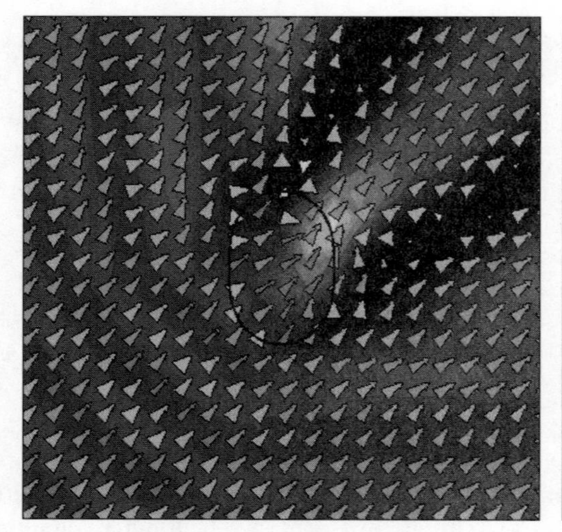

Figure 78: Left: time average of the Poynting vector field for the scattering of a plane wave at a dielectric cylinder with relative permittivity 2, oblique incidence. Right: distribution of the error function η along the boundary obtained when all six components of the field are matched.

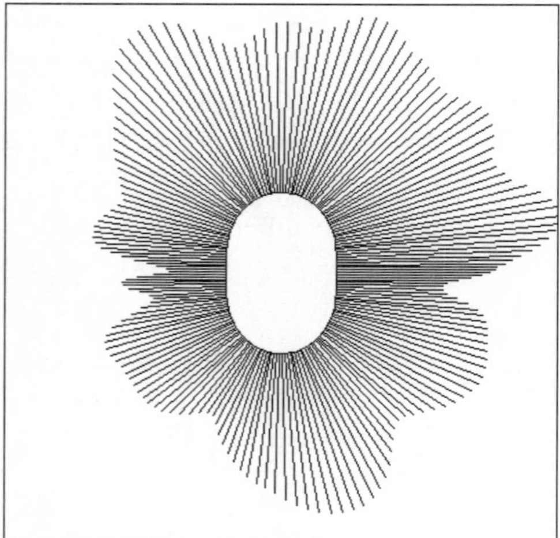

Figure 79: Error distributions of the same problem as in Figure 78, obtained with 'minimal complete' sets of boundary conditions: tangential and longitudinal components of the E and H fields (left), longitudinal components of the E and H fields and normal components of the D and B fields (right).

For statics, one can split the sets of electrodynamics into an electrostatic part (E and D fields) and a magnetostatic part (H and B fields). Although the resulting sets may be useful, the situation is more complicated, because the 'minimal complete' sets depend on the problem to be solved, especially on the dimension of the problem and on the type of boundary conditions. To illustrate the situation, let us consider electrostatics. Here, one has electrodes that can be modeled as PEC boundaries. Therefore, one can impose the boundary conditions for the tangential E components. The potential of such an electrode is free, i.e., one has a floating electrode. Often, the potential of an electrode is known and not free. To model such electrodes, one must impose a Dirichlet boundary condition for the scalar potential. In this case, the Dirichlet boundary condition is a 'minimal complete' set and the boundary conditions for the tangential E components can be omitted. Instead of this, one can use 'non-minimal' sets with both types of boundary conditions. In cylindrical problems, the electrostatic field does not vary along the z axis. It is purely transverse. Therefore, continuity conditions of longitudinal components or of derivatives with respect to z are automatically met and must be omitted. As a result, 'minimal complete' sets in 2D electrostatics consist of a single boundary condition. Similar statements hold for magnetostatics.

Error Minimization

Once the boundary conditions to be fulfilled are selected, one can define an error function along the boundary that shall be minimized. In order to obtain a linear system of equations, the error function should contain the squares of the field quantities in the series expansion (4.5). To outline the procedure, we consider the standard set of boundary conditions for electrodynamics. The following is a simple square error definition for this set on a boundary ∂D_{ij} between the domains D_i and D_j:

$$\begin{aligned}\eta_{ij}^2 &= (E_{ti}^0 - E_{tj}^0)^2 + (E_{\tau i}^0 - E_{\tau j}^0)^2 + (D_{ni}^0 - D_{nj}^0)^2 \\ &+ (H_{ti}^0 - H_{tj}^0)^2 + (H_{\tau i}^0 - H_{\tau j}^0)^2 + (B_{ni}^0 - B_{nj}^0)^2 ,\end{aligned} \tag{8.3}$$

where the superscript 0 indicates the approximation. The subscripts t and τ indicate the two tangential components and n indicates the normal component. With this error definition, numerical problems occur because the field components have numerical values that may be completely different. For example, when the SI or MKSA system of units is used, the D field is numerically much smaller than the E field. Therefore, the continuity condition of the D field has almost no influence on the error. In fact, adding up field components with different units makes no sense in physics. Therefore, the error definition (8.3) should be replaced by a sum of terms with identical units. Note that the terms in (8.3) have almost the form of energy densities. To obtain error densities, we must 1) multiply the E terms with the permittivity ε, 2) divide the D term by ε, 3) multiply the H terms with the permeability μ, and 4) divide the B term by μ. The main problem now is that ε and μ in the domains D_i and D_j may be different. Unfortunately, there is no optimal solution of this problem for all kind of applications. A reasonable approach is to use the geometric average of the material properties in the two domains and to add user-definable weights w. As soon as the material properties are introduced, it is reasonable to replace the D and B fields by E and D fields:

$$\begin{aligned}\eta_{ij}^2 &= w_{Et}^2 \varepsilon_{ij} (E_{ti}^0 - E_{tj}^0)^2 + w_{E\tau}^2 \varepsilon_{ij} (E_{\tau i}^0 - E_{\tau j}^0)^2 + \frac{w_{En}^2}{\varepsilon_{ij}} (\varepsilon_j E_{ni}^0 - \varepsilon_j E_{nj}^0)^2 \\ &+ w_{Ht}^2 \mu_{ij} (H_{ti}^0 - H_{tj}^0)^2 + w_{H\tau}^2 \mu_{ij} (H_{\tau i}^0 - H_{\tau j}^0)^2 + \frac{w_{Hn}^2}{\mu_{ij}} (\mu_i H_{ni}^0 - \mu_j H_{nj}^0)^2 ,\end{aligned} \tag{8.4}$$

where

$$\varepsilon_{ij} = \sqrt{\varepsilon_i \varepsilon_j}\,, \quad \mu_{ij} = \sqrt{\mu_i \mu_j}\,. \tag{8.5}$$

When the differences between the material properties in both domains are not very big, all user-definable weights may be set equal to one. Otherwise, one starts with all weights equal to one, analyzes the results and adjust the weights when the results are not balanced. For example, when one finds that the continuity condition of the normal components of the E field is met much less accurately than the continuity conditions of the other components, one increases the weight w_{En}.

The standard MMP implementation uses a slightly different error definition with an error that has the same dimension as the electric field. This is achieved when the right hand side of (8.4) is divided by ε_{ij}:

$$\begin{aligned} \eta_{ij}^2 = {} & w_{Et}^2 (E_{ti}^0 - E_{tj}^0)^2 + w_{E\tau}^2 (E_{\tau i}^0 - E_{\tau j}^0)^2 + w_{En}^2 (\frac{\varepsilon_i}{\varepsilon_{ij}} E_{ni}^0 - \frac{\varepsilon_j}{\varepsilon_{ij}} E_{nj}^0)^2 \\ & + Z_{wij}^2 \left[w_{Ht}^2 (H_{ti}^0 - H_{tj}^0)^2 + w_{H\tau}^2 (H_{\tau i}^0 - H_{\tau j}^0)^2 + w_{Hn}^2 (\frac{\mu_i}{\mu_{ij}} H_{ni}^0 - \frac{\mu_j}{\mu_{ij}} H_{nj}^0)^2 \right], \end{aligned} \tag{8.6}$$

where

$$Z_{wij} = \sqrt{Z_{wi} Z_{wj}} = \sqrt[4]{\frac{\mu_i \mu_j}{\varepsilon_i \varepsilon_j}}\,. \tag{8.7}$$

This error definition is reasonable when one compares the error on the boundary with the size of the electric field either near the boundary or of the incident plane wave in scattering applications.

On PEC and PMC boundaries, half of the boundary conditions are omitted, and one therefore has only three terms instead of six as in (8.6). For SIBC boundaries, one has only two boundary conditions, and in static applications one may have only one boundary condition. Although this simplifies the error definition, the procedure remains the same.

Once the error function is defined, one can obtain a matrix equation from the error minimization technique (see section on Error minimization techniques).

Generalized Point Matching (GPM)

It has been mentioned that the error minimization leads to a symmetric square matrix that can be considered as the product of two rectangular matrices (see chapter on 4 Generating matrix equations) when the integral contained in the matrix elements is approximated by a sum. This decomposition leads to the Generalized Point Matching (GPM) technique that is especially useful when the condition number of the matrix is high. Since MMP often deals with ill-conditioned matrices, we replace the error minimization technique by an appropriate GPM, i.e., we minimize the error defined in (8.6) implicitly by imposing the following six boundary conditions in a matching point P_k on the boundary ∂D_{ij} between the domains D_i and D_j:

$$w_{Et} \sqrt{a_k} (E_{ti}^0 - E_{tj}^0) = 0\,, \tag{8.8}$$

$$w_{E\tau}\sqrt{a_k}\,(E^0_{\tau i} - E^0_{\tau j}) = 0\,, \tag{8.9}$$

$$w_{En}\sqrt{a_k}\left(\frac{\varepsilon_i}{\varepsilon_{ij}}E^0_{ni} - \frac{\varepsilon_j}{\varepsilon_{ij}}E^0_{nj}\right) = 0\,, \tag{8.10}$$

$$w_{Ht}\sqrt{a_k}\,Z_{wij}(H^0_{ti} - H^0_{tj}) = 0\,, \tag{8.11}$$

$$w_{H\tau}\sqrt{a_k}\,Z_{wij}(H^0_{\tau i} - H^0_{\tau j}) = 0\,, \tag{8.12}$$

$$w_{Hn}\sqrt{a_k}\,Z_{wij}\left(\frac{\mu_i}{\mu_{ij}}H^0_{ni} - \frac{\mu_j}{\mu_{ij}}H^0_{nj}\right) = 0\,, \tag{8.13}$$

where a_k is a geometric weight associated to the matching point P_k. Note that the boundary is subdivided into areas of size a_k. The matching point P_k is placed in the center of the corresponding area. In 2D applications, the boundary and a_k are one-dimensional. In this case, a_k is essentially the average distance of the matching point P_k from its two neighbors.

On PEC, PMC, and boundaries, as well as in statics, one has fewer than six boundary conditions, but their construction remains the same. Each boundary condition is a product of 1) a user-definable weight w, 2) a geometric weight, i.e., the square root of an area a_k associated to the current matching point P_k, 3) a weight that guarantees equal units of all boundary conditions, and 4) the boundary condition itself.

In GPM, one usually works with overdetermined systems of equations. Typically, about twice or three times as many equations as unknowns, i.e., overdetermination factors of 2-3, are used. For example, in Figure 78 and Figure 79 the overdetermination factor is approximately equal to 2. In tricky situations, one may increase the overdetermination factor for safety reasons.

Without overdetermination, the errors of the continuity conditions imposed at the matching points are zero, but the continuity conditions that are not matched explicitly and the continuity conditions between the matching points are much higher and less balanced than with a reasonable overdetermination. This is illustrated in Figure 80.

The most important advantage of the overdetermination in the GPM is that it provides much more freedom in the placement of the matching points. Without overdetermination, the placement of the matching points is extremely critical and a slightly 'sub-optimal' placement may lead to completely wrong results in tricky situations. For more information, see [Hafner, 1990]. Note that the MaX-1 code uses a relatively robust, automatic setting of matching points that makes it hard to make mistakes with bad matching point settings. To get experience with badly placed matching points, older MMP versions [Hafner, 1990/S] must be used.

GPM allows one to select a finer discretization in critical areas, which results in a local reduction of the error distribution when appropriate MMP expansions are used. Figure 81 shows a typical example of electrostatics.

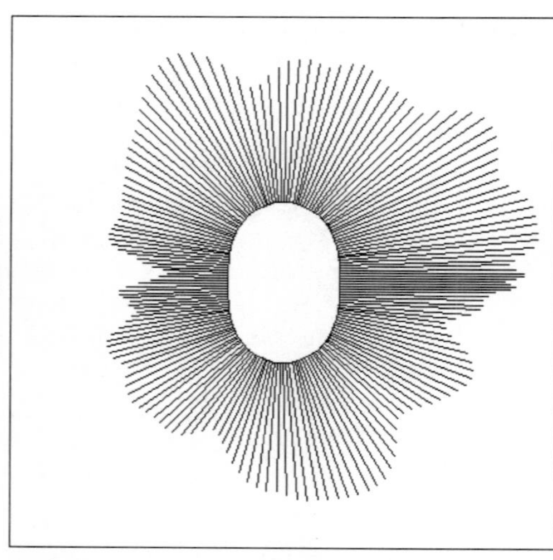

Figure 80: Error distribution for the same problem as in Figure 78, obtained without overdetermination (left). Since the maximum error is considerably higher than in Figure 78 and in Figure 79 and than in the picture on the right, the errors in the picture on the left were scaled by a factor of ½. Note that the error distribution is also quite unbalanced. On the right hand side, the error distribution obtained with a huge overdetermination is shown (overdetermination factor 24). As one can see, this error distribution is almost the same as in Figure 78, where a moderate overdetermination was used.

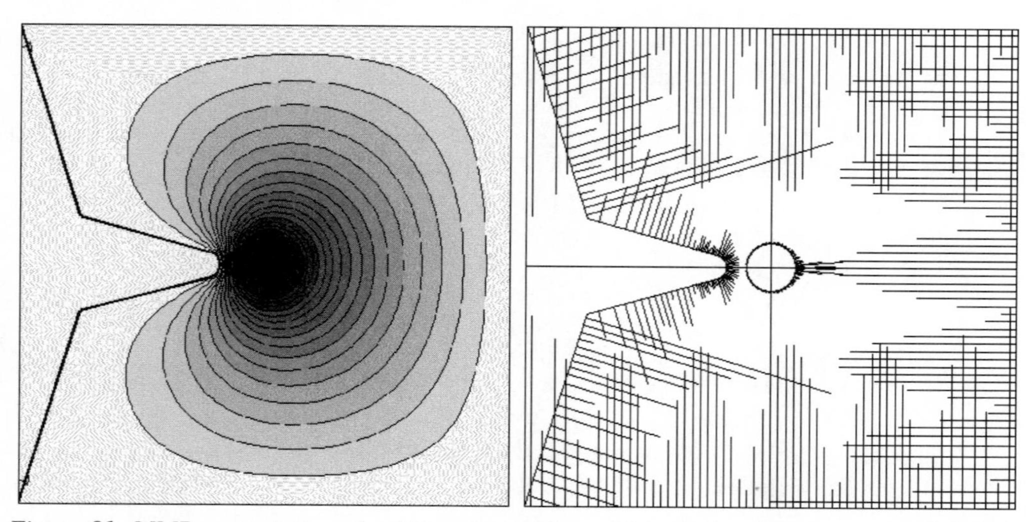

Figure 81: MMP computation of a 2D static problem with a tip in a box on potential 0 and a circular conductor on potential 1. Lines of constant potential (left) and error distribution (right). As one can see, the error has been reduced in the critical area by a fine discretization. Note that the computation is very accurate. The maximum error on the boundary is less than 0.026%. Note that this problem is similar to the problem in Figure 68 that was used with a combination of FD with conformal mapping.

Expansion-boundary correlation

Numerically, one works on the boundaries of the given domains. Here, the field of the expansions is evaluated for setting up the rectangular MMP matrices. To obtain some insight, it is useful to consider the behavior of different types of expansions on a given boundary. The Boundary Element Method (BEM) subdivides the boundary into parts and applies basis functions that are nonzero only in one part of the boundary. Plane wave expansions have the same amplitude on the entire boundary and therefore are a counterpart of the BEM basis functions. As we have seen, multipole expansions are concentrated around their origin, i.e., these expansions have a local behavior. When one observes the field of a multipole along a boundary, this field is non-zero almost everywhere, but it is concentrated in an area of the boundary that is close to the origin of the multipoles. Figure 82 illustrates this. Therefore, multipole expansions have a field that is dominant in some area like the BEM basis. Unlike the BEM basis, the multipole field is also non-zero further away.

Although the multipole expansions are not restricted to a finite area, it is convenient to define an 'area of biggest influence' for multipole expansions as illustrated in Figure 83. Since the multipole fields spread to infinity, the size of the area of biggest influence is not uniquely defined. From experience, one knows that a circle around the origin of the multipole with radius $r=c*d_{min}$, where d_{min} denotes the minimum distance of the multipole from the boundary and where $c = 1.2\ldots1.4$, is useful.

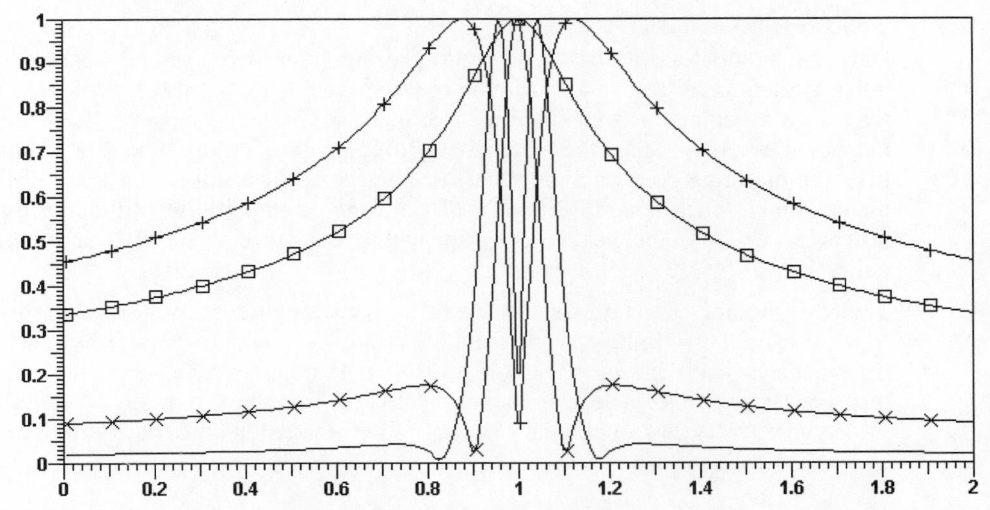

Figure 82: Amplitude of the multipole orders 0 (square markers), 1 (× markers), 2 (+ markers), and 3 (no markers) along a straight boundary from x=0 to x=2m. The multipole is placed at x=1m, y=0.1m. Note that there is some frequency dependence. The frequency for this plot is 300MHz.

Once the area of biggest influence is defined, one may distribute multipoles along the boundary of each domain in such a way that each boundary point is exactly within of one of the areas of biggest influence. This rule may be violated because the multipole functions are not restricted to their area of biggest influence.

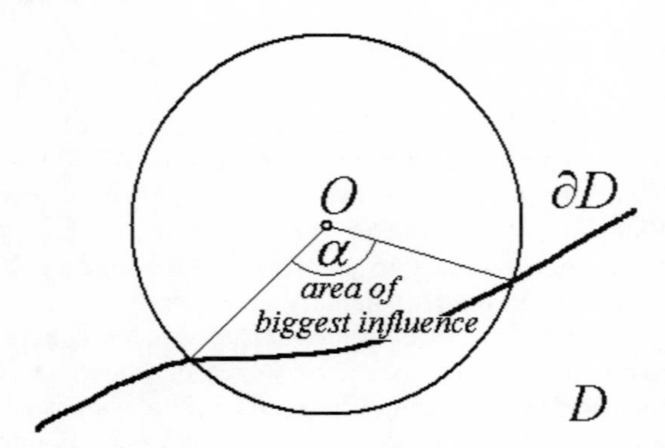

Figure 83: A multipole expansion is located at O. Its area of biggest influence is within a circle around O with radius r. The radius r is evaluated from the minimum distance d_{min} of O from the boundary ∂D of the domain D, according to $r=c*d_{min}$, where c>1 is a constant. The value of c is not uniquely defined. Usually, it is set to 1.2...1.4. Note that the origin O is outside D. From O one can see the part of ∂D within the area of biggest influence under a 'view angle' α

Since numerical dependencies occur when two multipoles are close together, the neighbor multipoles of a given multipole should be outside its area of biggest influence or at least close to its periphery.

Once the multipoles are placed, the multipole orders must be defined. Usually, one uses all integer orders from 0 up to a maximum order *m*. The higher *m*, the more accurate results one may expect within the area of biggest influence. This allows one to locally increase the numerical accuracy. Since the amplitude of the multipole orders should be computed mainly from the matching points within the area of biggest influence, one has a correlation between the maximum order *m* and the number of matching points to be set within the area of biggest influence. Usually, the matching points within this area should be placed more or less uniformly.

The angle α under which the part of the boundary within the area of biggest influence is seen from a multipole origin, is called the view angle. When this angle is small, numerical dependencies between the multipole orders may occur, which can cause ill-conditioned matrices (see section on Ill-conditioned matrices). Therefore, one should either avoid small view angles, or use not all of the multipole orders (see section on Condition number reduction).

Matrices

Rectangular matrix

The Generalized Point Matching (GPM) technique leads to a rectangular matrix M that can easily be interpreted. Each column of this matrix corresponds to a basis function of the series expansion and each row corresponds to a boundary condition imposed at a single matching point. In the MMP implementation, some expansions may define more than a single basis function and therefore occupy more than one column of the matrix. Moreover, in each matching point, several boundary conditions may be imposed. Therefore, a matching point can occupy several rows of M. The MaX-1 implementation of MMP works with automatic definition of matching points along the given boundaries, i.e., each matching point corresponds to a boundary. Figure 84 illustrates this.

In general, the matrix M is full. Whether it is dense or not depends on the structure of the domains of the given model. Since each matching point belongs to a boundary and each boundary belongs to a domain, it is easy to associate each row of the matrix to two domains. The correlation of columns and domain numbers is more difficult. Note that a single-domain expansion belongs to exactly one domain, whereas multi-domain expansions belong to several domains.

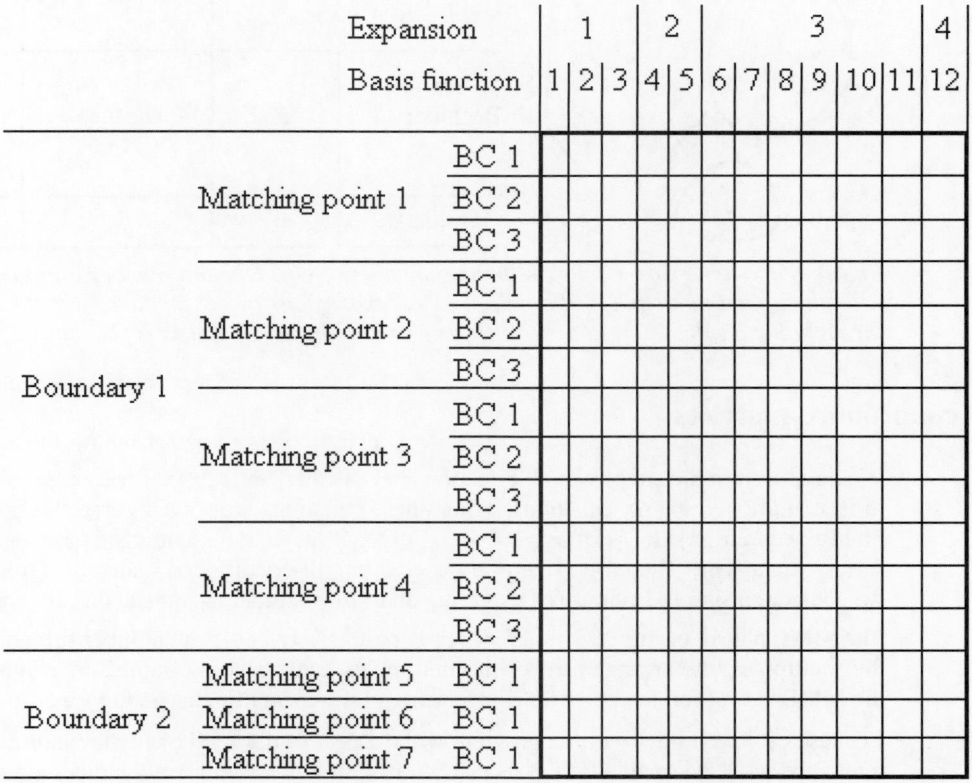

		Expansion	1			2		3						4
		Basis function	1	2	3	4	5	6	7	8	9	10	11	12
Boundary 1	Matching point 1	BC 1												
		BC 2												
		BC 3												
	Matching point 2	BC 1												
		BC 2												
		BC 3												
	Matching point 3	BC 1												
		BC 2												
		BC 3												
	Matching point 4	BC 1												
		BC 2												
		BC 3												
Boundary 2	Matching point 5	BC 1												
	Matching point 6	BC 1												
	Matching point 7	BC 1												

Figure 84: Rectangular MMP matrix R. Each column corresponds to a basis function and each row corresponds to a boundary condition. An expansion may contain several basis functions and a matching point may contain several boundary conditions.

The field in some special domains is not modeled by any expansion (zero fields inside perfect conductors and fields inside good conductors that are modeled by SIBC's). For all other domains, the field is modeled by at least one expansion. In general, one cannot easily subdivide the matrix into blocks that represent a domain because of the multi-domain expansions. When only single-domain expansions are used, one may obtain blocks by an appropriate arrangement of the matching points and expansions. Figure 85 illustrates this. Note that many zero blocks may be obtained when many domains are present, whereas only a few or even no zero blocks are obtained for simple problems with few domains. The sparsity of the MMP matrix depends on the number of the domains and on the geometric topology.

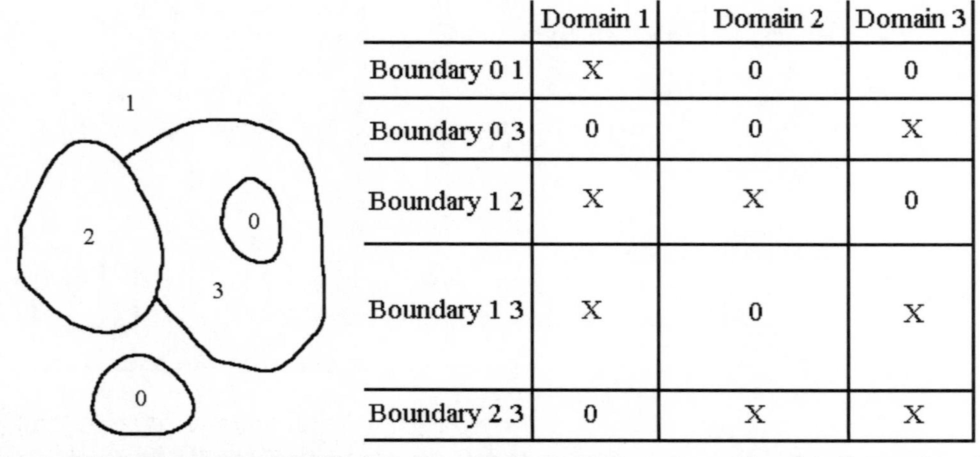

	Domain 1	Domain 2	Domain 3
Boundary 0 1	X	0	0
Boundary 0 3	0	0	X
Boundary 1 2	X	X	0
Boundary 1 3	X	0	X
Boundary 2 3	0	X	X

Figure 85: Geometric model with domain numbers (left). Domain number 0 indicates domains without explicit modeling of the field, i.e., no expansions are associated to these domains. On the right hand side, the resulting block structure of the MMP matrix is shown.

Ill-conditioned matrices

We have seen in the chapter on 2 Function analysis that generalized series expansions can lead to ill-conditioned matrix equations. When these equations are solved correctly, one can obtain highly accurate results [Hafner, 1993]. The MMP code is essentially based on generalized series expansions. Therefore, one often encounters ill-conditioned matrices. To illustrate this, we consider a simple plane wave scattering at a PEC cylinder as indicated in Figure 86.

The MMP matrix of this test model depends on the frequency, the number of matching points, the multipoles, the arrangement of the multipoles and matching points, etc. Figure 87 shows the MMP matrix for an appropriate arrangement of four multipoles on the y axis.

Because of the symmetry of the problem with respect to the x axis, only the half plane with $y \geq 0$ must be modeled explicitly. Let us distribute N multipoles uniformly along the y axis with the first multipole at y=0 and the N-th multipole at y=0.9m. With an increasing number of multipoles, the distance between neighbor multipoles is reduced. Therefore, we expect higher condition numbers for more multipoles. To get a fair comparison, we keep the number of unknowns fixed, i.e., we reduce the maximum orders of the multipoles when we increase the number of multipoles. Note that the number of unknowns per multipole is not constant. The first multipole has a much smaller area of biggest influence than the last multipole, i.e., the biggest portion of the matching points is close to the last expansion.

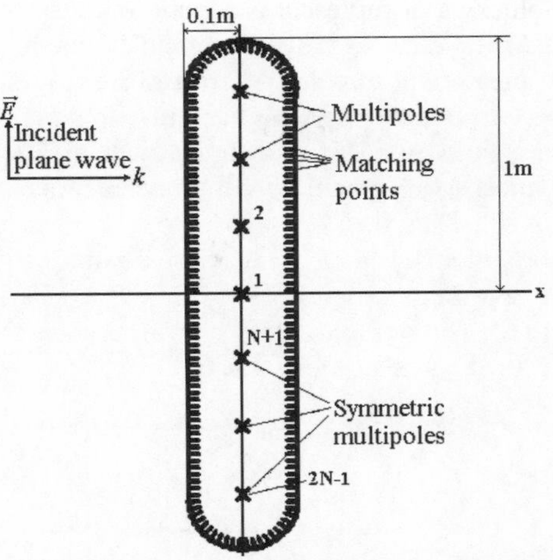

Figure 86: Cross section of a PEC cylinder illuminated by a plane wave. Because of the symmetry, only the upper half (matching points and multipoles) is explicitly modeled.

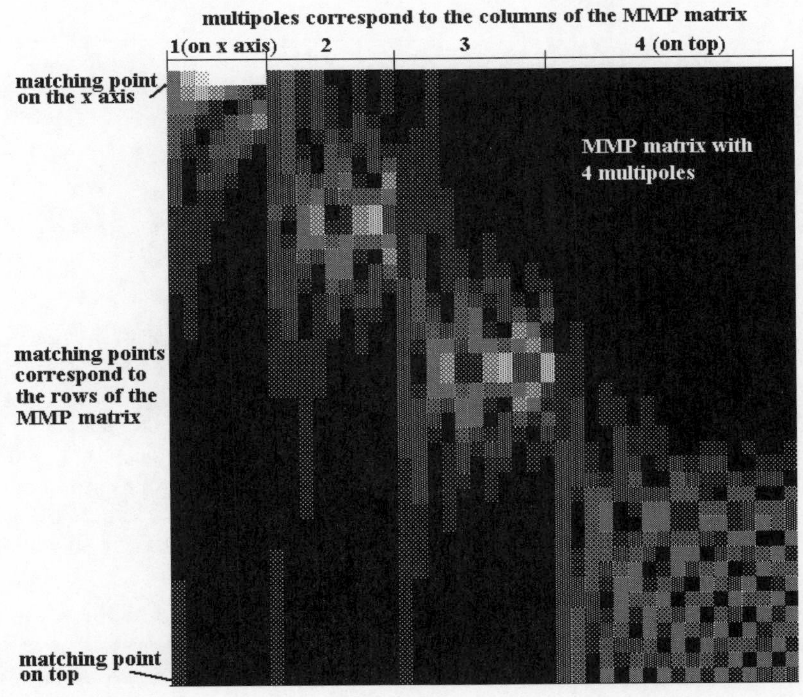

Figure 87: Typical MMP matrix for the test example with 2N-1=7 multipoles. Only the upper half of the matrix is shown because of its symmetry. The intensity corresponds to the size of the matrix elements. Note that the size of the matrix elements depends on the frequency. A more or less block-diagonal structure with four blocks corresponding to the four multipoles may be observed for all frequencies when the matching points are numbered appropriately.

The number of orders of a multipole is proportional to the number of unknown parameters, i.e., columns of the MMP matrix, and this should be more or less proportional to the number of matching points near the multipole. When we fix the number of columns, for example, to 44 as in Figure 87, the number of orders for each multipole is not uniquely defined. In the following, we compute the maximum order in such a way that it is approximately proportional to the area of biggest influence. Moreover, we keep the matching points and the number of rows of the MMP matrix fixed. The MMP matrix of the following computations has 82 rows and 44 columns, i.e., the overdetermination is approximately equal to 2. As one can see from Figure 88, the residual for all computations is quite low over a large frequency range when at least 5 multipoles are used. The optimal number of multipoles is nine, but the number of multipoles is not very critical for the size of the residual.

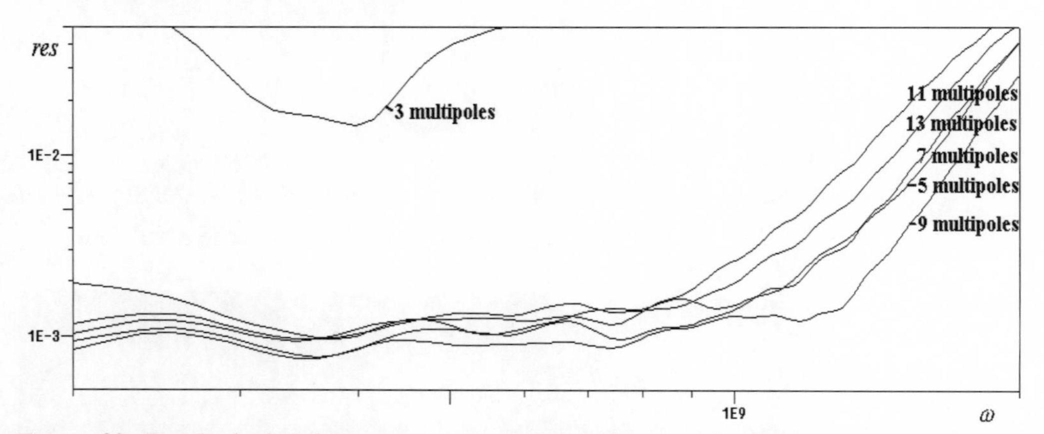

Figure 88: Residual of MMP computations versus frequency for different total numbers of multipoles for the test example shown in Figure 86. The residual shown here is the square root of the sum of the squares of the weighted residuals.

Figure 89 shows that the condition number of the MMP matrix increases very much with the number of multipoles. To estimate the condition number, the SVD has been used, but the Givens updating procedure has been applied for computing the unknown parameters of the MMP expansion. Of course, the condition number reduces the accuracy of the results, but this loss of accuracy plays no role as long as the numerical accuracy is much higher than the desired accuracy of the results. Since double precision was used, the numeric accuracy is approximately 13 digits. With a condition number of 10^7, we expect to lose about seven digits, i.e., we have still six digits of accuracy for the result of the matrix solver, which is much higher than the desired accuracy of three digits (residual 0.001).

When we would use normal equations, i.e., when we would multiply the MMP matrix with its adjoint matrix, we would obtain the symmetric square matrix of the error minimization technique. With this method we would lose approximately twice as many digits, i.e., 14 digits, and we would end up with completely wrong results. To obtain useful results, we therefore would have to select five or seven multipoles when we want to work with normal equations, whereas we can work with even more than 13 multipoles when we directly work on the rectangular MMP matrix. Since we considered a simple test example, the situation is not critical. As soon as we work on more complicated problems, it may become extremely difficult to set up the multipoles in such a way that the condition number of the normal equations remains sufficiently small and that a sufficient accuracy of the results is obtained.

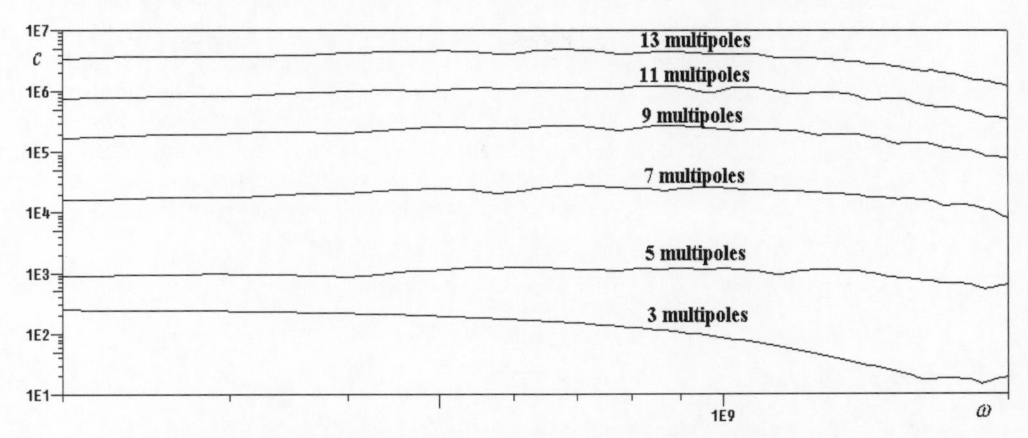

Figure 89: Condition number of the MMP matrix versus frequency for different total numbers of multipoles for the test example shown in Figure 86.

Condition number reduction

As long as one works with direct rectangular matrix solvers that are sufficiently robust (Givens, QR, or SVD [Golub, 1983]) typical MMP matrices of problems with less than 2500 unknowns can be handled without any technique that reduced the condition number of the MMP matrix. When larger problems are to be solved or when less robust matrix solvers are to be applied, such techniques may be of interest.

In order to obtain some idea, we consider the simple scattering problem defined in Figure 86 with only three multipoles at y=0, 0.3, 0.6 respectively. For an incident E wave, the z component E_z of the different basis functions (different orders of different multipoles) along the surface of the scatterer is plotted in Figure 90, Figure 91, and Figure 92.

The area of biggest influence of the first multipole should be the part of the boundary with y<0.15m, i.e., $0<s<0.15$ and $1.68<s<1.83$. As one can see in Figure 90, all multipole orders in this area behave like a cos function. The 'frequencies' of these cos functions increase with the multipole order, but the 'frequency' increases not as quickly as in a standard Fourier series. The simplest way to obtain a stronger increase of the 'frequency', is to omit some multipole orders. When one keeps the number of unknowns of the multipole constant, this means that the highest order is increased. For example, one can select the even orders 0, 2, 4, 6, 8 instead of 0, 1, 2, 3, 4, when one wants to have five unknowns.

Because of the symmetry decomposition, the analysis of the multipoles that are not on the x axis is more complicated. To obtain the desired symmetry, for each multipole a symmetric mirror multipole is added. This has the effect that the maximum values of the orders 0 and 2 of the second multipole at y=0.3m is at s=0 and s=1.83 rather than at s=0.3 and s=1.53. This means that the maximum values are even outside the area of biggest influence of this multipole. Especially the 0 order has a field similar to the lowest orders of the first multipoles, which causes some numerical dependencies and increases the condition number of the resulting MMP matrix. To obtain a lower condition number, one may select only the odd orders of this multipole.

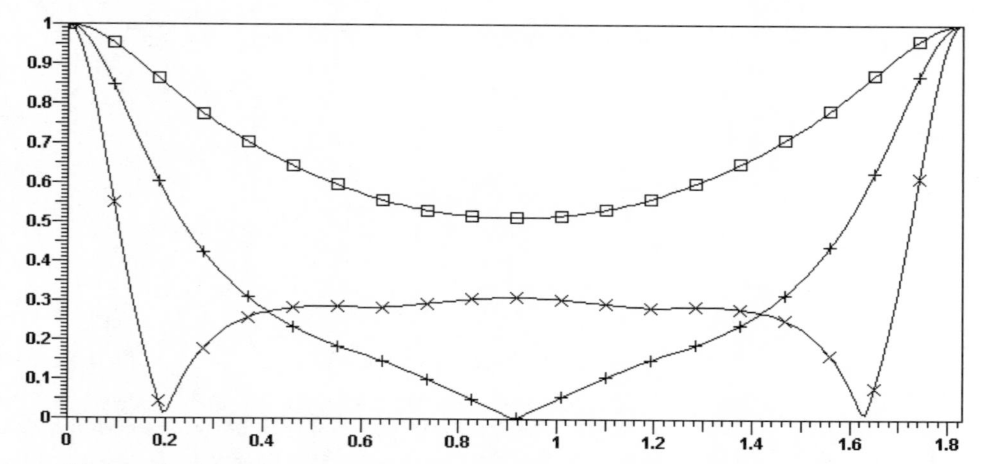

Figure 90: Amplitude of the orders 0 (square marker), 1 (× marker), 2 (+ marker) of the first multipole at y=0 along the boundary of the scatterer for the test example. The boundary parameter s starts at the point x=-0.1m, y=0 and runs clockwise around the scatterer. The lower half of the scatterer (s>1.83, y<0) is not shown explicitly because of the symmetry of the problem. The maximum of all amplitudes is scaled to 1.

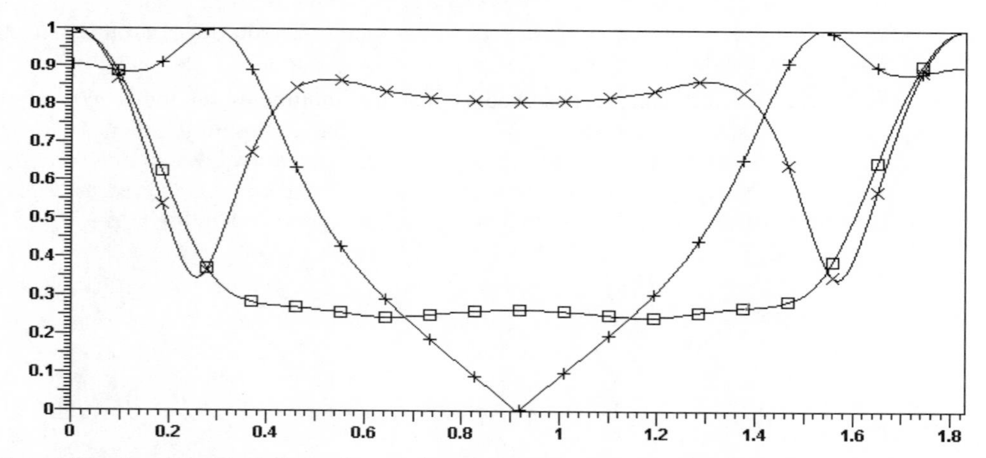

Figure 91: Amplitude of the orders 0 (square marker), 1 (× marker), 2 (+ marker) of the second multipole at y=0.3m (with symmetric mirror multipole at y=-0.3m) along the boundary of the scatterer for the test example. The boundary parameter s starts at the point x=-0.1m, y=0 and runs clockwise around the scatterer. The lower half of the scatterer (s>1.83, y<0) is not shown explicitly because of the symmetry of the problem. The maximum of all amplitudes is scaled to 1.

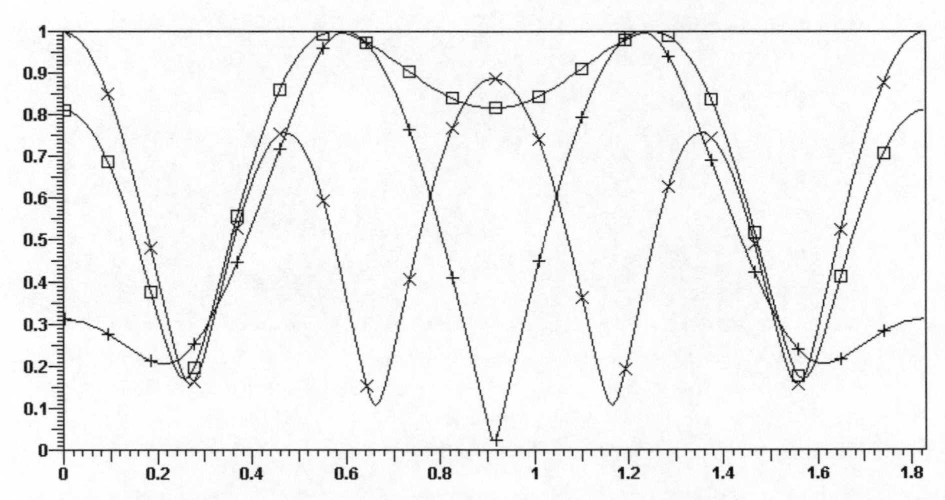

Figure 92: Amplitude of the orders 0 (square marker), 1 (× marker), 2 (+ marker) of the third multipole at y=0.6m (with symmetric mirror multipole at y=-0.6m) along the boundary of the scatterer for the test example. The boundary parameter s starts at the point x=-0.1m, y=0 and runs clockwise around the scatterer. The lower half of the scatterer (s>1.83, y<0) is not shown explicitly because of the symmetry of the problem. The maximum of all amplitudes is scaled.

The third multipole has the widest area of biggest influence ($0.45<s<1.38$). As one can see, the different orders of this multipole are quite different in the area of biggest influence, but the lowest orders are also similar to the orders of the first multipole in its area of biggest influence. Despite of this, it makes not much sense to omit many orders of the third multipole for reducing the condition number.

Another effect that can be observed is that the multipoles on the y axis of the scatterer 'illuminate' the back side ($x>0$) of the scatterer with the same intensity as the front side ($x<0$). This means that the area of biggest influence of these multipoles is not compact. It consists of two separate parts. To overcome this problem, one may move the multipoles away from the axis and select a 'front' set and a 'back' set. Figure 93 shows the resulting MMP matrix for a set of 10 off-axis multipoles. As one can see, one can obtain a quite nice matrix with maximum elements in blocks along the diagonal. To reduce the condition number, one can select, for example, only even multipole orders, or every third multipole order. This technique increases the contrast in Figure 93 considerably, i.e., the diagonal-dominance of the matrix becomes more pronounced and the condition number is reduced. At the same time, the error of the results is increased in most cases. Good results are obtained with every second order when the view angle (see section on Expansion-boundary correlation) of a multipole is approximately 90 degrees or less. For obtaining reasonable results with every third multipole order, the view angle should be 60 degrees or less.

Omitting some of the multipole orders allows one to reduce the condition number with a moderate increase of the errors when the matrix size is not changed. When the field along the boundary varies very rapidly, one might even obtain more accuracy, but the chance for increasing the accuracy with this technique is very low in most applications of practical importance.

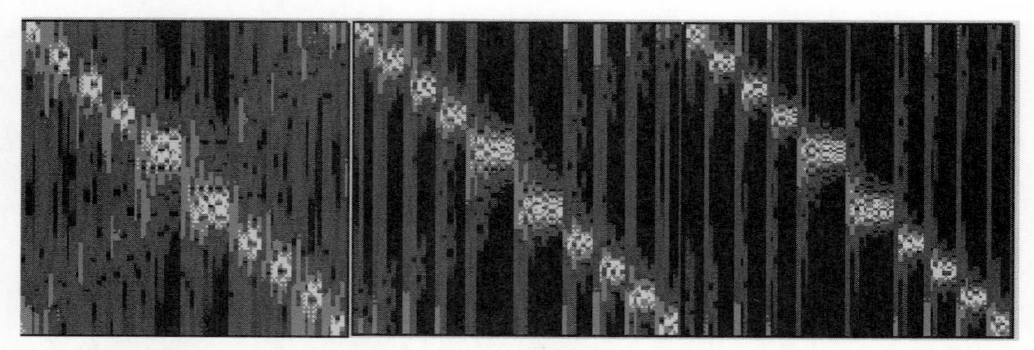

Figure 93: MMP matrices for the test example shown in Figure 86 for 10 off-axis multipoles. All multipole orders used (left), even orders used, odd orders skipped (middle), orders 0,3,6,... used (right). As one can see, omitting orders increases the contrast of the corresponding pictures, which indicates that the condition is improved, i.e., the condition number is reduced.

When one omits some multipole orders, one increases the maximum order without increasing the number of unknown parameters of the multipole. A more general method consists of the following two steps: 1) Increase the maximum order from m_0 to m. 2) Generate m_0 linear combinations of the multipole orders $0 \ldots m$, with predefined parameter sets a_{i0}, a_{i1}, ..., a_{im} with $i=0 \ldots m_0$. These linear combinations may be defined as connections and are used as new basis functions. In the special case, where one parameter is 1 and all other parameters of each set are 0, one obtains standard multipoles when $a_{ii}=1$, even order multipoles when $a_{i,2i}=1$ ($m=2m_0$), and so on. The appropriate choice of the parameter sets a_{ik} allows one to construct expansions that illuminate only a small part of the boundary. Such expansions are called multipole beams. As one can see from Figure 94, multipole beams can be focused on a small part of the boundary in such a way that an almost diagonal MMP matrix with a very low condition number is obtained.

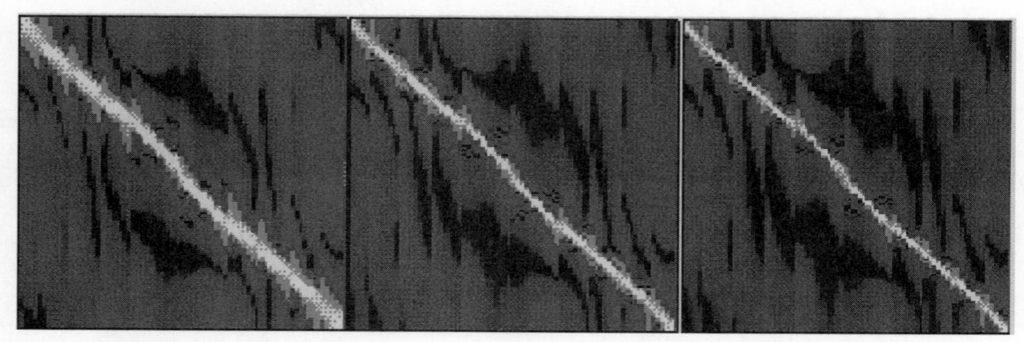

Figure 94: MMP matrices for the test example (see Figure 86) obtained with multipole beam expansions. Soft beam (left), medium beam (middle), sharp beam (right).

Although the multipole beam method seems to be very promising, it turns out that this method increases the final error much more drastically than the more simple method of omitting some multipole orders. To obtain useful results, the distance of the beam multipole from the boundary should be relatively large (several wavelengths). Therefore, the method is only attractive for very big scatterers. Since this method is very demanding, it has not been implemented in the MaX-1 code. A more sophisticated, complex multipole beam method has been presented for large problems [Boag, 1994].

Alternative methods to reduce the condition number of MMP matrices may be obtained from wavelet transforms and other appropriate transforms that replace the MMP matrix by a more sparse matrix. These methods always reduce the accuracy of the results. Therefore, condition number reduction should be applied with care. Such methods are interesting when the direct matrix solver may be replaced by a much faster iterative matrix solver or when the problem is so complicated that the condition number is too high for all available matrix solvers.

Direct matrix solvers

Since the MMP matrix may have a quite arbitrary structure, it is reasonable to use direct matrix solvers, except for big problems where it might be preferable to take advantage of the zero blocks. One then can either apply iterative or block-iterative matrix solvers [Kuster, 1992].

Note that most of the iterative matrix solvers and many direct matrix solvers were designed for square matrices. To apply such matrix solvers, one can solve the normal equations of the form (4.24). Remember that this causes numerical problems when the condition number of the matrix is not sufficiently low. Therefore, only matrix solvers based on the rectangular matrix equation should be applied. Otherwise, methods that reduce the condition number of the MMP matrix are required.

The standard MMP implementation uses the Givens updating procedure [Dongarra, 1979] which is a QR decomposition that updates an upper triangular matrix T. Since the rectangular matrix M must not be stored explicitly, this allows one to save memory, which is attractive when one works on small computers. For example, when the MMP matrix has n rows and m columns ($n>m$), one only stores $m*(m+1)/2$ matrix elements, which is approximately $2o$ times less than $n*m$, when o is the overdetermination factor. The Givens algorithm is very robust and well suited for typical MMP matrices. The more sophisticated Singular Value Decomposition (SVD) [Dongarra, 1979] is much more memory consuming (approximately $n*m+n*n+m*m$ elements, i.e., approximately $2*(1+o+o*o)$ times as many elements as Givens).

Note that the rows of the MMP matrix are weighted, but the columns are neither weighted nor scaled. In general, the square norm of each column depends on the expansion type, its location, and additional data. It may be far away from unity. When a user makes some mistakes in the modeling, it may even happen that a zero column is encountered which causes severe numerical problems – even for Givens and SVD. In order to obtain useful information on the condition number from the singular values, the column norm should be equal for all columns. Therefore, the MaX-1 code contains features that compute the column norm and allow one to scale the columns. This allows one to detect zero columns and to erase such useless columns. This results in a more robust code. When one analyzes the matrix for zero columns, one can also search and erase useless zero rows. Moreover, one may set small matrix elements equal to zero when one wants to increase the sparsity of the MMP matrix. This reduces the condition number, but it also reduces the accuracy of the final results.

Matrix analysis and tuning features require the explicit storage of the rectangular matrix and may be quite time consuming. The computation time of the MMP matrix with a direct matrix solver (Givens, QR decomposition, SVD, etc.) is proportional to $n*m^2$. The computation times for the detection of small elements, the elimination of zero rows and zero columns, and the scaling of columns are proportional to $n*m$. Therefore, such matrix tuning algorithms are much less time consuming than direct matrix solvers when the number of unknowns, i.e., the number of columns, m is big enough. To analyze linear dependencies between the basis functions, one may evaluate and analyze the scalar products of all columns of the matrix. This allows one to detect and remove misplaced multipoles, but the computation time for such an analysis is proportional to $n*m^2$. Therefore, it can considerably slow down the performance. Misplaced multipoles can be detected easily from the location of their origins. Moreover, the Givens algorithm does not crash even when two multipoles are placed – by mistake – at one and the same location, i.e., when a matrix contains several pairs of identical columns. Therefore, the automatic removal of identical columns is not worthwhile. Sophisticated matrix analysis and repair algorithms may be more time consuming than the matrix solvers. Despite this, such algorithms may be useful when a high robustness of the entire computation is required, for example, when the MMP algorithm is embedded in an optimization environment.

Iterative matrix solvers

Most of the well-known iterative matrix solvers have been designed for square matrices [Barrett, 1994]. A typical example is the Conjugate Gradient (CG) solver [Barrett, 1994]. CG and other iterative matrix solvers may be considered as a machine that takes a given input vector X^i and outputs a vector X^{i+1}, where X^i and X^{i+1} are approximations of the solution vector X of the given matrix equation MX=B. The iterative method is usually designed in such a way that it converges uniformly toward X for any start vector X^0. This means that X^{i+1} is closer to X than X^i. Note that uniform convergence is desirable but not necessary. However, CG converges uniformly for any start vector X^0. Therefore, one can simply start with a zero vector.

In the first CG publication, a variant for rectangular matrices was presented [Hestenes, 1952]. Although this is attractive for MMP, the first attempts with CG implementations in the MMP code were discouraging because the CG algorithm encountered severe problems with MMP matrices. First of all, the maximum number of CG iterations required for square matrices ($n=m$) is equal to the number of columns m. Since the computation time for one iteration is proportional to the number of matrix elements m^2, the maximum CG computation time is in the same order of magnitude as the direct matrix solvers. This would also hold when the number of CG iterations for rectangular matrices ($n>m$) would be equal to m. Unfortunately, the CG for rectangular matrices exhibits a staircase convergence that may be very slow for MMP matrices and that may require much more than m iterations. Note that staircase convergence is still uniform. The problem of such a convergence is that one never knows when the iterations can be stopped because one never knows whether a next step is coming that will considerably reduce the errors of the results. Therefore, the user must define a maximum error or residual that is admitted and CG iterations stop when this residual is reached. When this ‘barrier’ is too low, it might happen that the CG iterations never stop. Therefore, the number of CG iterations must be limited, for example, to m. When this limit is reached, it is more efficient to apply a direct matrix solver.

There are two different techniques to improve the CG performance. The first thing that comes to mind is to find a good start vector for the CG iteration. This is the idea that leads to the parameter estimation technique (see the section on Eigenvalue Estimation Technique). The second technique to improve the CG performance is well-known as preconditioning. The reason for matrix preconditioning is the fact that the convergence rate of CG (and other iterative matrix solvers) depends on the spectral properties of the matrix [Barrett, 1994]. Therefore, one can try to transform the given system of equations into another one with the same solution, but more favorable spectral properties. To illustrate the procedure, assume that the overdetermined system of equations (MX=B) is solved with the normal equations SX=M*MX=M*B=C, where M* denotes the adjoint matrix of the rectangular MMP matrix M. Therefore, S=M*M is a symmetric, square matrix that can be handled by standard CG. If the inverse of S were known, one could simply multiply SX=C with S^{-1} to obtain $S^{-1}SX=IX=S^{-1}C=D$, where I is a diagonal unit matrix. The solution of IX=D is trivial and the correct solution is obtained with a single CG iteration. This can be considered as an optimal preconditioning. When the inverse S^{-1} is unknown, one still can multiply SX=C by some matrix, which leads to a preconditioned system with nicer spectral properties that speed up the CG convergence. Usually, one multiplies the given matrix S both from the left and from the right with a left preconditioner P_l and a right preconditioner P_r:

$$P_l^{-1} S P_r^{-1} (P_r X) = P_l^{-1} C \quad \Rightarrow \quad \tilde{S}\tilde{X} = \tilde{C}. \tag{8.14}$$

With a good preconditioning, one can drastically reduce the number of CG iterations. Since the preconditioning requires both memory and computation time, one shifts a part of the work to the preconditioning. The 'perfect' preconditioning with S^{-1} is usually even more time-consuming than direct matrix solvers. Therefore, some compromise is required.

For the rectangular MMP matrix M, the situation is more complicated than for a square matrix S because the computation of the normal equations with S=M*M is to be avoided. Remember the remarks on the column scaling of the MMP matrix. The column scaling can be interpreted as a multiplication of the given matrix M with the inverse of a diagonal matrix D that contains the norms of the columns of M as diagonal elements. Therefore, MD^{-1} is the scaled matrix. With this scaling, the following normal equations are obtained:

$$(MD^{-1})^* MD^{-1}(DX) = D^{-1}M^* MD^{-1}(DX) = D^{-1}M^* B = D^{-1}C. \tag{8.15}$$

Note that D^{-1} is a real valued, symmetric matrix. Therefore, it is equal to its adjoint matrix. From (8.15) one can easily see that the column scaling of M corresponds to a preconditioning of the normal equations with $D=P_l=P_r$, which is known as the most simple, Jacobi preconditioning. The advantages of the Jacobi preconditioning are the simplicity of the evaluation of the matrix D and the small memory requirement for storing a diagonal matrix. Moreover, one can apply CG iterations directly to the scaled, rectangular matrix equation

$$MD^{-1}(DX) = \tilde{M}\tilde{X} = B. \tag{8.16}$$

As soon as the scaled system of equation is solved, the unscaled parameter vector X is easily obtained from

$$X = D^{-1}\tilde{X}. \tag{8.17}$$

Note that the inversion of a diagonal matrix is almost trivial and very quick. The diagonal elements of the inverse matrix are equal to $1/d_{ii}$, where d_{ii} are the diagonal elements of D. Experience shows that Jacobi preconditioning is useful for MMP matrices that are to be handled with CG, but this precondition is not as robust as desired. Therefore, the CG solution of MMP matrices may still fail, especially when relatively high condition numbers are encountered.

Special techniques

Having considered the main MMP procedure, some additional techniques will be outlined. These techniques are not applied to all kinds of problems, but they may be helpful for solving some special problems or for obtaining a better performance in special situations.

Symmetry decomposition

The geometry of applications of technical interest exhibits no symmetry in general, but for reducing the numerical effort, one usually works with some idealizations that have a high symmetry. Note that the geometric symmetry must not coincide with the symmetry of the field. When the symmetry of the field coincides with the symmetry of the geometry of the problem, the procedure is usually straightforward. This happens for example in electrostatics when symmetric Dirichlet boundary conditions are present. A typical example is a circular coaxial cable. In the following, we focus on the much more demanding situation, where the field has a lower symmetry than the geometry. A typical example is higher order modes in a circular coaxial cable.

There are many different types of symmetry that may be handled with various techniques. In general, a symmetry is described by some symmetry operations that leave the given geometry unchanged. For example, symmetry operations of rotational symmetry are 1) the rotations around a given symmetry axis with an arbitrary angle and 2) reflections at any plane that contains the symmetry axis. Similarly, the symmetry operations of cylindrical symmetry are 1) translations parallel to a given axis and 2) reflections at any plane perpendicular to the symmetry axis. A high symmetry means that there are many independent symmetry operations. The symmetry operations of the examples above are not independent because the results of each of the symmetry operations 'rotation' or 'translation' may be obtained by combinations of two appropriate reflections. Therefore, it is sufficient to study reflective symmetries in principle. This may be done with the so-called representation theory of groups, which is quite demanding. For simple symmetries, such sophisticated theories may be replaced by more intuitive procedures. For example, the z separation is appropriate to handle problems of computational electromagnetics with cylindrical symmetry, where z is the cylinder axis (see section on 2D: cylindrical structures, z separation). Similarly, a separation of the angular coordinate φ may be used for rotationally symmetric problems, which leads essentially to Fourier series in φ. Such symmetries are usually inherent in a specific implementation of a numerical method. For example, the 2D MMP codes are designed for 3D problems with cylindrical symmetry. Other implementations of MMP were designed for rotationally symmetric problems [Klaus, 1985].

Reflective symmetries with respect to a few symmetry planes are often used in numerical codes. For simplicity, we consider reflective symmetries in 2D codes for cylindrically symmetric 3D problems. As mentioned before, such problems have reflective symmetry with respect to all transverse planes that are perpendicular to the z axis. The z separation and the corresponding 2D MMP expansions take these symmetries into account. Now, it may happen that the geometry is also symmetric with respect to a symmetry plane parallel to the z axis. By a proper selection of the Cartesian coordinate system we can assume that the xz plane is a symmetry plane. In 2D MMP, we only model the transverse xy plane explicitly. Therefore, we can say that the 2D geometry (in the xy plane) is symmetric with respect to the x axis. What does this mean for the electromagnetic field? First of all, we can decompose any scalar field $f(x,y)$ in the xy plane into a symmetric part f_s and an anti-symmetric part f_a:

$$f(x,y)=c_s f_s(x,y)+c_a f_a(x,y)\,,\quad f_s(x,y)=f_s(x,-y)\,,\quad f_a(x,y)=-f_a(x,-y), \tag{8.18}$$

where c_s and c_a are constant factors that are obtained from the excitation of the problem, i.e., the right hand side of the matrix equation to be solved. A special situation is obtained when the right hand side, i.e., the excitation is missing. This happens in eigenvalue problems. The solutions of eigenvalue problems may then be decomposed in even modes with c_s=1, c_a=0 and odd modes with c_s=0, c_a=1. To get an idea how the c_s and c_a may be computed when the right hand side is known, assume that you want to compute a geometrically symmetric structure with two electrodes, one on potential V_1 and one on potential V_2. From (8.18) one obtains the following two equations when one inserts a point on the first electrode and its symmetric counterpart on the second electrode:

$$\begin{aligned} V_1(x,y)&=c_s f_s(x,y)+c_a f_a(x,y)\\ V_2(x,-y)&=c_s f_s(x,y)-c_a f_a(x,y)\,. \end{aligned} \tag{8.19}$$

Now, the symmetric and anti-symmetric parts may be scaled to 1 et the given point (x,y) and one obtains

$$c_s=(V_1+V_2)/2\,,\quad c_a=(V_1-V_2)/2\,. \tag{8.20}$$

We therefore can split the entire computation into a symmetric model with potential c_s on both electrodes and an anti-symmetric model with potential c_a on the first electrode and $-c_a$ on the second electrode. As soon as we have found these two solutions, we obtain the desired solution from the superposition (8.18). Obviously, such a symmetry decomposition is only reasonable when the computation time for each of the two parts is less than half of the computation time of a direct solution without symmetry decomposition.

How can we model the symmetric and anti-symmetric problems? First of all, the symmetry operation (reflection at the x axis) allows us to easily construct the solution for y<0 when y>0 is known. Therefore, we may explicitly model and compute the field only in the upper half plane. This allows us to reduce the number of rows and columns of the MMP matrix by a factor of 2 for both the symmetric and the anti-symmetric solution. This results in two MMP matrices with $n/2$ rows and $m/2$ columns instead of a single n by m matrix. When direct matrix solvers are applied, the computation time is proportional to nm^2. Therefore, the total computation time may be reduced to ¼ when the symmetry decomposition is used.

To model a symmetric or an anti-symmetric solution, we must apply symmetric or anti-symmetric basis functions. It may happen that some of the basis functions are either symmetric or anti-symmetric with respect to the x axis, but most of the basis functions are neither symmetric nor anti-symmetric in general. For example, a monopole is symmetric with respect to the x axis only when its origin is on the x axis.

To construct a symmetric expansion, we apply the symmetry operator to a given basis function $f_k(x,y)$. This generates an implicit basis function $f'_k(x,y)$. For example, when we explicitly define a monopole expansion $f_k(x,y)$ with origin at (x_o,y_o), the basis function $f'_k(x,y)$ is the field of a monopole at $(x_o,-y_o)$. When both the explicit and implicit expansions have identical amplitudes, $f_k(x,y)+f'_k(x,y)$ is a symmetric and $f_k(x,y)-f'_k(x,y)$ an anti-symmetric basis function. Note that usual MMP expansions are characterized not only by an origin, but also by an orientation. The symmetry operations must be applied to both the origin and the orientation. Figure 95 illustrates this.

When two symmetry axes are present, the construction may be repeated for both symmetry axes, which results in a 'corona' of four expansions that are geometrically symmetric to each other. The superposition of the corresponding fields allows one to obtain four different types of basis functions that are either symmetric or anti-symmetric with respect to both symmetry axes. Figure 95 illustrates this construction.

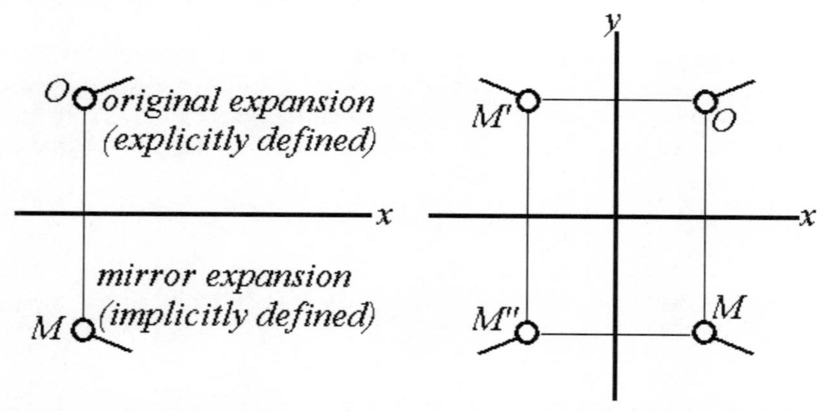

Figure 95: Construction of symmetric expansion for an asymmetric original expansion when one symmetry axis (left) or two symmetry axes (right) are present. Note that both the origin and the orientation of the original expansion are reflected at the symmetry planes.

Note that one and the same corona of expansions may be used for constructing all four different types of basis functions. Figure 96 shows two examples for problems with two symmetry axes. When only expansions with the correct symmetry with respect to the x and y axes are used, only the first quadrant must be defined explicitly. The boundary, the mirror expansions, and the field in the remaining quadrants may than be constructed with symmetry operations. Therefore the original n by m matrix is decomposed in four symmetric parts with $n/4$ rows and $m/4$ columns. This allows one to reduce the computation time to 1/16.

Unfortunately, one cannot split the field into even and odd parts with respect to each symmetry axis as soon as more than two axes are present. In this case, the symmetry composition becomes quite difficult. To find the correct symmetry decomposition for arbitrary diedric symmetries with an arbitrary number of symmetry axes, or for other types of symmetry, the representation theory of finite groups should be applied [Stiefel, 1992], [Hafner, 1981].

The symmetry decomposition of non-scalar fields is slightly more difficult, when a single or two perpendicular symmetry axes are present. Since the E field is the (negative) gradient of the scalar potential in electrostatics, one can easily recognize how this vector field is decomposed. When the scalar potential is symmetric with respect to a symmetry axis, the gradient and the E field are tangential to the axis, and when it is anti-symmetric, the gradient and the E field are perpendicular to the axis. From the E field one obtains the D and j fields when one applies appropriate material properties. Now, one has to recognize that the geometry and the material properties are symmetric with respect to each symmetry axis, at least for isotropic material. Therefore, the symmetry of the D and j fields is the same as the symmetry of the E field. In electrodynamics, the H field is obtained from the curl of the E field. The curl operator associates a pseudovector field to a vector field. Therefore, the H field is tangential to the symmetry axis when the E field is normal and vice versa. To obtain the B field, one can again use the material properties, i.e., the B field behaves like the H field. Finally, the vector potential A and the B field are linked with the curl operator. From this, it follows that A has the same behavior as the E field. Briefly, all vector fields are decomposed like the E field, whereas all pseudo-vector fields are decomposed like the H field.

Figure 96: Construction of basis functions that are symmetric with respect to the x axis and anti-symmetric with respect to the y axis. Lines of constant scalar potential obtained with four static monopoles (left) and four static dipoles with an orientation of 30 degrees with respect to the x axis.

To obtain scalar fields from vector fields, one can also consider the components of the vector fields in an appropriate coordinate system. One then finds that the following components have the same symmetry as the scalar potential V: E_{tang}, D_{tang}, j_{tang}, A_{tang}, H_{norm}, B_{norm}, where the subscript tang indicates the components tangential to the symmetry axis and norm indicates the components perpendicular to the symmetry axis. All other components have opposite symmetry.

Note that the symmetry decomposition into even and odd components can easily be extended to 3D geometry with up to three symmetry planes that are perpendicular to each other. For more complicated 3D symmetries, the representation theory of finite groups may be applied.

When one observes the symmetric and anti-symmetric parts of the different fields with respect to a symmetry plane, one can recognize that a plane PEC wall may be inserted in the symmetry plane when V, E_{tang}, D_{tang}, j_{tang}, A_{tang}, H_{norm}, B_{norm} are symmetric and E_{norm}, D_{norm}, j_{norm}, A_{norm}, H_{tang}, B_{tang} are anti-symmetric. For the opposite symmetry, a plane PMC wall may be inserted. PEC and PMC walls therefore play the role of two different kinds of mirrors for the electromagnetic field. Such walls are often introduced in numeric codes for taking symmetries into account without implementing symmetry decompositions.

The drawback of the replacement of symmetry planes by PEC and PMC planes is that these planes must be discretized, whereas the symmetry planes are not explicitly discretized. When a symmetry decomposition is implemented in the code, it is therefore preferable to work with the symmetry decomposition. In this case, one should even replace models with PEC planes that are perpendicular to each other by models with symmetry planes. Figure 97 shows a typical electrostatic example with an electrode consisting of two perpendicular PEC walls. To model the PEC walls explicitly, additional multipole expansions in the second, third, and fourth quadrant would be required. Moreover, only a part of the PEC walls could be discretized, because these boundaries extend to infinity. This would result in bigger MMP matrices and in less accurate results.

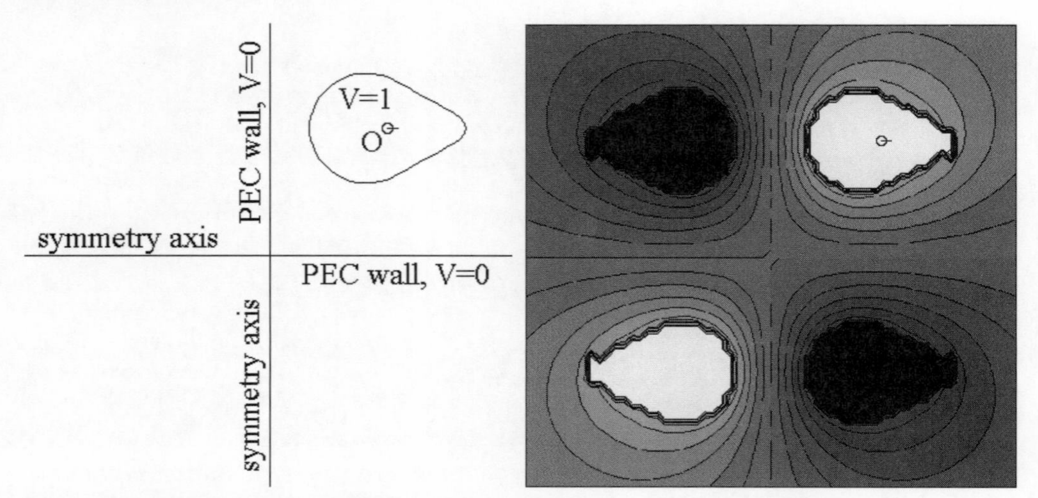

*Figure 97: Electrode on potential V=1 and two perpendicular PEC walls on potential 0 that may be replaced by symmetry axes (left). Resulting MMP solution obtained with a single multipole (indicated by a circle) in the electrode with appropriate symmetry decomposition. The field in the first quadrant is a good approximation that matches the boundary conditions well. The matrix size is very small, 20*10.*

Conductors (PEC and SIBC)

We have seen that PEC planes may be replaced by symmetry planes in some cases. Since PEC's are often used to approximate good conductors, codes for computational electromagnetics should contain features to handle arbitrary PEC's. The modeling of PEC's is very simple, because the electromagnetic field inside a PEC is zero and the scalar potential is constant. Moreover, the surface currents and charges on the PEC surface reduce the number of boundary conditions to be imposed.

In dynamics, only the tangential components of the E field and the normal component of the H field must be matched. Moreover, these components are zero on the PEC surface. The normal component of the H field may be omitted, because it is automatically fulfilled when the tangential components of the E field are zero everywhere. In the standard MMP implementation, all three boundary conditions are imposed for obtaining higher reliability.

In electrostatics, a PEC is an electrode with constant scalar potential. There are two types of electrodes: 1) electrodes of fixed, known potential and 2) floating electrodes. On floating electrodes, only the boundary conditions for the electric field may be imposed, whereas the known potential must be matched on the electrodes fixed potential. In this case, one may omit the boundary conditions for the electric field. Note that mixing boundary conditions for the potential and for the E field can cause weighting problems. For more details, see [Gnos, 1997].

Imperfect conductors may be modeled exactly as any other material, i.e., one approximates the field inside the conductors by an appropriate series expansions and one matches the same continuity equations on the boundaries as on the boundaries of dielectric domains. For good conductors, this causes numerical problems. On such conductors, one observes a strong skin effect, i.e., the amplitude of the field decays almost exponentially with the distance from the surface and it oscillates rapidly. Therefore, many terms in the series expansions are required for obtaining accurate results. Moreover, severe cancellations may occur in the evaluation of the series expansion which cause inaccurate results. Fortunately, an exact knowledge of the field inside good conductors is not required in most cases. Therefore, the PEC approximation may be reasonable. From the PEC approximation, one may compute surface currents that approximate the integral of the currents near the boundary. With simplified skin effect models, this allows one to estimate losses caused by heating of the imperfect conductor.

A simpler, and more accurate computation of the losses caused by an imperfect conductor is obtained when Surface Impedance Boundary Conditions (SIBC) are imposed on the surface of the conductor (see section on Surface Impedance Boundary Conditions (SIBC)). In this case, the field inside the conductor is not computed as in the PEC model, but the PEC boundary conditions are replaced by SIBC's. Figure 98 illustrates the differences between a PEC and a SIBC model of a simple grating with circular conductors.

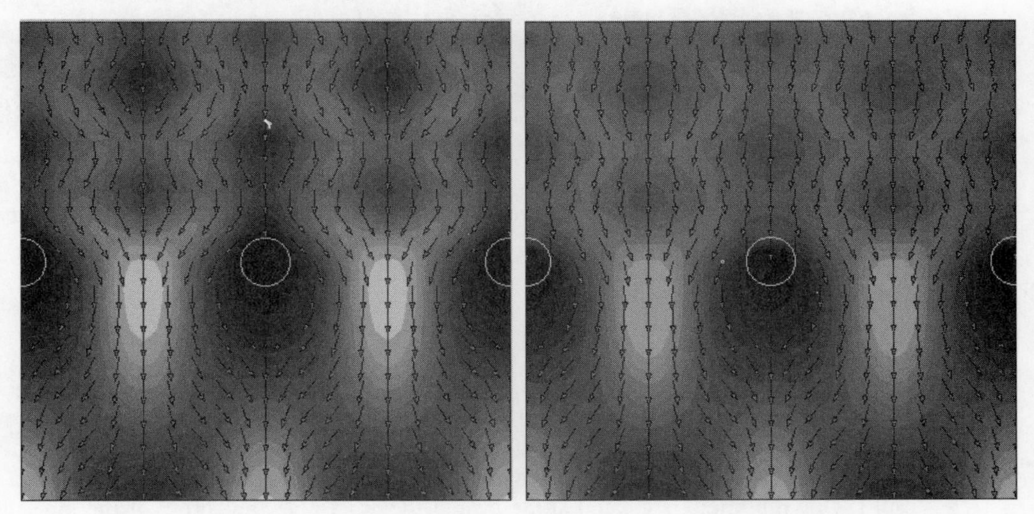

Figure 98: Time average of the Poynting field for a plane wave that illuminates a grating consisting of lossy conductors. PEC approximation (left) and SIBC approximation (right).

The main difference is that the time-average of the Poynting field does not penetrate the PEC conductor, i.e., the Poynting vector is tangential to the PEC surface, but not tangential to the

SIBC surface. Therefore, one can compute the absorbed power in the conductor from the flux of the time average of the Poynting field through the surface of the conductor, which is obtained from a simple integration.

Both the PEC and SIBC models are approximations of real conductors that are more effective the more pronounced the skin effect is. A comparison with accurate computations of imperfect conductors shows that SIBC approximations may be quite accurate, even when the skin effect is not pronounced. Figure 99 illustrates this. Note that the SIBC's are derived from a simplified model with a plane boundary. Therefore, these boundary conditions are most accurate when the curvature of the boundary is low. As long as one is interested in integral quantities such as the absorbed energy in the conductor and not in the field close to the conductor, one may obtain accurate SIBC results also when the curvature is very high at some points. For more details, see [Hafner, 1995].

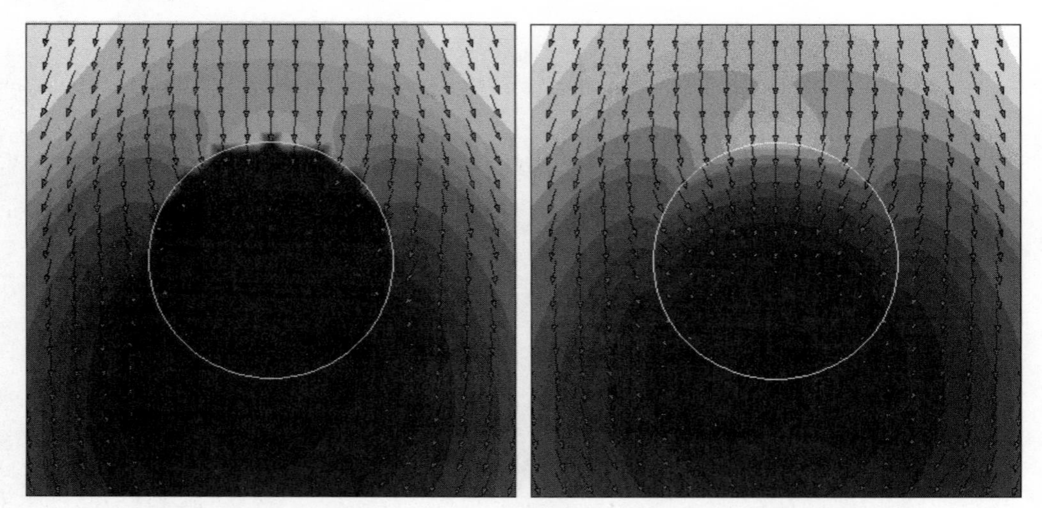

Figure 99: Same as in Figure 98, zoomed closer to the conductors. SIBC approximation (left) without modeling of the field in the conductors and correctly modeled field in the conductor. To make the differences visible, a relatively bad conductor with weak skin effect is considered.

Fictitious boundaries

The field outside a PEC or outside an electrode in electrostatics is usually modeled by a couple of multipoles placed inside the PEC. This causes problems with the placement of the multipoles when the PEC is thin. To illustrate this, we consider the electrostatic computation of a rectangular coaxial cable (see Figure 100). As long as the height/width aspect ratio of the inner conductor is close to zero, two multipole expansions are sufficient to obtain accurate results. One multipole is located at the center, the other near the corner of the inner conductor in the first quadrant. Because of the symmetry with respect to the x and y axes, only the first quadrant is modeled explicitly. When the aspect ratio is close to zero, many multipoles must be placed along the axis. The distance between these multipoles is approximately equal to half of the height of the inner conductor.

When the height of the inner conductor tends to zero, the number of multipoles in the inner conductor tends to infinity, which leads to huge series expansions and MMP matrices. To overcome these problems, one could use distributed multipole expansions, such as line multipoles. This leads to time-consuming integration. Instead, one should increase the space where the multipoles may be placed. This can be done by inserting a fictitious boundary, i.e., by subdividing the domain between the conductors into two subdomains with identical material properties. Now, two sets of multipoles are required for modeling the field in each of the domains. These multipoles may be placed anywhere outside the corresponding domain, i.e., the multipoles for modeling the field in domain 1 near the inner conductor may be placed not only inside the inner conductor, but also somewhere in domain 2. The multipoles outside the outer conductor can again be replaced by a Bessel expansion in the center. Since there now are two domains, one Bessel expansion for each of the domains are required.

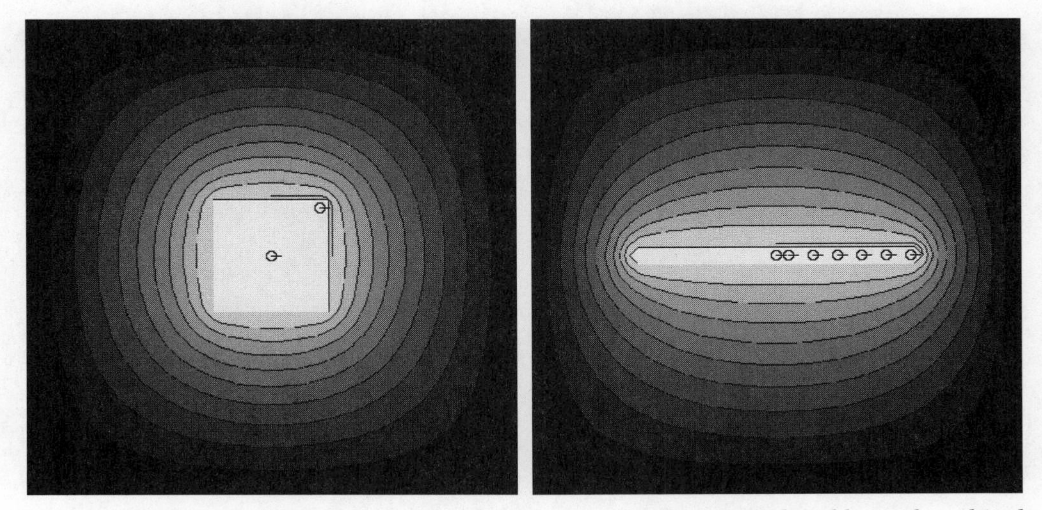

Figure 100: Lines of constant potential in a rectangular coaxial cable and multipole expansions used to approximate the field. Note that only the first quadrant is modeled explicitly because of the symmetry. Multipoles are indicated by circles. A Bessel expansion in the center replaces all multipole expansions outside the outer conductor. Only two explicit multipole expansions are required for a square inner conductor (left), whereas six multipole expansions are required for the flat inner conductor (right).

Note that splitting the domain into subdomains can reduce the symmetry of the problem. When the fictitious boundary is defined along the x axis, this axis is no longer a symmetry axis. The y axis remains a symmetry axis. Therefore, only the right half space must be modeled explicitly. As one can see in Figure 101, excellent results can be obtained for an arbitrarily thin inner conductor with a reasonable number of multipoles. Since the electric field is singular near the corner of the inner conductor, it is reasonable to place higher order multipoles close to this critical area. For an infinitely thin conductor, i.e., for a strip, the singularity is described by a monopole located on the edge of the strip.

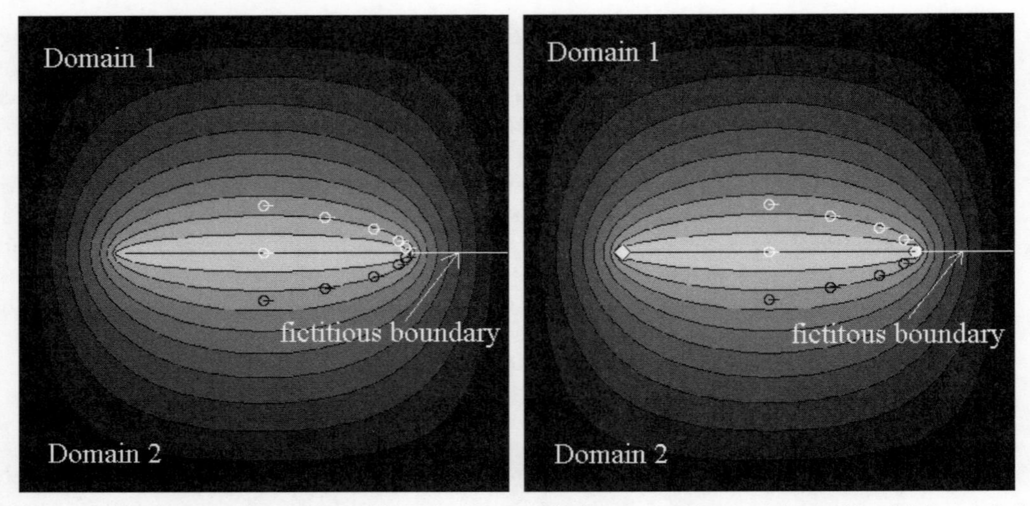

Figure 101: Same as in Figure 100, with a very thin inner conductor. To avoid a high number of multipoles in the inner strip, the space between the conductors is split into two domains by a fictitious boundary. The multipoles may then be located outside the strip. Black circles indicate multipoles for expanding the field in domain 1, white circles indicate multipoles for expanding the field in domain 2. To obtain an accurate solution near the corner of the inner conductor, multipoles should be placed very close to the corner (left). For an infinitely thin conductor, a monopole may be placed directly on the edge (right) to simulate the singularity.

Fictitious boundaries may be introduced not only to obtain more freedom in the multipole setting, but also to subdivide geometrically complicated domains into more simple subdomains that can be modeled with more simple expansions. Obviously, this is a step toward a domain method such as FE [Hafner, 1991].

The MMP code minimizes the mismatching error along the boundaries. The error of the field in each domain is typically maximum near the boundaries. The fictitious boundaries therefore have a considerable influence on the error distribution in the domains. With appropriate fictitious boundaries, one may achieve a more balanced error distribution or one may insert higher errors in areas where high errors may be tolerated. This allows the code to reduce the errors in critical areas.

Periodic problems

Gratings and antenna arrays are periodic structures that exhibit a special symmetry. For simplicity, we only consider structures that are periodic in one direction, as illustrated in Figure 102. As one can see, the symmetry operation of such a structure is given by a translation with a vector $\vec{d}$. The periodic structure may be separated in three different parts or layers by use of fictitious boundaries. Although fictitious boundaries may have an arbitrary shape, it is reasonable to use straight lines. The top and bottom layers are two homogeneous domains, whereas the middle layer is inhomogeneous and contains the periodic structure. Obviously, one cannot explicitly model such an infinite structure. To overcome this problem, one may try a symmetry decomposition using the representation theory of symmetry groups. This is quite demanding. Therefore, a more intuitive procedure is preferable at least for engineers.

Assume that the x axis is parallel to the translation vector $\vec{d}$. Then a Fourier decomposition in the x direction is what certainly comes to mind. This idea leads to the so-called Floquet theory [Collin, 1991], [Yeh, 1988]. We have already seen in the section on Dependent expansions - Rayleigh expansions, that the scattered field in the top and bottom layer for an incident plane wave may be approximated by Rayleigh expansions or Floquet modes. These expansions are plane waves (that may be evanescent) with a direction that is obtained from the data of the incident plane wave and the vector $\vec{d}$. The same expansions may also be used to approximate the field in the top and bottom layers radiated by an array of antennas contained in the middle layer, provided that the feed of each antenna is correlated with the feed of the neighbor antenna to the left by a complex factor C. What is missing is the modeling of the field in the middle layer. To expand this field, one may implement 'periodic' expansions like the Rayleigh expansions. For example, one can use an infinite array of multipoles along the x axis, generated by repeated translations with the vector $\vec{d}$. When the amplitude of each multipole is equal to the amplitude of the multipole to its left – multiplied by the complex factor C – the multipole array has the correct symmetry. Thus, one can model the field in the middle layer by several periodic multipole arrays. Unfortunately, the numerical evaluation of such multipole arrays is very time-consuming because the convergence is quite bad. To overcome these problems, one may find special techniques to speed up the convergence, for example, for monopole arrays [Tayeb, 1991]. Since this requires a careful analysis and implementation of all types of basis functions, this procedure is not attractive for MMP.

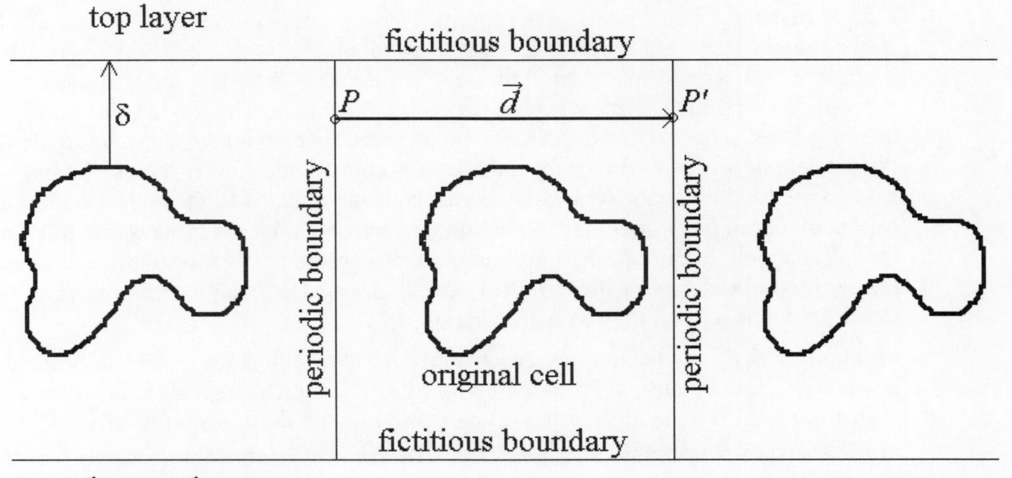

Figure 102: Separation of three layers of a periodic structure by straight fictitious boundaries. The middle layer is subdivided by a special class of fictitious boundaries, i.e., periodic boundaries. This allows one to isolate a single cell that must be modeled explicitly.

Note that the situation is similar to the computation of a thin conductor that was outlined in the previous section. There, a fictitious boundary was introduced instead of a distributed multipole.

Now, we can introduce a fictitious boundary in the middle layer that connects the fictitious boundaries of the top and bottom layers. For simplicity, we select a straight line in the y direction. Note that its location on the x axis is arbitrary. The optimal location is in an area where we expect the field to be smooth. Now, the symmetry operations generate an array of fictitious boundaries that separate geometrically identical cells. The field in each cell is identical to the field in the left neighbor cell multiplied by the complex factor C. Therefore, it is sufficient to compute the field in one single cell. When we compare the field along the left and right boundary of such a cell, we see that the field on the right boundary is equal to the field on the left boundary multiplied by C. This is a periodic boundary condition. Because of the symmetry operations, only a single cell – called the original cell – and only a single fictitious boundary (in the y direction) on one side of the cell must be modeled explicitly. Moreover, only the parts of the fictitious boundaries (in the y direction) between the top and bottom layers and the original cell must be modeled explicitly. Therefore, only a finite number of boundaries of finite lengths must be modeled. The field inside the original cell can be modeled with the usual expansions that must not be periodic at all. For example, one can apply standard multipoles.

As one can see from Figure 98, excellent results can easily be obtained for periodic structures, when fictitious and periodic boundaries are inserted for isolating a single cell. It is important to note that the locations of these boundaries are not unique. One may move them to different places. Moving the fictitious boundaries affects the length of the periodic boundary. One might assume that the discretization of the shortest possible periodic boundary is numerically most efficient because the number of matching points increases with the length of the boundary to be discretized. This assumption is wrong, because the discretization depends not only on the length, but also on the complexity of the field. At those locations where the field varies rapidly, a high matching point density is required. The highest complexity of the field is usually encountered near true boundaries between domains with different material properties. Therefore, one should move the fictitious boundaries between the top and bottom layers and the original cell at a sufficient distance from the boundaries inside the original cell. This increases the length of the periodic boundaries, but it allows one to apply a less fine discretization of the fictitious boundaries. Moreover, the higher order Rayleigh terms that expand the field in the top and bottom layer are evanescent, i.e., the corresponding field decays exponentially in the y direction. When the fictitious boundary is close to the true boundaries of the original cell, high order Rayleigh terms are required to obtain accurate results. As soon as the fictitious boundary is moved away, the high order Rayleigh terms become small on the fictitious boundary and may be omitted [Hafner, 1995].

Moving the fictitious boundaries far away from the true boundaries of the original cell makes not much sense because this increases the discretization of the periodic boundaries without allowing a further reduction of the discretization of the fictitious boundaries. When the size d of the translation vector $\vec{d}$ is bigger than half the wavelength, the distance δ of the fictitious boundaries from the true boundaries of the original cell should be approximately a quarter wavelength. When d is small compared with the wavelength, δ should be in the same order of magnitude as d.

Eigenvalue solver

Most of the problems in computational electromagnetics lead to an inhomogeneous matrix equation of the form MX=B, where B is the inhomogeneity. Such matrix equations may be easily solved with many different matrix solvers. The inhomogeneity contains essentially the excitation in electrodynamics. There are two important types of applications, where the excitation is not modeled, namely resonators and guided waves. Resonators are energetically closed objects. It is assumed that the field in the resonator had been excited a long time ago. Waveguide structures are cylindrical structures that can be split into a longitudinal and a transverse problem. Along the z axis, one observes wave propagation, whereas the transverse problem is a 2D resonator. Here, the excitation is assumed to be somewhere at infinity in the z direction. Since the excitation is modeled, a homogeneous matrix equation of the form MX=0 is obtained. Such a matrix equation has obviously the trivial solution X=0. When M is a square matrix, one may also find non-trivial solutions with X≠0 that are of physical interest. These solutions are the resonator modes or the modes propagating along a cylindrical structure, i.e., waveguide modes. To obtain non-trivial solutions, the determinant of the matrix M must be zero. For general eigenvalue problems in electromagnetics, the elements of M depend on an eigenvalue e. For resonators, the eigenvalue is the resonance frequency ω, whereas the eigenvalue of guided waves is the propagation constant γ. Therefore, one has

$$\det(M(e))=0. \tag{8.21}$$

In general, the determinant of M is a non-linear function of e. Usually this is a transcendent function and finding analytic solutions is impossible, except in simple situations. There are many numerical algorithms for finding zeros of functions [Press, 1992]. Even when (8.21) is real valued, the search for all zeros may be difficult and time consuming.

First of all, the condition number of M may become high, especially near the zeros of det(M). This reduces the numerical accuracy and generates some noise on the function det(M(e)). This noise may heavily disturb the algorithm to find the zeros.

Several solutions may be close to each other, i.e., the difference e_1-e_2 of two modes may be very small or even zero when the modes are degenerate. This may also disturb the algorithm that searches the zeros.

The search algorithm proceeds iteratively and it may be hard to find out when to stop the iteration for the following reasons. Assume that the algorithm has found a zero of (8.21) after k iterations, i.e., an approximation e^k of some eigenvalue e. When the correct value e would be known, one could compute the error $\varepsilon^k=|e-e^k|$. This error does not tend to zero for $k\to\infty$ because of inaccuracies of the model. Instead, it tends toward some positive value ε_∞. Good algorithms may estimate $\varepsilon^k-\varepsilon_\infty$ and can stop, for example, when $|e^k-e_\infty|/e_\infty$ is guaranteed to be smaller than some relative error defined by the user. First of all, this error is misleading, because the true relative error $|e^k-e|/e$ may be much bigger than $|e^k-e_\infty|/e_\infty$. When the search process is stopped after k iterations, the parameter vector X^k is evaluated and from this, an approximation of the field F^k of the mode is obtained. The relative errors $\|F^k-F\|/F$ and $\|F^k-F_\infty\|/F_\infty$ of the field may be much larger or also much smaller then the errors $|e^k-e|/e$ and $|e^k-e_\infty|/e_\infty$. It may turn out that many fewer than k iterations would have been enough to obtain the desired accuracy of the field. In this case, one has wasted computation time. It may also turn out that k iterations are not enough for the desired accuracy of the field; also the desired accuracy of the eigenvalue has been reached. In this case, one has to resume the eigenvalue search.

The determinant of square matrices is usually computed by Gauss and similar algorithms that transform the original matrix into a triangular matrix T with det(M)=det(T)=$\Pi(t_{kk})$, where $\Pi(t_{kk})$ denotes the product of the diagonal elements of T. When T is an upper triangular matrix, one can see from the last row of the matrix equation TX=0 that either the last element x_K of the parameter vector X or the last diagonal element t_{KK} of the matrix T must be zero. Therefore, one can replace the eigenvalue condition (8.21) by

$$t_{KK}(e)=0. \tag{8.22}$$

It is obvious from (8.22) that the last parameter and the corresponding basis function plays a prominent role for the eigenvalue computation. Good results are obtained when the last parameter is relatively big, i.e., when the corresponding basis function is important for the description of the field of the mode to be searched. When a 'poor' basis function is at the last position, it may happen that the desired mode cannot be found with (8.22). This happens when the scalar product of the field of the mode with the field of the basis function is zero, for example, when the mode has even symmetry and the basis function has odd symmetry with respect to some axis. Note that this 'mode suppression' can also be desirable, especially when one is searching for a specific mode of a waveguide with many different modes. With an appropriate selection of the last basis function, one can suppress undesired modes, which may result in a simpler eigenvalue search.

For MMP, the most important drawback of the method outlined above is that the matrix M must be square. In this case, MX=0 has only a trivial solution and the determinant of M is not defined. Of course, one can obtain a square matrix when one uses the normal equations M*MX=0. From this, one obtains

$$\det\,(M^*(e)\cdot M(e))=0. \tag{8.23}$$

Remember that these equations should be avoided because of the problems caused by high condition numbers. Therefore, one has to find a way to compute eigenvalues with a rectangular system of equations. The normal equations minimize the square norm of the error vector E of MX=E. Since the matrix elements are functions of the eigenvalue e, the square norm of E is also a function of e. In general, this function is a positive, real number. When an analytic solution is found, it is zero. Therefore, it makes sense to analyze $\eta(e)=\|E(e)\|^2$.

First of all, it is important to recognize that it may happen that a minimum of η corresponds to an almost zero solution, i.e., an almost trivial solution that is not desired. Secondly, one can multiply det(M(e)) with an arbitrary function $f(e)$ that has no zeros. This will have no influence on the locations of the zeros of det(M(e)). Therefore, the scaling of rows or columns of M has no influence on the eigenvalue that is found. This is not the case for the minima of $\eta(e)$. Scaling of the rows and columns of the rectangular matrix M may affect the positions of the minima. Does this mean that the idea of a minimum search of $\eta(e)$ is useless? On the contrary! The sensitivity of the location of the minima and the shape of $\eta(e)$ near the minima allow one to obtain information on the accuracy of both the eigenvalue and the eigenvector or the corresponding field.

Before a useful method is obtained, the problem with the almost trivial solutions must be solved. The key to that is the observation that the amplitude of the solution is not explicitly defined. Once the amplitude A is defined, it is also a function of e. Since almost trivial solutions are characterized by a small square norm of the amplitude, one may define η as follows:

$$\eta(e)=\|E(e)\|^2/A^2(e). \tag{8.24}$$

With a proper definition of the amplitude, one can avoid almost trivial solutions. Note that the locations of the minima of $\eta(e)$ depend also on the definition of the amplitude. Again, the robustness of the results when the definition of the amplitude is modified gives one useful hints on the accuracy. There are many ways to define the amplitude. For transmission lines, one may use the total current in a wire or the voltage between the wires, i.e., line integrals over the H or E field. In waveguides, the total power flux along the z axis may be more reasonable. In resonators, the total energy seems to be a natural amplitude definition, and so on. Note that power flux and energy densities have units that are different from the units in the error vector E. Remember that the MMP matrix is scaled in such a way that each row has the unit either of the electric field or of the square root of an energy density. Therefore, one should define the amplitude as an integral over a field (E or H) when one wants to work with (8.24). Instead of this, one can also replace A^2 by A and define the amplitude as an integral over energy density or over the Poynting flux.

Since the amplitude definition contains the field $F(e)$, the procedure for finding minima of (8.24) is the following:

```
While stopping criteria are not met:
Increase iteration number k
Select eigenvalue e^k from the previous search data
Solve M(e^k)X^k=E^k in the least squares sense
Compute the field F^k from X^k and integrate to obtain A^k
Compute η^k from E^k and A^k
Analyze the values η^1,η^2,...,η^k to obtain stopping conditions
```

Obviously, two steps in this procedure must be clarified: 1) the selection of the k-th eigenvalue approximation e^k and 2) the stopping criteria. The former means that an algorithm is defined that selects e^k from the previous search data (namely η^j, e^j for $j<k$) in such a way that e^k is as close as possible to the desired eigenvalue e. This is nothing else than a minimum search routine for a real valued function $\eta(e)$ [Hafner, 1985], [Hafner, 1990/1]. Note that the argument of this function is real for loss-free problems and complex for lossy problems.

Real eigenvalue search for loss-free problems

The eigenvalues that correspond to guided waves of loss-free waveguides or modes of loss-free resonators are real numbers. Therefore, the eigenvalue search is performed on the real axis of the complex e plane. Note that evanescent waves of waveguides, i.e., modes below cutoff, are characterized by a purely imaginary propagation constant. If such waves are of interest, one has to search along the imaginary axis. For the search procedure, this does not make any substantial difference. Therefore, we focus on the search along the real axis.

First of all, the search interval on the real axis can be restricted for physical reasons. When the waveguide consists of several dielectrics, the propagation constant of all guided modes is smaller than the maximum wave number (in the most dense dielectric). When no conductors are present, the propagation constant of all guided modes is also bigger than the minimum wave number. If there is an outer conductor, the lower limit of the propagation constant is zero. Therefore, one has always a finite search interval. Within this interval, $\eta(e)$ is a smooth function with several maxima and minima. Each minimum corresponds to a mode.

Theoretically, the function $\eta(e)$ may be differentiated with respect to e. Then, each mode is described by a zero of the derived function $\eta'(e)$. There are many useful and efficient algorithms for detecting zeros of a function [Press, 1992]. Note that some of the zeros of this function correspond to maxima that do not represent any mode. Moreover, $\eta'(e)$ has to be evaluated numerically, which causes numerical problems and inaccuracies. Therefore, it is better to use algorithms that directly search for minima of $\eta(e)$; also minimum search is more demanding than zero finding. One can distinguish two categories of minimum search algorithms [Press, 1992]: 1) algorithms that can start at any point and 2) algorithms that start in an interval that must contain a minimum. The former seem to be simpler, but for our purpose, such algorithms are rather useless because the minimum that will be found depends on the start point and it may be hard to find all minima of interest with such algorithms (see also the section on Non-linear optimization). The latter usually start with three points $e_1<e_2<e_3$ and the condition

$$\eta(e_1)>\eta(e_2)<\eta(e_3) \tag{8.25}$$

must be met. These algorithms refine the interval that contains the minimum and can stop when 1) the maximum number of iterations is reached, 2) the interval is sufficiently short, i.e., when the eigenvalue has been found with the desired accuracy, or 3) the function $\eta(e)$ is so flat within the search interval that an accurate eigenvalue computation is impossible. Note that scaling, amplitude definition, etc. affect the shape of $\eta(e)$. When a 'flat minimum' is detected, the location of this minimum would be shifted considerably when another scaling or another amplitude definition is selected. Therefore, an accurate eigenvalue evaluation would be wasting computation time because the eigenvalue estimate must be inaccurate. To obtain more accurate results, one then has to improve the initial model until a less flat function $\eta(e)$ is obtained.

Very flat minima may even be artificial solutions that do not correspond to any mode. This is easily verified when the amplitude definition is exchanged. It is useful to distinguish the following four types of minima:

1) Flat minimum ($\eta(e_1)\approx\eta(e_2)\approx\eta(e_3)$) with high function value ($\eta(e_2)$ big). Such a minimum indicates a high risk of being an artificial one. The entire model should be refined to obtain acceptable results and the amplitude definition should be modified. When the minimum disappears, it was an artificial one and can be ignored.

2) Flat minimum ($\eta(e_1)\approx\eta(e_2)\approx\eta(e_3)$) with low function value ($\eta(e_2)$ small). The low function value indicates that the electromagnetic field can be accurately computed; also the eigenvalue itself might be very inaccurate. This is typical for the computation of transmission line modes at low frequencies. In this case, it is recommended to evaluate the eigenvalue from the electromagnetic field and not from the minimum search (see section on Transmission lines).

3) Sharp minimum ($\eta(e_1)>>\eta(e_2)<<\eta(e_3)$) with high function value ($\eta(e_2)$ big). The high function value indicates that the field computation is inaccurate, therefore the minimum search should be refined. Since the minimum is sharp, a few additional iterations will considerably

reduce $\eta(e_2)$. When one is only interested in the eigenvalues and not in the field, this is not necessary.

4) Sharp minimum ($\eta(e_1)>>\eta(e_2)<<\eta(e_3)$) with low function value ($\eta(e_2)$ small). An accurate computation of both the eigenvalue and the electromagnetic field has been achieved.

The minimum search algorithm should be designed in such a way that it is quick, at least when the search is sufficiently close to the desired minimum. The function $\eta(e)$ may be approximated quite well by a parabolic curve in the vicinity of the minimum. Therefore, a parabolic approximation of $\eta(e)$ at the three points e_1, e_2, e_3 is appropriate. The minimum e_p of the parabolic approximation is then a good estimate of the minimum. To refine the search, one drops e_1 when $\eta(e_2)>\eta(e_p)<\eta(e_3)$ holds and one replaces the three points e_1, e_2, e_3 by e_2, e_p, e_3. Otherwise, one drops e_3 and one replaces the three points e_1, e_2, e_3 by e_1, e_p, e_2. This procedure may be repeated. Numerical problems occur when the triplet of points e_1, e_2, e_3 becomes very unbalanced, i.e., when the center point is very close to one of the other points. To avoid such problems, one can replace the parabolic interpolation by a geometric interpolation that defines a new point e_p in such a way that the triplets e_1, e_p, e_2 and e_2, e_p, e_3 become more balanced.

Note that the original interval of the eigenvalue search is usually too big for the parabolic search routine mentioned above – except when the user has a good estimate from a previous computation or from experience. Therefore, the eigenvalue search typically starts with a rough search, where $\eta(e)$ is computed in a predefined set of N points $e_1<e_2<\ldots<e_n<\ldots<e_N$ that are uniformly distributed over the entire search interval. The rough search detects a minimum near e_n when $e_{n-1}<e_n<e_{n+1}$ holds. It then may pass the triplet e_{n-1}, e_n, e_{n+1} as input values e_1, e_2, e_3 to the fine search routine. The main problem of the rough search routine is that it is time-consuming when N is high. Otherwise, there is some risk that the routine skips sharp minima or minima that are close together.

To illustrate the procedure, we consider a simple, symmetric cable consisting of two circular PEC wires with radius 2mm embedded in a circular dielectric with relative permittivity 4 and radius 8mm. The distance between the axes of the wires is 8mm. The cable has two symmetry axes. The transmission line mode on such a cable has odd symmetry with respect to one axis and even symmetry with respect to the other. If we take these symmetries into account, we reduce the number of modes described by the function $\eta(e)$. Usually, such a cable is used at frequencies where no higher order modes propagate. To observe higher order modes, we start at a very high frequency where several higher order modes with the same symmetry as the transmission line mode propagate. Figure 103 shows $\eta(e)$ for different definitions of the amplitude and for two different expansions in the last position. The normalized propagation constant γ / k_0 is used as eigenvalue e. For waves with a field that is almost entirely within the dielectric, γ / k_0 is close to the square root of the relative permittivity of the dielectric, i.e., 2. When the field is almost entirely outside the dielectric, γ / k_0 is close to 1. Therefore, the search interval is 1…2. Modes near cutoff have γ / k_0 close to 1, whereas the lowest order mode (transmission line mode) has the highest γ / k_0.

As one can see, the shape of $\eta(e)$ depends very much on the expansion in the last position when no amplitude is defined. When a high multipole order is the last expansion, the minima become very sharp and the transmission line mode and the lower order modes are suppressed or at least hardly detectable. As soon as a prominent expansion is placed at the last position, $\eta(e)$ becomes much nicer and all minima can easily be detected. The same happens when the amplitude is well defined. On transmission lines, one usually works with voltages and currents. The voltage can be defined by a line integral over the electric field from one wire to another one. At high frequencies, the result depends on the path of the integral. In the computations illustrated in Figure 103, the path is a straight line on the symmetry axis of the cable. Similarly, the current can be defined by a closed line integral over the magnetic field along a path around one wire. The result also depends on the shape of the path. Here, the path coincides with the surface of one wire. In Figure 103, one can observe artificial minima when the current is used to define the amplitude. No such minima are observed when the time average of the Poynting flux along the z axis is used to define the amplitude. Since the cable is an open structure, the numeric integration should run over the entire xy plane, which is not easy. Fortunately, a very rough approximation is sufficient. For the computation in Figure 103, the Poynting vector has been computed at 25 points, uniformly distributed over a square area of the xy plane. Note that all minima are sharp and deep. Therefore, both the eigenvalues and the field of the modes can be computed accurately. With a total number of 50 unknowns, i.e., with a very small MMP matrix, at least six digits of the propagation constant of the transmission line mode were obtained with 13 iterations. The maximum error of the field computation was below 0.3%.

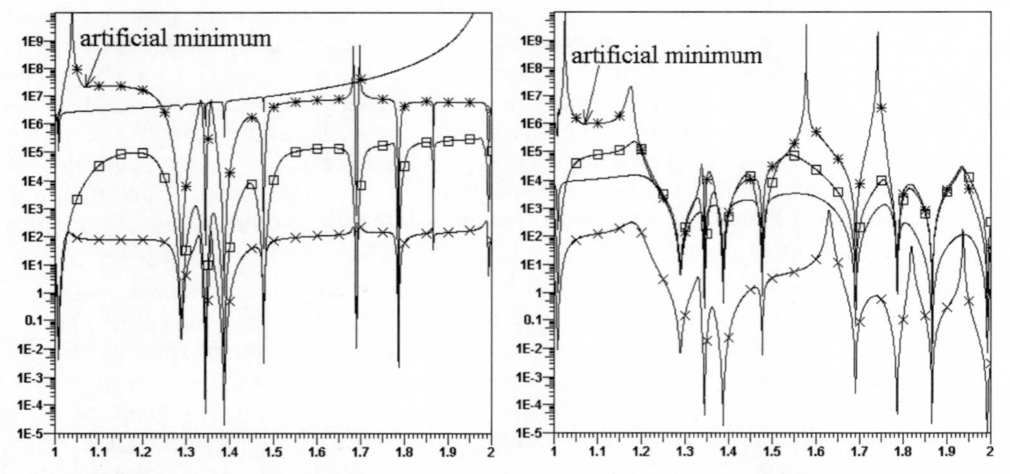

Figure 103: Minimum search function $\eta(\gamma / k_0)$ for a symmetric cable with two circular conductors embedded in a circular dielectric at 30GHz. Curves without markers: no amplitude definition, × markers: amplitude defined as voltage, star markers: amplitude defined as current in a wire, square markers: amplitude defined as energy flux in the z direction. Left hand side: last expansion is a high multipole order, right hand side: last expansion is the zero order multipole inside a wire.

When the frequency is reduced all minima shift to lower values. When a minimum reaches the lower limit of the search interval, the cutoff frequency of the corresponding mode is reached. Below cutoff, the mode no longer propagates. When we reduce the frequency from 30GHz to 3GHz, all higher order modes are cutoff and $\eta(e)$ exhibits a single, nice minimum, as one can see in Figure 104. At this frequency, the minimum is deeper, which indicates that the field can be computed even more accurately than at 30GHz. This is no surprise because the number of unknowns required typically increases with the size of the boundary with respect to the wavelength. At the same time, we observe that the minimum is less sharp, i.e., the eigenvalue can be computed less precisely. This trend is continued when we further reduce the frequency. At 30MHz, the field can be computed very accurately, but the minimum is already quite flat. Therefore, no accurate eigenvalue computation is achieved. At 3MHz, the minimum is so flat that it is impossible to estimate the eigenvalue. At the same time, the value of the minimum is higher than at 30MHz, which indicates that also the field is computed less accurately than before. Despite this, one can evaluate the field very accurately, when one sets any value for γ / k_0. This means that one can omit the eigenvalue search and set, for example, γ / k_0=1.5 to evaluate the field. As soon as the field is computed, one can accurately compute the propagation constant from the field.

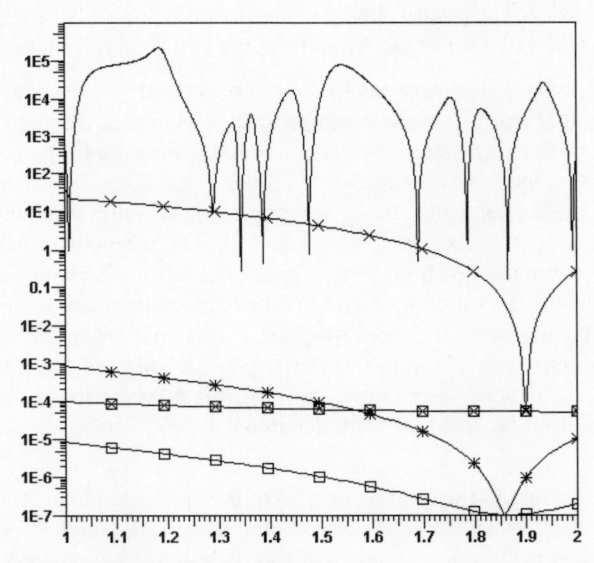

Figure 104: Minimum search function $\eta(\gamma / k_0)$ for a symmetric cable with two circular conductors embedded in a circular dielectric at different frequencies. Curve without markers: 30GHz, × markers: 3GHz, star markers: 300MHz, square markers: 30MHz, square and × markers: 3MHz. Amplitude defined as energy flux in the z direction, last expansion is the zero order multipole inside a wire.

It is interesting to see that the field of the transmission line mode can be computed accurately also when a completely wrong propagation constant is set first. When one applies static computations to obtain the electric and magnetic fields, one replaces the Helmholtz equations by Laplace equations, which means that one sets the transverse wave number κ=0 in all domains. This means that one implicitly sets γ / k_0=1 in the magnetostatic computation. For the electrostatic computation of the field outside the dielectric one also sets γ / k_0=1, whereas one sets γ / k_0=2 for the field inside the dielectric. Although this is theoretically not consistent, it allows one to approximate the field very accurately.

Note that the skin effect and losses caused in the imperfectly conducting wires have been ignored in our model of a cable with PEC wires. The PEC assumption implies that the currents are on the surface of the wires. As one knows, the currents in the wires are almost homogeneous at sufficiently low frequencies. Therefore, the computation becomes increasingly inaccurate with decreasing frequency. To obtain more reasonable results, one has to replace the PEC wires by more realistic conductors that may be characterized by a complex permittivity. As a consequence, the propagation constant becomes complex and the eigenvalue search must be performed in the complex plane.

Complex eigenvalue search for lossy problems

Theoretically, lossy eigenvalue problems are not well understood, because a correct eigenvalue definition is based on energetically closed models. The eigenmodes of resonators do not exchange any energy with the exterior of the resonator. Similarly, guided modes on cylindrical structures are energetically closed systems in the transverse plane. When losses are introduced, energy is converted into heat. Therefore, the structure is no longer energetically closed.

It is well known that a resonator is characterized in practice by some function $f(\omega)$ of the frequency that exhibits some peaks near the resonance frequencies ω_n that are computed from idealized models. The shape of $f(\omega)$ near the resonance frequencies ω_n is important in practice, but not contained in the idealized models. In fact, to measure $f(\omega)$, one has to open the resonator and to connect it to some equipment. When one models an opened resonator with its input and output ports, one obtains a scattering problem rather than an eigenvalue problem. When the resonator has high losses, $f(\omega)$ exhibits no sharp peaks and it is questionable whether it makes sense to compute discrete resonance frequencies in such a case. However, once one has defined a search function $\eta(e)$ for a loss-free resonator, one can easily generalize this function to one with a complex argument, i.e., a complex eigenvalue e. A complex frequency ω describes a damped oscillation and its imaginary part is the damping constant. Similar statements hold for lossy waveguides that can be described by a complex propagation constant γ.

The numeric search for minima of $\eta(e)$ in the complex plane is much more demanding than the search in the loss-free case. As long as the losses are small, it is quite obvious that the minima of $\eta(e)$ are close to the real axis and the damping constant $\mathrm{Im}(e)$ is very small compared with the real part $\mathrm{Re}(e)$. In this case, a very high accuracy of the computation of e is required to guarantee a sufficient accuracy of $\mathrm{Im}(e)$. Therefore, it is reasonable to use a loss-free model first and to apply some perturbational method to obtain the damping constant $\mathrm{Im}(e)$ (see the section on Perturbations).

For higher losses, the perturbational method becomes inaccurate. In this case, a complex eigenvalue makes sense. The main problem now is that there is no well defined search area in the complex plane. When one uses the time dependence $\exp(-i\omega t)$ for resonators, a damped oscillation is described by a complex frequency either in the second or in the fourth quadrant of the complex plane. In both quadrants one finds identical solutions. Therefore, one can restrict the search to the fourth quadrant, but this is still an infinite search space. Similarly, one can restrict the search for the propagation constant of waveguides to the first quadrant of the complex plane, when the z,t dependence $\exp(i\gamma z - i\omega t)$ is assumed. Since extremely damped modes are of no interest in practice, one can reduce the search area to a strip along the real axis. One might be tempted to restrict it to a rectangle with a baseline on the real axis that coincides with the search interval that is used for the corresponding loss-free model. This is not recommended, because the real part of a lossy mode can be bigger than the maximum value of the corresponding loss-free mode. To illustrate this, we consider a so-called Sommerfeld wire, i.e., a single, circular wire in free space. Such wires were sometimes used as antenna feeds. The fundamental mode on such a wire is rotationally symmetric. Since the total electric current of this structure through the *xy* plane is not zero, this is no transmission line mode. The 'return current' is a pure displacement current, i.e., the electric field of this mode is not purely transverse – except the idealized, loss-free case. Figure 105 illustrates the fundamental mode on a lossy Sommerfeld wire. The normalized propagation constant γ / k_0 of an idealized Sommerfeld wire is equal to one. Losses in the wire shift γ / k_0 into the complex plane. As one can see from Figure 105, both the real part and the imaginary part are increased. The same effect can be observed for the Harms-Goubau wire, which is a Sommerfeld wire with a dielectric coating. The search interval for such a loss-free wire is $1 \ldots \sqrt{\varepsilon_r}$, where ε_r is the relative permittivity of the dielectric coating. When losses are present, $\mathrm{Re}(\gamma / k_0) = \sqrt{\varepsilon_r}$ may be observed.

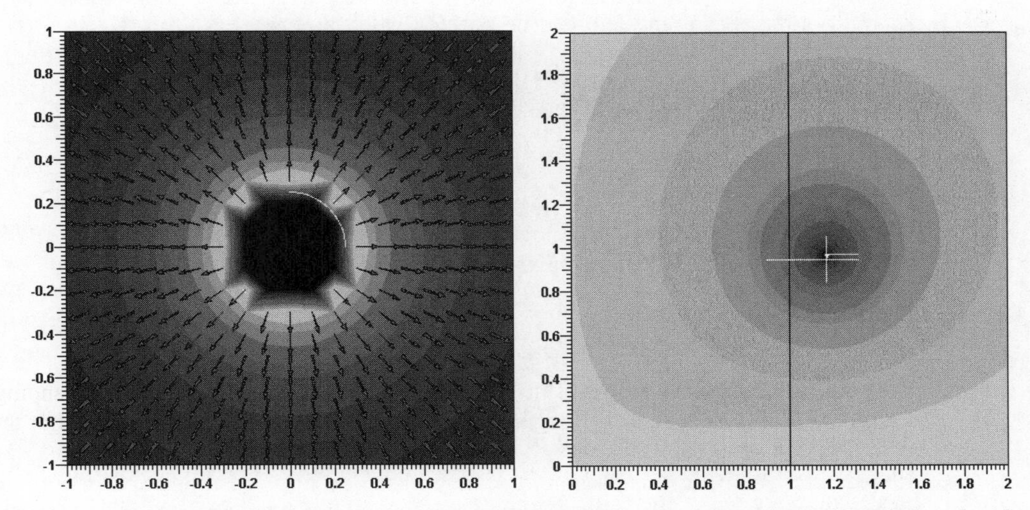

Figure 105: Fundamental mode on a Sommerfeld wire (lossy conductor). Left hand side: Electric field in the xy plane. The baseline of the arrows is proportional to the z component of the field. Right hand side: search function $\eta(\gamma / k_0)$ in the complex plane. The black line indicates the path of the initial, rough search, whereas the fine search is described by white lines.

When the losses are not too big, one can first evaluate the corresponding loss-free mode and search afterwards for a minimum of $\eta(e)$ along a line parallel to the imaginary axis in the complex plane. Of course, this line will not hit the minimum in the complex plane, but the minimum of $\eta(e)$ on the line will be closer than the eigenvalue of the loss-free computation. Once the minimum on the line is detected, one may switch to another line that is parallel to the real axis and search for the minimum on this line, and so on. Figure 105 illustrates this complex minimum search procedure.

There are many other minimum search procedures [Press, 1992] that might be applied for searching complex eigenvalues. Unfortunately, $\eta(e)$ is not as nice in the complex plane as on the real axis. For example, one encounters cuts in the complex plane that are hard to analyze and disturb such a procedure. Moreover, it may be difficult to identify the modes, i.e., to find out which of the minima corresponds to which of the modes. Finally, since lossy modes are characterized by curves $f(e)$ that are not very sharp, it can be difficult to clearly separate different modes. To overcome such problems in the case of guided waves, it is recommended to study the frequency dependence of the modes and to apply the method outlined in the following section.

Eigenvalue Estimation Technique (EET)

When computing the propagation constant γ of a guided wave on a cylindrical structure, one is often interested in its frequency dependence, i.e., in the function $\gamma(\omega)$ within some frequency range. Of course, one can repeat the entire search procedure for a set of frequencies $\omega_1<\omega_2<\ldots<\omega_n<\ldots<\omega_N$ distributed over the entire frequency range. Remember that the rough search is time-consuming and can be omitted when the user has a good guess for the triplet e_1,e_2,e_3, i.e., $\gamma_1(\omega_n),\gamma_2(\omega_n),\gamma_3(\omega_n)$, for the fine search. Such a triplet can be obtained from previous computations of γ for other frequencies. Therefore, one can estimate $\gamma_1(\omega_n),\gamma_2(\omega_n),\gamma_3(\omega_n)$ from $\gamma_1(\omega_{n-1}),\gamma_2(\omega_{n-1}),\gamma_3(\omega_{n-1})$. The main problem of this estimate is that the triplet $\gamma_1(\omega_n),\gamma_2(\omega_n),\gamma_3(\omega_n)$ that starts the fine search must fulfill the condition (8.25), i.e.,

$$\eta(\gamma_1(\omega_n))>\eta(\gamma_2(\omega_n))<\eta(\gamma_3(\omega_n)). \tag{8.26}$$

Assume that the search for $\gamma(\omega_{n-1})$ has started with the triplet $\gamma_1(\omega_{n-1}),\gamma_2(\omega_{n-1}),\gamma_3(\omega_{n-1})$ and that it ended at some value $\gamma_{n-1}(\omega_{n-1})$. In this case, a first (zero order) guess for $\gamma_1(\omega_n),\gamma_2(\omega_n),\gamma_3(\omega_n)$ might be:

$$\gamma_2(\omega_n)=\gamma_{n-1}(\omega_{n-1}),\quad \gamma_1(\omega_n)=\gamma_2(\omega_n)-\delta(\omega_{n-1}),\quad \gamma_3(\omega_n)=\gamma_2(\omega_n)+\delta(\omega_{n-1}), \tag{8.27}$$

where $\delta(\omega_{n-1})=(\gamma_3(\omega_{n-1})-\gamma_1(\omega_{n-1}))/2$ is the length of the initial search interval. When the frequency step $\omega_n-\omega_{n-1}$ is sufficiently small, there is a good chance that the condition (8.26) holds and that the fine search can be started without a preliminary rough search. Of course, the rough search must be performed for the first frequency ω_1 when no estimation of the triplet is available. Note that some estimate may also be available for the first frequency from previous computations of similar cylindrical structures. This more advanced type of eigenvalue estimation is not detailed here.

Remember that there are several methods to obtain an extrapolation of a function $f(t)$ that is known in $N-1$ points $t_1, t_2,\ldots,t_{N-1}$. To obtain a k-th order extrapolation of $f(t_N)$, one can select $k+1$ or more points t_n and approximate $f(t)$ by some series expansion of the form

$$f(t)=f^0(t)+error(t)=\sum_{i=0}^{I}a_i basis_i(t)+error(t)\cdot \tag{8.28}$$

The parameters a_i are then optimized in such a way that the square error norm at the points t_n is minimized. Finally, $f^0(t_N)$ is the desired extrapolation of $f(t_N)$. Obviously, the quality of this extrapolation depends on the location of the points t_n, on the order I of the extrapolation, and on the selection of the EET basis functions $basis_i(t)$. Of course, one will preferably select those points t_n that are close to t_N. Since high order extrapolations require the information on many points, some of these points will not be close to t_N. This may reduce the accuracy of high order EET. Remember that overdetermined systems of equations are preferred in the MMP code.

Such systems are also obtained for the EET when more than k+1 points are used for the extrapolation. Here, weighting plays an important role. It is obvious that points that are closer to t_N should have a higher weight, but up to now, no reasonable weighting procedure has been found that leads to better extrapolations than k-th order EET with k+1 points.

What happens when the minimum search routine finds that the condition (8.26) is violated? In this case, one can increase the interval size δ and restart the fine search until (8.27) holds. Instead, one can also replace the minimum search procedure that can capture a minimum outside the initial search interval. Note that such a procedure is also used in the complex eigenvalue search that was outlined in the previous section. Of course, EET can also be applied to complex eigenvalue problems. In this case, $f(t)=\gamma_2(\omega)$ is a complex function with a real argument. Instead of using (8.28) with complex basis functions, one can also split $\gamma_2(\omega)$ into real and imaginary parts and extrapolate both parts with the same (real) EET basis.

The Eigenvalue Estimation Technique (EET) outlined above estimates the center γ_2 of the search interval from the previous result as a zero order extrapolation and leaves the size 2δ of the original search interval fixed. As soon as the propagation constant has been evaluated for more than one frequency, one can apply higher order extrapolation techniques as outlined in the section on Approximation, interpolation, extrapolation. This allows one to find very accurate estimates of γ_2. Even an extremely accurate estimate does not reduce the computation of the fine search algorithm, because the number of iterations to be performed in this algorithm depends on the size of the initial interval and on the desired accuracy. Therefore, it is reasonable to adapt δ, i.e., to reduce the initial interval size when the estimate of γ_2 is accurate. This means that one estimates not only the eigenvalue itself, but also an additional system parameter that has a considerable influence on the performance of the search routine. Of course, this method can also be extended to the extrapolation of other important parameters (see next section).

Remember that the quality of extrapolations depends very much on the set of basis functions that is applied for the extrapolation. When standard power series are applied, first order (linear) extrapolation is usually much better than zero order extrapolation, but higher order extrapolations are very often worse than first order extrapolation. To improve the quality of higher order extrapolations, the power series basis must be replaced by a basis that is more appropriate for a certain class of problems. Finding a good set of basis functions for the extrapolation is very demanding. Such an adapted set may be found by evolutionary strategies, especially by GP, GGP and similar algorithms (see the chapter on 3 Optimization). The optimized EET can be much more efficient than a linear EET, but only within a restricted range of application, i.e., the extrapolation basis is well adapted to some class of eigenvalue problems, whereas it can be inefficient for other problems. Since the optimization procedure is extremely time consuming, the optimized EET is only useful for those who frequently solve eigenvalue problems within a well defined area, for example, similar types of waveguides. However, simple, linear EET is a powerful tool for eigenvalue problems that may be applied to MMP and any other numerical code.

Figure 106 illustrates the application of the EET for the computation of the frequency dependence of modes of an optical fiber with a lossy aluminum cladding. Note that the material properties of the cladding are strongly frequency dependent. Such fibers are applied in Scanning Optical Nearfield Microscopes (SNOM). As one can see, the curves are smooth within a big range, which makes EET very efficient. In some area (near a cut in the complex plane) the curve changes direction very quickly. This is quite demanding for the EET. One of the curves ends in the cut, another is bent and runs along the cut, whereas the third runs out of the first quadrant of the complex plane. For a detailed discussion, see [Novotny, 1994].

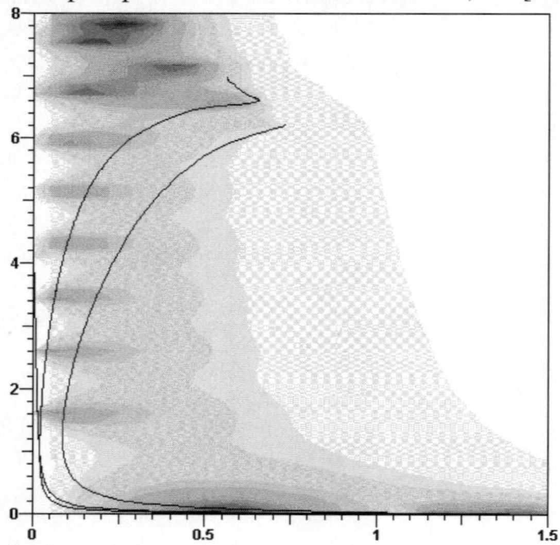

Figure 106: Search function $\eta(\gamma / k_0)$ in the complex plane for an optical fiber with a lossy cladding. Three 'propagating' modes with small losses are represented by minima near the real (horizontal) axis. Several 'evanescent' modes are characterized by the minima near the imaginary (vertical) axis. The black lines indicate the traces for the three 'propagating' modes when the frequency is decreased. EET has been used to compute these lines.

Parameter Estimation Technique (PET)

Obviously, we can extend the EET concept to any set of parameters that are important for the solution of a problem, as long we are interested in a sequence of solutions of the problem rather than in a single solution. The computation of the frequency dependence is a typical example of such a sequence of computations. In this case, the frequency may be considered as a variable. Another variable might be the angle of incidence for scattering problems, or some geometric variable, such as the radius and location of a wire, and so on. In general, each parameter p_k of a set of K parameters may be considered as a function of J variables v_j: $v_j \rightarrow p_k(v_1, v_2 \ldots v_j \ldots v_J)$. Each of these multi-dimensional functions may be analyzed with an appropriate set of basis functions that allow some extrapolation. Once the parameters have been evaluated for several sets of variables, these data may be used to obtain extrapolations for a given set of variables. This is a very general formulation of the PET, which also includes the EET as a special case.

Non-eigenvalue problems are characterized by an inhomogeneous matrix equation of the form MX=B. Here, the unknown vector X is a natural set of parameters that describes the solution. Therefore, it is natural to apply the PET to the parameter vector X. For simplicity, we consider only a single variable, for example, the frequency ω. Consequently, all elements x_k of X are functions of ω. Now, we assume that we have computed X(ω) for a finite number of frequencies $\omega_1, \omega_2 \ldots \omega_{n-1}$ and that we now want to compute $X(\omega_n)$. Now, we can obtain a guess $X^0(\omega_n)$ from the known solutions $X(\omega_1), X(\omega_2) \ldots X(\omega_{n-1})$. The maximum order of this guess is n-2. For this guess, an extrapolation of each component x_k of X is required. Theoretically, one could use different sets of basis functions for the extrapolation of each component, but this would result in too high a degree of freedom. Therefore, one best starts with a single set of basis functions for extrapolating all components, i.e., all parameters. As in the EET, one can use a simple power series basis, more sophisticated sets of basis functions, or even an optimized basis for a certain class of problems. This means that each component $x_k(\omega)$ of $X(\omega)$ is expanded by a series of the form

$$x_k(\omega) = x_k^0(\omega) + error_k(\omega) = \sum_{i=0}^{I} a_{ik} basis_i(\omega) + error_k(\omega) \cdot \tag{8.29}$$

To take advantage of a good guess $X^0(\omega_n)$ of the desired solution $X(\omega_n)$ of the given problem, an iterative matrix solver must be applied to the matrix equation MX=B (see the section on Iterative matrix solvers). Direct matrix solvers are useless, because they do not allow one the prior knowledge, i.e., $X^0(\omega_n)$. Although it is quite evident that the number of iterations required for solving MX=B can be reduced when a good starting vector $X^0(\omega_n)$ is known, the procedure is less trivial than one might expect.

First of all, to compute the first solution $X(\omega_1)$, no starting value is available. Therefore, it is reasonable to evaluate this solution with a direct matrix solver. For the second solution $X(\omega_2)$ one then can only apply a zero order extrapolation, which is not very accurate in general, and so on. For a higher order PET one therefore has a starting phase with successively increased order extrapolation. During the starting phase, the quality of the extrapolation is somehow reduced. Therefore, the steps of the variable (frequency) should be small and may grow during this phase. Obviously, a proper selection of the sequence $\omega_1, \omega_2 \ldots \omega_{n-1}$ is not trivial. When the frequency is the PET variable, one knows that the complexity of the solution increases with the frequency. Therefore, the functions $x_k(\omega)$ are typically smooth at low frequencies, which makes it reasonable to start at the lowest frequency and to increase the frequency afterwards, i.e., to select a sequence of frequencies with $\omega_1 < \omega_2 < \ldots < \omega_n < \ldots < \omega_N$.

Since the field at highest frequencies is most complex, it makes sense to model at the highest frequency first. A model that is sufficiently accurate at the highest frequency will also allow one to obtain sufficiently accurate results at lower frequencies. Therefore, one might prefer to start the PET at the highest frequency and to run downwards. In this case, it may be reasonable to adapt the step size, i.e., to start with relatively small frequency steps and to increase the steps when the extrapolation is so good that the iterative matrix solver stops after the first iteration. When too many iterations are required, one should reduce the step size.

Although very encouraging results have been obtained [Hafner, 1994/2], [Fröhlich, 1997] with the PET, this technique has not been extensively studied because even fast PCs are not sufficiently fast for the demanding optimizations of the PET basis functions. Moreover, MMP works with different types of basis functions. It seems that different sets of PET basis functions are optimal for the extrapolation of MMP parameters that correspond to different MMP basis functions. Note that the MMP basis functions are multipoles, Bessel expansions, harmonic expansions, etc., whereas the PET basis functions in (8.29) are simple functions of a single variable.

Obviously, the PET could be applied not only to MMP, but also to any numerical code that is applied to compute the frequency dependence or some other dependence of the field or of any derived quantity. Incidentally, the time domain procedure in FD and other codes may be considered as a very simple, zero order PET. It is assumed that PET is one of the important keys to efficient handling of sequences of computations that occur in the optimization of devices.

Optimal and automatic multipole setting

The high degree of freedom in the setting of multipole expansion usually causes problems for inexperienced MMP users. Moreover, one sometimes observes that the setting of the multipoles is quite critical, i.e., that small variations of the origins of the multipoles can have a strong impact on the accuracy of the solution. Therefore, some optimal or sub-optimal pole setting strategy seems to be desirable. Early attempts at optimal pole setting for 2D electrostatic problems by Leuchtmann [Leuchtmann, 1987] were discouraging and the situation in electrodynamics is even more complicated. Therefore, no optimal pole setting routines have been implemented in the MMP codes and in the correponding graphical tools [Hafner, 1990/S], [Hafner, 1993/S], [Regli, 1992].

To get some insight into the problems of optimal pole setting, we consider the scattering at a circular PEC cylinder of 1m radius. Because of the rotational symmetry, it is reasonable to model the scattered field by a set of multipoles that are located on a concentric circle with a radius that is smaller than the radius of the cylinder. Moreover, it is reasonable to distribute the multipoles uniformly on that circle and to use equal multipole orders for all multipoles. Therefore, one only has two integer and two real parameters that characterize the MMP expansion. The first integer parameter N is the number of multipoles and the second integer parameter M is the multipole order. The first real parameter r is the radius of the circle. This parameter must be in the interval 0…1m. The second real parameter φ_1 indicates the angle of the first multipole with respect to the x axis. When a polar coordinate system with origin in the center of the cylinder is introduced, the locations of the n multipoles are described by $r_n=r$, $\varphi_n=\varphi_1+2\pi(n-1)/N$. Obviously, one can restrict φ_1 to the interval $0\ldots 2\pi/N$. Let us first set $\varphi_1=0$ and M=0 (only monopole expansions). Thus, we only have a real parameter r and an integer parameter N and we can easily explore what happens when these parameters are varied. Figure 107 shows the average error obtained for a plane wave incident on the PEC cylinder for N=4 and N=8 as functions of r for different frequencies. As one can see, there are very sharp and deep minima near r=0 for sufficiently low frequencies. For r=0, all monopoles are placed at the center of the cylinder and simulate a higher order multipole, i.e., a multipole placed at the center of the cylinder is optimal for this problem. Note that this effect is much less pronounced at higher frequencies, i.e., the minimum is much less sharp and deep.

Now, let us replace the incident plane wave by a monopole source located at x=-1.5m on the negative x axis. Is the multipole expansion at r=0 still optimal? As one can see in Figure 108, one again observes a sharp and deep minimum at low frequencies, but the minimum is far away from r=0. The reason for that is as follows: The analytic solution of a monopole in front of a circular PEC cylinder in electrostatics is obtained with a 'mirror' monopole inside the PEC. The mirror point depends on the location of the source point and of the radius. In the test example, it is near x=-0.65m on the negative x axis. For r=0.65m one of the monopoles happens to be placed at this location. Again, the minimum is much less sharp and deep for higher frequencies. Although a multipole at r=0 is not optimal at all, it still allows one to obtain acceptable results when the order is sufficiently high.

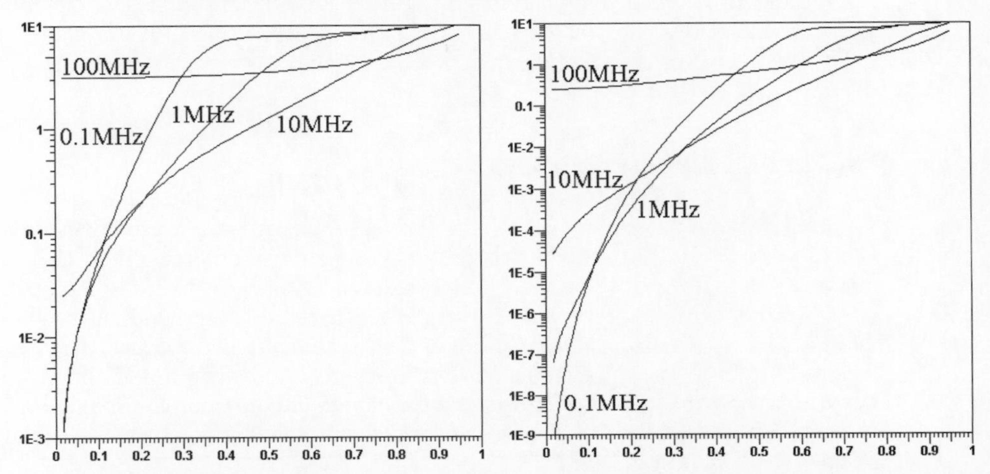

Figure 107: Scattering of a plane wave at a circular PEC cylinder of 1m radius, electric field polarized parallel to the axis of the cylinder. Scattered field modeled with a set of monopoles uniformly distributed on a circle. Average error as a function of the distance of a set of monopoles from the origin. Left hand side: four monopoles on the x and y axes, right hand side: eight monopoles on the x and y axes and on the diagonals.

When we rotate the monopoles around the point r=0, i.e., when we increase φ_1, we observe a similar behavior for the incident plane wave, but a completely different behavior for the monopole excitation. Figure 109 shows that the minima near r=0.65m vanish because no monopole is set near the mirror point when φ_1 is not properly set. Obviously, one now has very flat minima and a multipole expansion at r=0 is as good as a monopole set with r=0.65m.

To increase the accuracy, we can increase either the number N of multipoles or the multipole orders M. Figure 109 illustrates what happens when we increase the orders from M=0 to M=1. The error can be considerably reduced. Again, we observe very flat minima, but the flat area is shorter and the optimal solutions are obtained for r near 0.5m. Moreover, the function is very noisy near r=0. The reason for this effect is that the condition number of the MMP matrix tends to infinity for small r. This indicates that one should not place several multipoles close together. Such multipole clusters should be replaced by a single multipole of higher order.

We only have considered an extremely simple example here, but similar observations may be made in general. Since the degree of freedom of the arbitrary MMP expansions is extremely high, the error function has many variables and cannot be easily illustrated. First of all, sharp minima may be observed sometimes, but only for a specific excitation of a given problem. Searching for such minima does not make much sense because the search procedure is usually much more time consuming than an overdiscretization, i.e., an MMP solution gives more higher order multipoles than are required. Such an overdiscretization allows one to easily obtain extremely accurate results because of the excellent convergence properties.

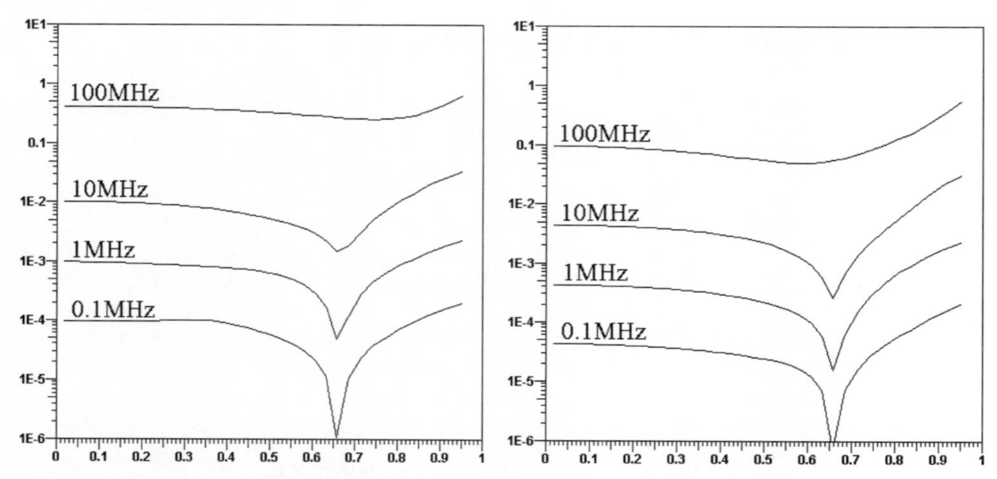

Figure 108: Same as in Figure 107, but with a monopole source at x=-1.5m, y=0.

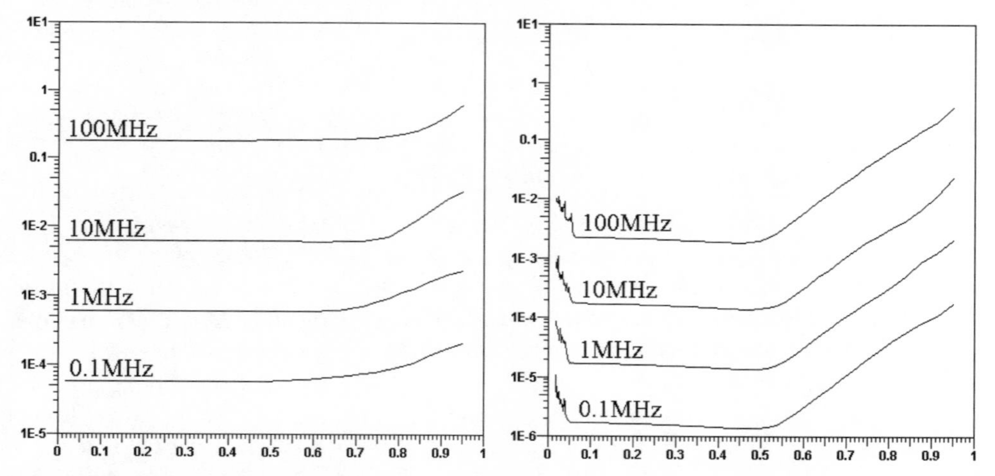

Figure 109: Same as Figure 108 with another set of expansions. Left hand side: the eight monopoles are rotated around the center of the cylinder with an angle of 30°. Right hand side: the monopoles are replaced by first order multipoles (monopole + dipole expansion).

Secondly, an optimal MMP expansion for a specific excitation must not even be close to an optimal one as soon as the excitation is modified. Finally, as soon as a sufficient number N of sufficiently high order multipoles is set in such a way that multipole clusters are avoided, the error functions exhibit large flat areas, which indicates that the multipole setting is robust rather than critical and that a sub-optimal but good multipole distribution can easily be found.

For these reasons, an optimization of the multipole placement makes no sense. Automatic pole setting routines should be based on simple geometric rules [Hafner, 1987], [Hafner, 1990] and constructions rather than on sophisticated and time-consuming techniques.

Error analysis and adaptive multipole setting

Once a MMP solution is obtained, the analysis of the error distribution along the boundary allows one to adaptively improve the multipole expansions. Note that MMP minimizes the square norm of an error function that is defined on the boundary and contains the weighted mismatching of the boundary conditions. The MMP code allows one to visualize this error function. Figure 110 shows a typical error distribution of an MMP model of a planar structure with a rib. The biggest errors are obtained in the area $x<-1$. Assume that this area is not critical. Therefore, it has been discretized relatively roughly and the errors in this area may be tolerated. For $x>0$, the biggest errors are observed near $x=0.5$, $y=0.5$. These errors may be reduced by increasing the orders of the multipoles in this area. Note that the multipole near $x=0.4$, $y=0.4$ is quite close to the boundary. A more balanced multipole expansion is obtained when this multipole is moved a little further away from the boundary. Figure 111 shows that the error in the critical area is considerably reduced when the orders of the two multipoles in this area are increased only by two. As a result, the number of columns of the MMP matrix grows from 307 to 315, which causes an increase of the computation time of about 8%.

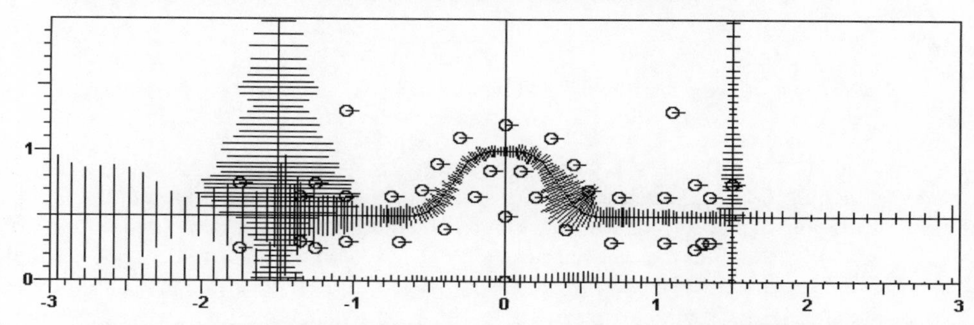

Figure 110: Error of an MMP model of a rib structure. Multipoles and other types of expansions are indicated by circles; the errors are proportional to the lengths of the lines perpendicular to the boundaries.

The error analysis feature of the MMP code and the fact that the multipole expansions have a local behavior makes adaptive multipole setting easy. This allows one to obtain balanced error distributions with maximum errors that are not much higher than the average error. If desired, one can also locally reduce the error in areas where a small error is desired. The ability of a local control over the error distribution is very desirable in complicated models.

Automatic matching point setting

Older MMP versions [Hafner, 1990/S], [Hafner, 1993/S] define a model mainly by a set of MMP expansions and by a set of matching points. This is quite general, but also dangerous, because wrong results may be obtained when the matching point density on some part of the boundary is too low. Setting the matching points is relatively easy when highly overdetermined MMP matrices are used, but it is also a permanent source of mistakes of inexperienced MMP users. Note that the minimal distance between neighbor matching points depends not only on the shape (curvature) of the boundary, but also on the orders and locations of the multipoles that model the field along the boundary. When a high order multipole is set close to a boundary, a high matching point density is required near the multipole. Although the rules for setting the matching points for a given MMP expansion are quite simple, setting the matching points appropriately is tedious and annoying for the user. Moreover, the matching point must be modified as soon as some multipole location and order is modified. Thus, adaptive multipole setting is only easy when the matching points are defined automatically.

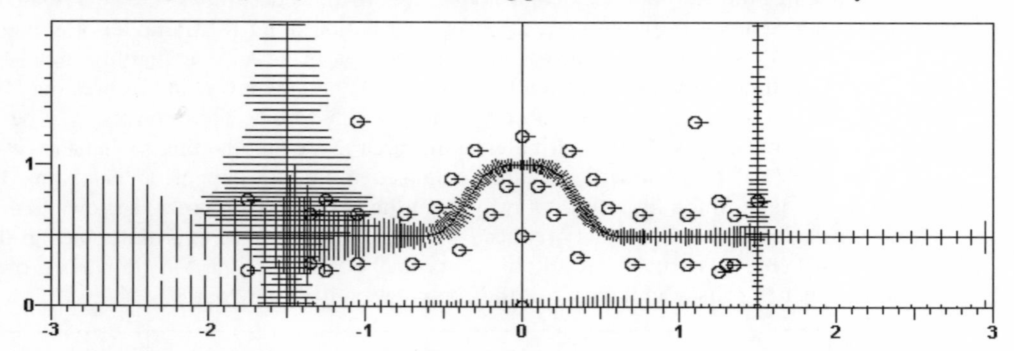

Figure 111: Orders of the two multipoles near the 'critical area' near x=0.5, y=0.5 are increased by 2. The corresponding numbers of parameters are increased from 8 to 12. Moreover, the lower of these multipoles is moved a little further away from the boundary.

The MaX-1 code uses a parametric definition of the boundaries and automatically sets the matching points in such a way that useful results are obtained as soon as a reasonable set of multipole expansions is defined. Figure 112 illustrates the automatic matching point setting for a simple example of a circular boundary. The matching point refinement is rough and simple. This is sufficient when the MMP matrix is slightly overdetermined with an overdetermination factor of ≈ 2. A sophisticated and time-consuming matching point definition is not required.

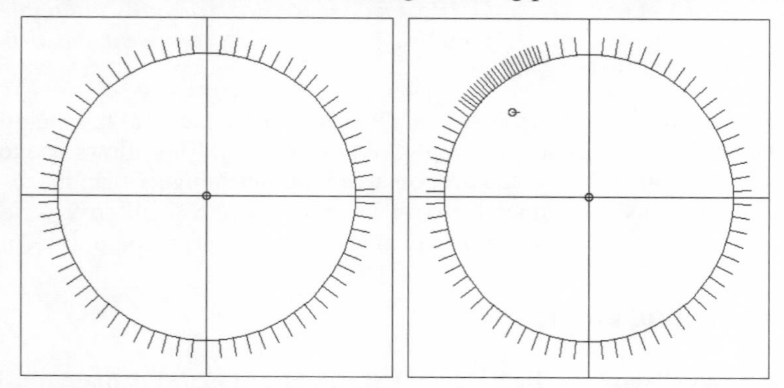

Figure 112: Matching point distribution on a circular boundary for a single multipole in the center (left) and an additional multipole closer to the surface (right). The multipoles are indicated by circles.

9 Applications

General remarks

The aim of this chapter is to outline specific properties of the most prominent types of applications in computational electromagnetics. The details of the modeling of these applications depend on the numerical method that is applied. Such details are not discussed here. Moreover, specialists working in one of those areas define specific terms and methods to analyze the results obtained from computer codes. They prefer function plots of quantities that are usually obtained from the electromagnetic field by integration, for example, the Radar Cross Section (RCS), Specific Absorption Ratio (SAR), reflection coefficients, and so on. These terms are often not well known to beginners and also to specialists working in a different area. Moreover, for getting started, it is more convenient to visualize the field rather than derived quantities. Therefore, mainly field plots are shown in this book. Those who intend to work special applications should read specialized books (see Bibliography for some examples) that define and discuss the corresponding terms.

Classifications of the applications of codes of computational electromagnetics often represent different teaching areas, such as high voltage technique, electrical machines, microwave electronics, electromagnetic compatibility, antenna design, radar technology, and so on. Although typical applications in these areas may be considerably different from the practical point of view, there are often no differences from the computational point of view. For example, for a computer code, there is no essential difference between a power line, a transmission line for data transfer, a micro strip, etc. In this chapter, a classification is used that affects the numerical modeling rather than the practical application.

Most of the numerical methods presented in this book might be applied to analyze all types of applications. The examples presented in the following have been computed with the MMP feature of the MaX-1 code [Hafner, 1998/S]. For a more detailed description of the MMP modeling of all these applications, for animated visualizations of the results, and for many more examples, see the MaX-1 tutorial and the MaX-1 movie description.

Electrostatics

Electrostatic computations are often used to simulate high voltage problems, where the accurate evaluation of the maximum field strength and of electric forces is most important. Especially domain methods tend to underestimate the maximum field strength. Therefore, boundary methods such as the Charge Simulation (CS) method [Singer, 1974] are preferred in high voltage simulations. Another method that is often applied is the Monte Carlo simulation that is not detailed in this book. Note that CS is a special case of MMP.

Electrostatic simulations are also used to compute the capacity per unit length on transmission lines that may be operated at quite high frequencies. The quality of such computation depends on the distance between the conductors with respect to the wavelength. Good results are obtained when the maximum distance is considerably smaller than one wavelength. Note that such computation cannot be used to obtain higher order modes and the even mode that may propagate at arbitrarily low frequency. The even mode is a mode where the currents in all wires flow in the same direction. The return current is a pure displacement current. Even modes are strongly excited when a plane wave or some other wave hits a transmission line. Therefore, even modes are important in EMC (electromagnetic compatibility). In EMC, even modes are usually called common modes. Such modes are computed as indicated in the section on Guided waves.

In electrostatics one usually has two types of materials, electrodes and dielectrics. An electrode consists of conducting material and is free of electric field. Therefore, the electric scalar potential on an electrode is constant. On the surface of an electrode, one has electric surface charges that are usually unknown. Therefore, the continuity condition for the normal component of the D field is usually used to compute the surface charge density. From that, one can also compute the mechanical forces on the electrode.

In practice, one may distinguish two different types of electrodes, fixed potential electrodes and floating electrodes that have a potential that is initially unknown. To model fixed potential electrodes, one usually works with the scalar potential and imposes the given, fixed potential on the electrode, i.e., the boundary condition (5.323). On floating electrodes, the scalar potential is also constant, but unknown. For such electrodes, one has an additional condition, for example, the condition that its total charge is zero. As a consequence, the derivative of the scalar potential in the tangential direction is zero on the surface of a floating electrode, i.e., one obtains from (5.322):

$$\frac{\partial \phi}{\partial t} = 0, \quad \frac{\partial \phi}{\partial \tau} = 0. \tag{9.1}$$

When one is working with the scalar potential in electrostatics, the corresponding continuity conditions (5.322) for the derivatives along a boundary between two dielectrics may be used. Instead of working with the continuity conditions for the derivatives of the scalar potential, one may also work with the continuity conditions for the electric field.

Remember that the simultaneous use of boundary conditions that contain the potential and conditions that contain the derivatives of the potential can cause weighting problems. To overcome such problems, one can model dielectric boundaries, floating electrodes, and also fixed potential electrodes with the continuity conditions of the E and D fields and impose the boundary conditions for the scalar potential of the fixed potential only in a few, highly weighted points.

The scalar potential in a dielectric that does not contain any fixed potential electrode is only defined up to an arbitrary constant. To uniquely determine this constant, one may impose continuity conditions for the potential at some points on the boundaries of such a dielectric. This improves illustrations of the scalar potential, but it has no influence on the electric field. For a more detailed discussion, see [Gnos, 1997].

Matrix methods for electrostatics end up with a matrix equation of the form MX=B. The inhomogeneity B contains the potential values of the fixed potential electrodes or the charges of the floating electrodes. Once the matrix equation is solved, the primary field may be computed. If one works with the scalar potential as primary field, one may use the definition of the scalar potential to compute the electric field. Finally, the capacity may be computed from integrals of the electric field of the form (5.119).

Figure 113 shows an example with two fixed potential electrodes, a floating electrode and two dielectrics. On the fixed potential electrodes, the boundary condition for the potential is imposed. On the floating electrode, the boundary condition for the tangential electric field is imposed and the expansion is defined in such a way that the total charge is zero. In the MMP code, the latter is obtained when no electric charges, i.e., no zero order multipole expansions, are set inside this electrode. On the boundary between the two dielectrics, both the scalar potential and the normal components of the D field are matched.

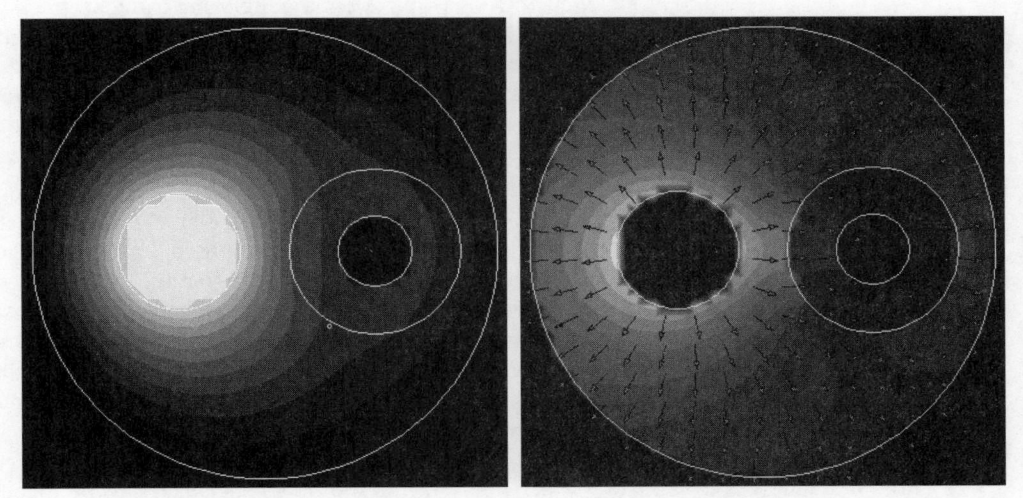

Figure 113: Electrostatic computation of a system of three circular electrodes and two different dielectrics. The outer electrode is on potential 0, the inner electrode to the left is on potential 1, the inner electrode to the right is floating. Its total charge is zero. The floating electrode is coated by a dielectric with relative permittivity 4. The second dielectric fills the space between the left electrode and the outer electrode. Its relative permittivity is 1. Left hand side: scalar potential, right hand side: electric field.

Magnetostatics

The differential equations that describe magnetostatics are very similar to those that describe electrostatics. Therefore, there are many similarities, but there are also some important differences. First of all, one usually works with the magnetostatic vector potential $\vec{A}$ that is more complicated than the electrostatic scalar potential. Note that the vector potential in cylindrical applications has only a non-zero z component. The B field is then the gradient of the z component of the vector potential, rotated by an angle of 90 degrees in the transverse plane. Therefore, the lines of constant A_z coincide with the magnetic field lines. This makes the use of the vector potential attractive. At the same time, the vector potential is not as important as the electric scalar potential in electrostatics, because there is no 'electrode' in magnetostatics that is described by a constant vector potential.

The magnetostatic object that corresponds to the electrode in electrostatics is a conductor that carries electric currents. In electrostatics, the charges are on the surface of the electrodes, whereas the currents in magnetostatics are not necessarily on the surface of the conductors. When one considers the magnetostatic computation of the inductance per wavelength of a transmission line, one has two special cases, the so-called low-frequency and high-frequency limits. In the high-frequency limit, surface currents are good approximations. Therefore, PEC models and the same procedure as in electrostatics may be used. At low frequencies, the current distribution in the conductors is almost homogeneous. Therefore, the current distribution is assumed to be known in the low-frequency computation. When all currents are embedded in a single non-conducting domain, one may apply the Ampère integral (5.6) to directly compute the magnetic field, i.e., no solution of a matrix equation is required. The low-frequency modeling of conductors in the presence of different materials with different magnetic properties is more difficult. Assume that any bounded magnetic domain contains conductors with a known current distribution. Then, the magnetic field in such a domain is a superposition of an unknown field and the field obtained from the Ampère integral over the known current distribution. The unknown field may be expanded by some series and the parameters of this series expansion are computed in such a way that the field equations and boundary conditions hold. This leads to an inhomogeneous matrix equation MX=B, where the inhomogeneity B contains essentially the Ampère integrals.

In transmission line computations one rarely encounters materials with relative permeability different from one, i.e., the high-frequency computation consists of a model with PEC cylinders in free space and the low-frequency solution is directly obtained from the Ampère integral. To illustrate the procedure, we consider an academic example with two circular wires within a circular magnetic material with relative permeability 4 (see Figure 114). The high-frequency computation of the magnetic field is dual to the electrostatic computation. For the low-frequency approximation, the field inside the magnetic material is modeled by a superposition of a series expansion and the solution of the Ampère integrals. For the free space outside the magnetic material, a conventional series expansion is applied.

Once the magnetic field is known, one can proceed as in electrostatics, i.e., one may compute the inductance from (5.120).

Scattering

Electromagnetic scattering contains many different applications. Some of them are treated in separate sections in the following. In general, an arbitrary geometric structure consisting of several different materials is illuminated by a known, incident wave. When matrix methods are applied, this always leads to inhomogeneous equations of the form MX=B, where the inhomogeneity contains the excitation, which is often assumed to be a plane wave.

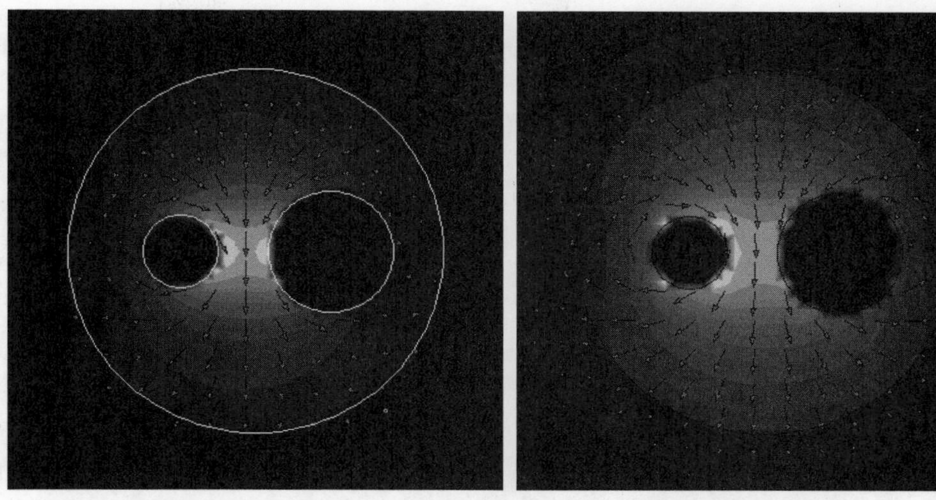

Figure 114: Magnetostatic field of two circular conductors embedded in a circular magnetic material with relative permeability 4. Left hand side: high-frequency computation with a PEC model of the conductors, right hand side: low-frequency computation with uniform current distribution inside the conductors. Note that the field inside the conductors of the low-frequency computation is not zero, but it has not been evaluated in this plot.

One may distinguish two several different types of scattering problems depending on the time-dependence of the incident field (time-harmonic, pulsed, etc.), the material properties (loss-free, lossy, isotropic, anisotropic, and so on), and the distance where the scattered field is to be computed. In radar applications, one is often interested in the far field. This causes some trouble when domain methods are applied which may only discretize a limited portion of the space. Therefore, near-to-far field transformations have to be added to domain methods. Such transformations may be based on multipole expansions or on Green's function formulations. Some implementations of Method of Moment (MoM) codes are specialized in such a way that only the far field may be computed with a sufficient accuracy. In far field computations one usually computes the scattered field only and one reduces the entire field information to a single scalar quantity, such as the Radar Cross Section (RCS) on simple function plots of the radiation pattern. The main reason for this is that the object is illuminated by some antenna placed far away. The antenna field is not a plane wave at all, but near the object it may be approximated by a plane wave. The scattered field is then received by another antenna that is also far away from the object. This antenna mainly detects the scattered far field.

Usually, one is interested in the scattered field produced by plane waves with varying angle of incidence. Therefore, one must solve $MX=B_k$ with several right hand sides B_k, $k=1,2,\ldots,K$. To save computation time, matrix solvers that can handle multiple right hand sides simultaneously are therefore preferred. Instead of such algorithms, one may also apply the Parameter Estimation Technique (PET) when no such matrix solver is available.

In microwave heating, bio-electromagnetics (illumination of organisms by antenna), and other applications the near-field computation is important. This provides no problems to both domain and boundary methods that are not specialized for far-field computations. The illumination of a small dielectric lens is a typical example, where the near field is of interest. Figure 115 shows that an incident plane wave may be focused on an area with a size that is in the same order of magnitude as the wavelength. Moreover, a monopole source is converted into a field behind the lens that is approximately a plane wave.

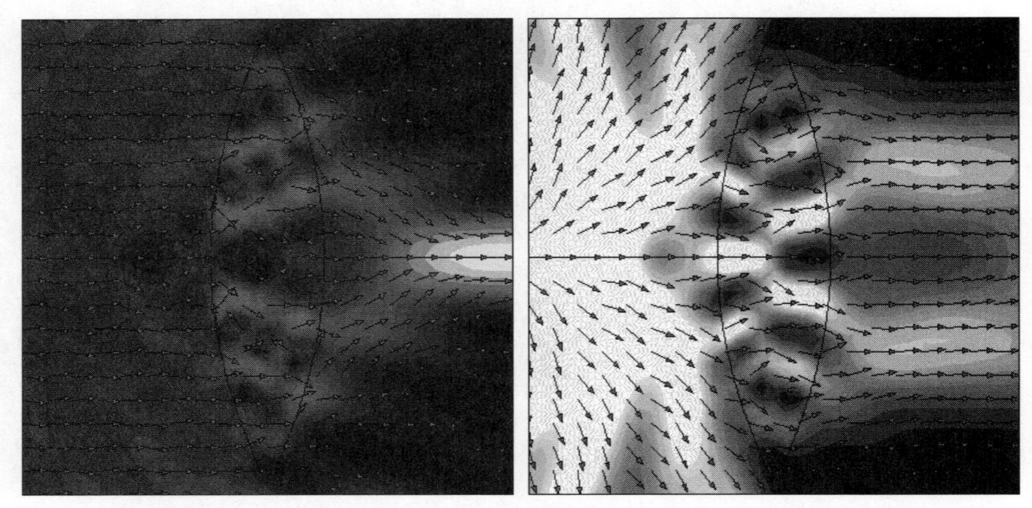

Figure 115: Time average of the Poynting vector field for a wave incident on a cylindrical lens with relative permittivity 4. Side length of the plot area is equal to 4 wavelengths. Incident E wave. Left hand side: incident plane wave in the x direction, right hand side: incident monopole wave with origin 2.2 wavelengths to the left of the center of the lens.

Figure 116 shows that a slot between two PEC cylinders may cause a near field that has some similarity to the field near the focus of a dielectric lens. Since the slot may be small compared with the wavelength, one may achieve a much smaller focus area than with a lens. This is one of the main ideas behind Scanning Nearfield Optical Microscopes (SNOM) [Novotny, 1996]. When a high optical resolution is desired, SNOM may replace traditional optical microscopes based on dielectric lenses.

Gratings

Gratings [Loewen, 1997], [Petit, 1980], [Turunen, 1997] can be considered as a special case of electromagnetic scattering at a periodic, cylindrical structure. This causes difficulties for both domain methods and boundary methods because both the domain and the boundary are infinite. To overcome these difficulties, the periodicity must be taken into account in such a way that only a single period of the grating must be explicitly modeled. This is usually achieved under the assumption that the incident wave is a plane wave that is also periodic but can have a periodicity that differs from the periodicity of the grating. For more detailed information on the modeling of gratings, see the subsections on Periodic problems.

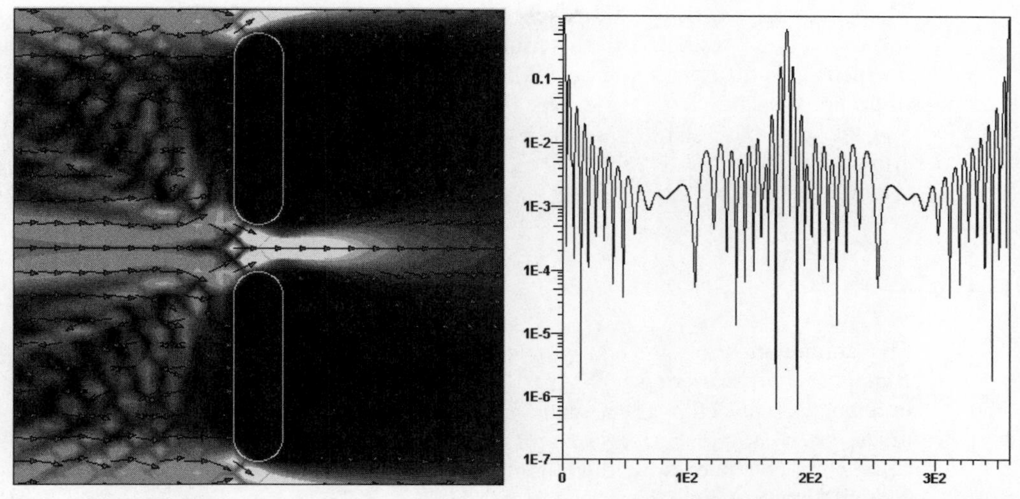

Figure 116: Plane E wave incident in the x direction on a structure consisting of two PEC cylinders. Left hand side: Time average of the Poynting vector field near the structure in a cross section (xy plane) of the cylinders. Right hand side: radiation pattern of the scattered field (time average of the radial component of the Poynting vector field far away from the structure as a function of the radiation angle – the maximum is normalized to 1).

Since gratings extend to infinity, such structures separate also the far field into two regions, one with reflected waves and one with transmitted waves. The reflected and the transmitted waves are pure plane waves propagating in several directions that are obtained from the periodicity of the grating and from the x component of the wave vector of the incident plane wave, where x denotes the direction of the grating. This means that the radiation pattern of a grating is not a continuous function of the radiation angle as for standard scattering problems. One now has a discrete spectrum with Dirac peaks for a finite number of angles. Therefore, the radiation pattern is replaced by the amplitudes and the directions of the peaks, i.e., plane waves. Note that there is an infinite number of evanescent plane waves that do not radiate and a finite number of radiating plane waves. The number of radiating plane waves depends on the periodicity with respect to the wavelength and on the angle of incidence. For sufficiently low frequencies, i.e., long wavelengths, only one reflected and one transmitted plane wave may be propagating. In special cases of non-transparent gratings, the transmitted waves may be missing.

As in the far-field scattering problems, the experimental setup for grating measurements is different from the theoretical model [Loewen, 1997]. In practice, the plane wave is an approximation of a laser beam or of an antenna field and the grating is finite. From this, one immediately can recognize that one will measure a continuous radiation spectrum as in standard scattering setups rather than a discrete spectrum with peaks. When the grating is large enough, the discrete spectrum is a good approximation and carries the most important information of practical interest, with one exception. For the practical use of gratings, it is important to know how sharp the peaks are. Note that not only the size of the grating, but also the manufacturing tolerances may reduce the sharpness of these peaks. The numerical modeling of finite, non-periodic gratings leads to a standard scattering computation.

There are many codes that are specialized to the simulation of gratings [Petit, 1980]. Such codes are often restricted to the computation of special classes of gratings without arbitrary geometry and material properties. When standard scattering codes are used for the computation of gratings, such severe restrictions are usually missing. For such codes the material properties and the geometric shape play no essential role. Figure 117 shows two completely different types of gratings. Both of them were computed with the MMP feature of the MaX-1 code [Hafner, 1998/S].

Antenna

The computation of antenna problems [King, 1956], [Lee, 1984] is very similar to the computation of scattering problems. The only difference is that the incident wave in scattering problems is replaced by the antenna feed, which may be a waveguide mode, a transmission line mode, or a simple current source. As in scattering problems, an inhomogeneous matrix equation is obtained. Now, the inhomogeneity contains the feed. Figure 118 shows an antenna computation that includes a scatterer in the near field of the antenna. The position of the scatterer may heavily affect the antenna pattern in the far field and causes reflections in the feed.

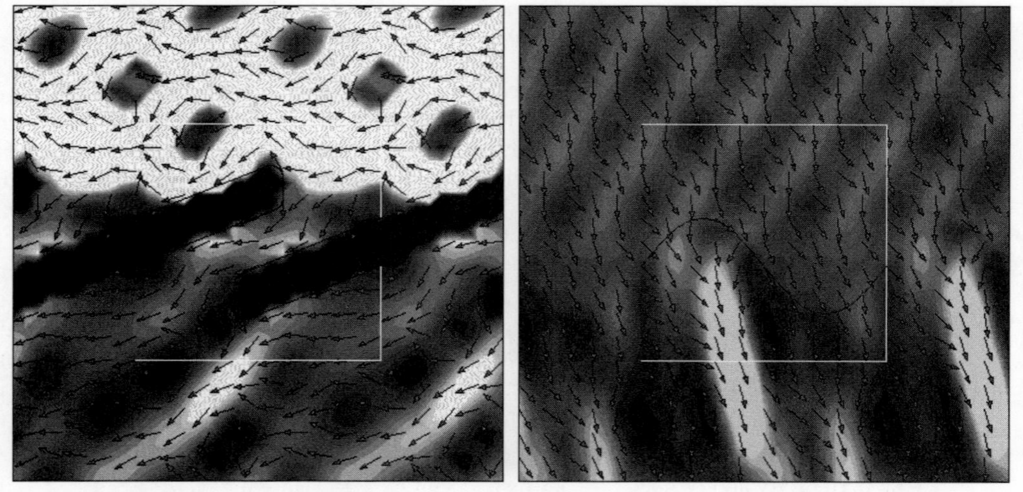

Figure 117: Time average of the Poynting vector field for two different gratings. The side lengths of both plots are equal to four wavelengths in free space. Left hand side: grating consisting of PEC lamella, H wave illumination in the –y direction. Right hand side: sinusoidal dielectric grating, relative permittivity of the dielectric is 4, H wave incident, angle of the incident wave vector –60° with respect to the x axis. Black lines indicate the physical boundaries of the gratings, whereas white lines indicate fictitious boundaries that separate a single cell of the grating. Only the field inside this cell is modeled explicitly.

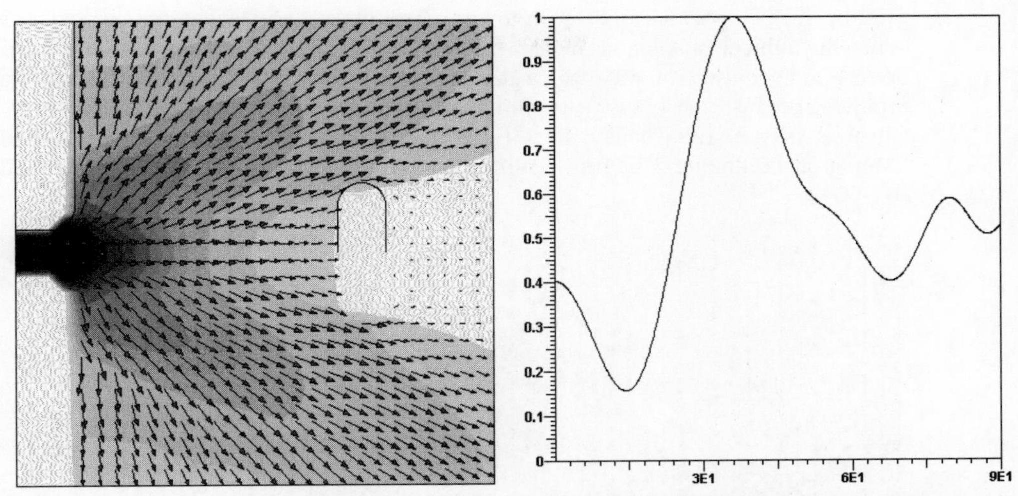

Figure 118: Slot antenna illuminating a PEC cylinder. H wave computation. Left hand side: Time average of the Poynting vector field. Side length of the plot area is four wavelengths. Dark indicates high field strength. Right hand side: radiation pattern of the entire structure, maximum normalized to one.

Note that many antenna design codes are restricted to certain classes of antenna, for example, wire antennas. Such codes provide many user-friendly features and allow one to obtain important antenna properties such as the radiation pattern, the impedance of the antenna, and so on. The influence of obstacles near the antenna on the impedance and on the radiation pattern is often not studied with these codes, because explicit modeling of the feed structure is omitted.

Antenna arrays

The step from standard scattering problems to gratings is the same as the step from usual antenna to periodic antenna arrays. The periodicity may be taken into account exactly as for gratings and the radiation pattern of such antenna arrays has the same properties as the scattered far field of gratings, which consists of a couple of plane waves traveling in different directions. The directions of the plane waves generated by periodic antenna arrays depend on the periodicity and on the phase shifts between neighbor antenna feeds.

When the periodicity is sufficiently short, one only obtains a single radiating plane wave. The direction of this plane wave may then be tuned with the phase shift between the antenna feeds, i.e., one may electronically modify the antenna pattern. Figure 119 illustrates this.

Note that more than a single plane wave are radiating at higher frequencies. The phase shift is a single number that affects all radiating waves simultaneously, i.e., one cannot tune each plane wave separately. Of course one has many more tuning parameters when one allows arbitrary phase shifts rather than a constant phase shift between neighbor antenna, but this destroys the periodic symmetry. As a result, one obtains a superposition of continuous radiation patterns that results again in a continuous pattern rather than in a pattern with discrete peaks that correspond to pure plane waves. This is not really a drawback because periodic antenna arrays must be finite in practice and for finite arrays one always obtains continuous radiation patterns. When the number of antennas in such a finite array is big, one obtains many tuning parameters. Setting them in such a way that a desired pattern is obtained is a demanding optimization problem, because the brute force computation of a radiation pattern of an array with many antennas is very time consuming. Therefore, additional techniques such as the Parameter Estimation Technique (PET) and features like the connections in the MMP code are desirable.

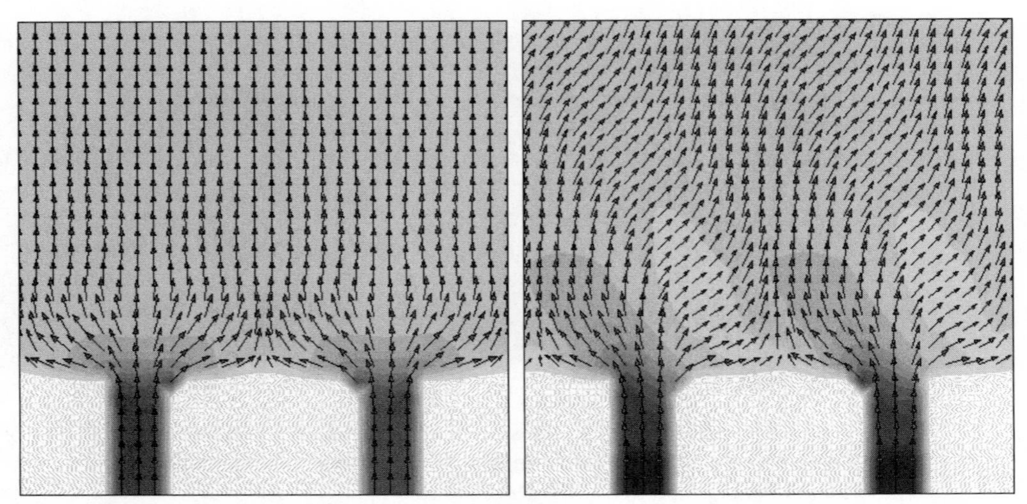

Figure 119: Time average of the Poynting vector field for an array of slot antennas producing H waves. Left hand side: no phase shift between the feeds, right hand side: phase shift between the feeds. The distance between the antenna feeds is equal to two-thirds of a wavelength.

Resonators

In practice, cavities or resonators [Ramo, 1965] are structures that are almost energetically closed. Such structures are often analyzed with an energetically completely closed model. As a consequence, there is no incident wave as in scattering problems and no feed as in antenna problems. One simply assumes that energy is in the resonator and has always been there. Thus, the right hand side of the resulting matrix equation is zero, i.e., the matrix equation is homogeneous and has the form MX=0. Such an equation has always trivial solutions X=0 that are of no interest because X=0 means that no field is in the resonator. Non-trivial solutions are only obtained at the resonance frequencies that play the role of eigenvalues. In the section on the Eigenvalue solver of the MMP code, methods for finding eigenvalues were outlined. These methods are similar for most matrix methods, at least for those working in the frequency domain.

Since one is searching for frequencies, working in the frequency domain is most natural, but it should be mentioned that one can also apply time domain methods. For such methods one often prefers a model of the resonator with an input port and an output port that is also used for measurements in practice. From the theoretical point of view, such open models belong to the category of waveguide discontinuities that are outlined below. Of course, one can also work with open resonator models when a frequency domain code is applied. The advantage of this procedure is that the model is closer to the measurement and this allows one to compute the frequency dependence that will be measured rather than the resonance frequencies that correspond to the peaks in the frequency dependence.

Note that the situation in the computation of resonators is very similar to the computation of gratings and periodic antenna arrays. In all cases, a continuous function that may be measured is idealized by a sequence of Dirac peaks that are obtained from the idealized models.

A resonator is energetically not completely closed when some ports are inserted that allow the resonator to exchange energy with the neighborhood. When no input port is present, where the resonator receives energy, the resonator continuously loses energy, i.e., one obtains damped oscillations that may be described by a complex frequency, where the imaginary part of the frequency is the damping constant. One may distinguish two types of output ports of a resonator. The first type is some opening in the resonator that causes radiation losses. When a waveguide is connected to an opening, the 'radiation' may be some waveguide mode. The second type of port converts electromagnetic energy inside the resonator into another type of energy. The most prominent example is the Ohmic losses, i.e., conversion into thermal energy. For the resonator, all kind of output ports have the same effect. They increase the damping constant. From the analysis of the complex frequency one cannot distinguish between the different types of losses. To obtain Ohmic losses, one can integrate the power loss density that is obtained from the electromagnetic field and the material properties, whereas the radiation losses are obtained from integrals of the time average of the Poynting vector field over the openings in the resonator, i.e., the power flux through the openings. Thus, the electromagnetic field must be analyzed when one is interested in the design of high quality resonators with low losses.

There are two different types of resonators depending on the materials and effects that are used to enclose the energy. For this purpose, essentially a total reflection at a wall is required. This can be achieved either with perfectly conducting walls or with boundaries between materials with different refractive index. Perfectly conducting walls allow one to enclose the energy in a finite volume, whereas the field of a resonator that uses the total reflection at a boundary between materials with different refractive index extends to infinity. Obviously, the computation of resonators of the second type is more demanding and especially causes problems for domain methods. Such resonators are often used in lasers. Note that perfect conductors do not exist. Realistic conductors cause losses. Sufficiently good conductors are only available at sufficiently low frequencies. At sufficiently high frequencies, all conductors become very lossy or even transparent. This is the main reason why lasers are built with resonators of the second type [Erni, 1996]. For masers, i.e., the microwave equivalent of lasers, one prefers metallic cavities.

A problem that often occurs in the numeric evaluation of eigenvalue problems is the detection of so-called spurious modes or non-physical modes. Whether or not such modes occur depends on the numerical method. Usually, it is not difficult to detect such modes, but this may be time-consuming. Therefore, methods that do not exhibit such modes or methods that allow one to suppress them are desirable.

Figure 120 shows a simple example of a loss-free rectangular cavity with PEC walls. Note that the visualization of such 3D fields is quite demanding. In this figure, the electric and magnetic fields are shown on three of the six walls and on two of the three symmetry planes. Perspective projection is used and the direction of the field vectors is indicated by triangles. The length of the baselines of these triangles is proportional to the field component that points to the viewer. Note that the electric field of the mode that is shown is purely vertical. When parallel projection is applied, this field would be represented by triangles with baselines of zero length, i.e., straight lines. The non-zero baselines stem from the perspective projection.

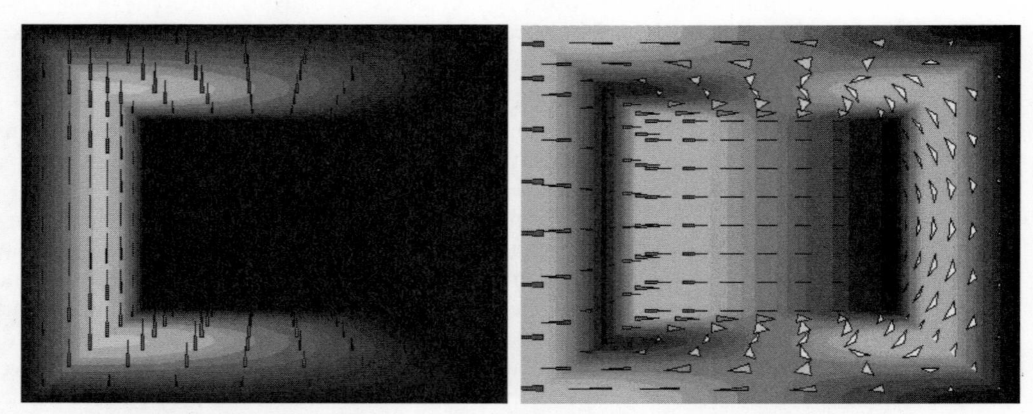

Figure 120: Electric field (left) and magnetic field (right) of an E_{011} mode in a loss-free rectangular cavity. The plane to the left and the bottom planes are symmetry planes; all other planes shown in these plots are PEC planes. The little triangles indicate the direction of the field.

Guided waves

Guided waves [Collin, 1991] that propagate along arbitrary cylindrical structures may be decomposed into a propagating part (in the z direction) and a transverse resonance problem. The z,t dependence of the field is described by the function $\exp(i\gamma z - i\omega t)$. Note that slightly different formulations are often used, for example, $\exp(i\gamma z + i\omega t)$ or $\exp(\gamma z + i\omega t)$. With the notation $\exp(i\gamma z - i\omega t)$, the real part of the propagation constant γ describes the propagation of waves in the z direction and the imaginary part is the attenuation constant. Waves traveling in the $+z$ direction are characterized by a propagation constant with positive real and positive imaginary parts. Obviously one has a harmonic time dependence. Therefore, it is natural to work in the time domain, exactly as in the computation of resonators. Since the transverse resonance problem is the core of the computation of guided waves, one essentially has a 2D resonance problem that can be solved exactly as any other resonance problem. The main difference from the resonators considered in the previous section is that the frequency is now given by the generator that excites the guided modes. The generator is assumed to be at $z=-\infty$ and is not modeled explicitly. The propagation constant γ now plays the role of the eigenvalue to be searched. The remaining procedure is the same as for standard resonator problems.

Since one often uses waveguides at different frequencies, ω is a modeling parameter rather than a constant. When one is interested in the frequency dependence of the propagation constant, one has to compute $\gamma(\omega_k)$ for a sequence of frequencies ω_k. To reduce the computation time, it is recommended to apply methods such as the Eigenvalue Estimation Technique (EET).

As for resonators, one can distinguish two different types of waveguides depending on the effect that causes the total reflection required in the transverse resonance problems. Waveguides with an outer PEC boundary as well as shielded micro strips belong to the first type and are often used in microwave technology [Booton, 1992], whereas unshielded transmission lines and dielectric waveguides (optical fibers [Haus, 1984], [Kapany, 1972], [Marcuse, 1991]) belong to the second type with a field that extends to infinity.

Figure 121 shows the electric field of two different modes on an optical fiber with aluminum cladding. Such fibers are used in Scanning Nearfield Optical Microscopes. The cladding is used to suppress the field of the fiber outside the core. As one can see, aluminum is not a good conductor at optical frequencies. Therefore, the field in the cladding is not zero and must be modeled explicitly. The electric field in the cladding causes high Ohmic losses that cause a high damping constant, i.e., a big imaginary part of the propagation constant. For a more detailed analysis of cladded optical fibers and the frequency dependence of such modes, see [Novotny, 1994].

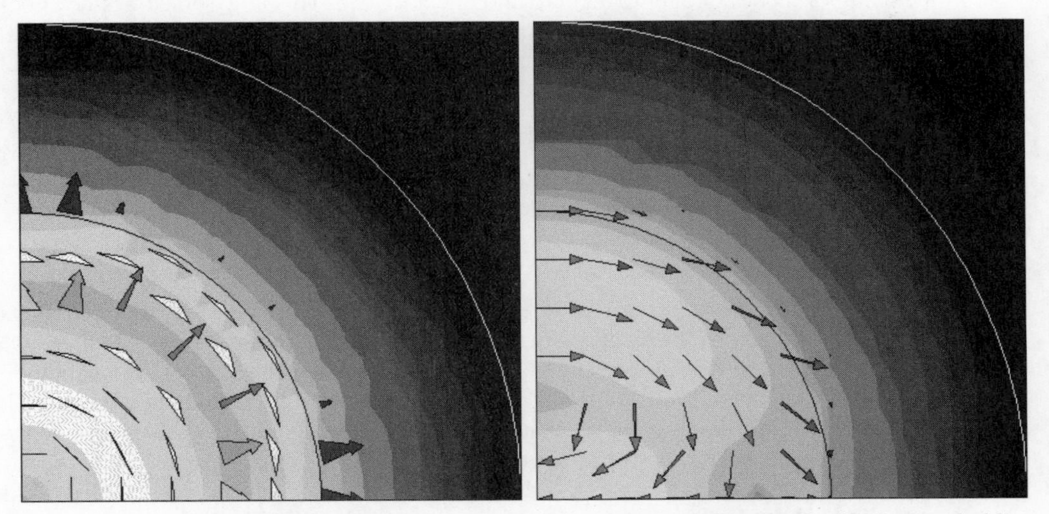

Figure 121: Electric field distribution of modes on an optical fiber with aluminum cladding used for SNOM. Only the first quadrant is shown because of the symmetry. The baselines of the arrows indicate the strength of the longitudinal components. The intensity of the field is indicated by a logarithmic gray scale. Right hand side: rotationally symmetric TM core mode, left hand side: higher order HE core mode.

Most of the modes on a waveguide propagate only for frequencies that are higher than the cutoff frequency. A proper definition of the cutoff frequency is only possible when no losses are present. Below the cutoff frequency, the modes are evanescent, i.e., the propagation constant is purely imaginary. Near the cutoff frequency, the eigenvalue search procedures may exhibit problems. Special techniques to determine the cutoff frequency are therefore sometimes derived. The cutoff condition depends on the type of the waveguide. For example, for a waveguide of the first type with an exterior PEC boundary, cutoff is reached when the propagation constant is zero. Optical fibers and other waveguides of the second type reach cutoff when the propagation constant is equal to the wave number of the medium with the lowest refractive index.

To search the cutoff frequency numerically, one can use the following procedure. First, the propagation constant is set according to the cutoff condition. Then, the frequency plays the role of the eigenvalue and one searches for the frequency exactly as for standard resonators. For more information on the search strategies that are useful for guided waves and for additional examples, see the tutorial of the MaX-1 code.

Coupling

Waveguide modes with a field that extends to infinity are disturbed by any object outside the waveguide. Usually, the exterior field decays exponentially with the distance from the waveguide. Therefore, strong effects are only observed when an object is close to the outer boundary of such a waveguide. When two waveguides are close together, one observes coupling of modes of one waveguide into the other one. This may sometimes be undesired and sometimes desirable. In practice, the coupling area where two waveguides are parallel to each other has a finite length. The entire structure has therefore no cylindrical symmetry and would have to be modeled with a 3D model which would be extremely demanding for all kinds of numerical methods. In principle, such a coupler of waveguide modes can be modeled as a waveguide discontinuity as outlined in the next section. When the coupling area is sufficiently long, one can obtain useful information from an idealized, cylindrical model of the coupling area. This is the main idea of coupled mode theory [Haus, 1984], [Marcuse, 1991], [März, 1995]. Since this theory is quite difficult it is not discussed here. Note that it always leads to more or less approximate descriptions of realistic waveguide systems.

Assume that we have several parallel dielectric waveguides and that the modes that would propagate in each of the waveguides in the absence of the others are known. Obviously, thinking of the waveguides separately is not correct when the waveguides are close to each other. With an appropriate code that can analyze arbitrary waveguide systems, one is free to consider each of the waveguides separately or the system of all waveguides as a new entity, as a new waveguide with its own modes. These modes are called super modes. When one feeds one of the waveguides at $z=0$ with a mode it would have in the absence of the others, one actually feeds simultaneously several of the super modes. When all of the excited super modes propagate with the same propagation constant in the z direction, one would observe a single field pattern in all cross sections. Usually, the propagation constants of the super modes are different. Therefore, the field pattern of the superposition of the super modes changes along the z axis. When the field pattern at $z=a$ is decomposed into modes of the separated waveguides, one obtains not only the initial mode that was excited at $z=0$, but also other modes on all waveguides, i.e., the initial mode is coupled with the other modes. From this, it is quite evident that the differences between the propagation constants of the modes on each of the waveguides and the propagation constants of the super modes are very important for obtaining information on the coupling.

There are different ways to compute the coupling coefficients between different modes on different waveguides in a system of two or more waveguides. Often, approximate solutions are used that work without the explicit computation of the super modes. In these solutions, integrals over the field of the modes are involved. Instead of this, one can directly obtain the coupling coefficients from the propagation constants of the modes on all waveguides and of the super modes of the entire system.

To illustrate the procedure, a system of two circular dielectric waveguides (fibers) with different diameters (0.14mm and 0.1mm) is shown in Figure 122. The relative permittivity of both dielectrics is 4, the exterior is free space. The fundamental mode of such waveguides is the HE_{11} mode. Its electric field at 1000GHz is shown in Figure 123 for both fibers. As one can see, most of the field is inside the bigger fiber, whereas it is much weaker inside the smaller fiber at this frequency. This can also be recognized when one considers the propagation constants. For the mode of the bigger fiber one has γ / k_0=1.494, whereas the mode of the smaller fiber is characterized by γ / k_0=1.165. For γ / k_0=1, the field would propagate entirely outside and for γ / k_0=2 (2 is the square root of the relative permittivity of the dielectric) it would propagate entirely inside the fiber.

In the system of two fibers one has four HE_{11} super modes. Two of them are mainly polarized in the y direction. The other ones are polarized in the x direction. These modes are shown in Figure 122.

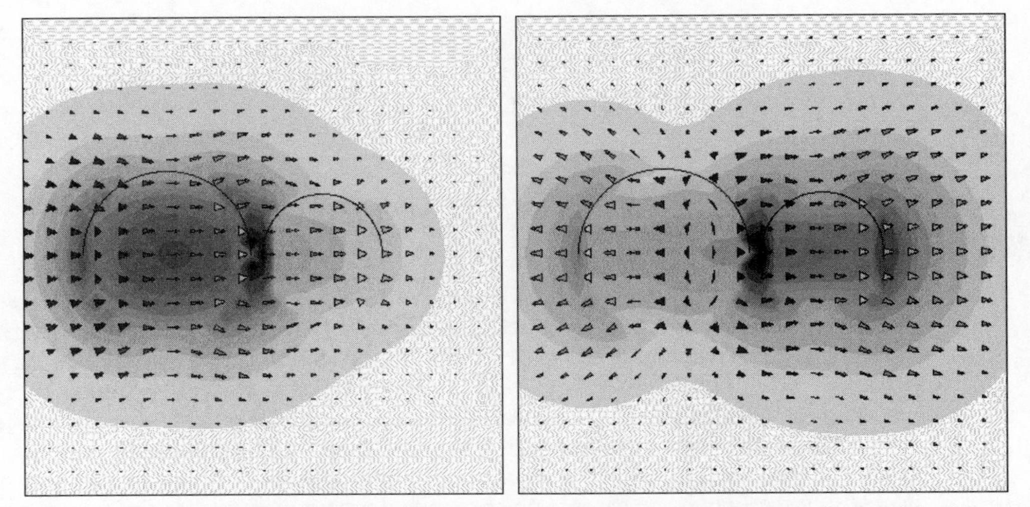

Figure 122: Electric field of HE_{11} super modes of a system of two optical fibers with different diameters, polarized in the x direction. Left hand side: even mode, right hand side: odd mode. Dark areas indicate high field strengths. The size of the baselines of the arrows indicates the strength of the longitudinal components.

The corresponding normalized propagation constants are γ / k_0=1.530 and γ / k_0=1.220. As one can see, the field of the even HE_{11} mode of the system is similar to the field of the HE_{11} mode of the bigger fiber and its propagation constant is also closer to the propagation constant of the bigger fiber. Obviously, the odd mode has also a field inside the smaller fiber, although not as strong as in the bigger one. From this, one can expect that the odd mode couples primarily with the mode in the bigger fiber. Even if the field were not known, one could expect this from the fact that the propagation constant of the odd mode is close to the propagation constant of the mode in the bigger fiber, i.e., one can obtain the coupling constants of the modes either from the analysis of the field or from the propagation constants. Similar statements may be made for the odd mode. For more information on fiber coupling, see [Marcuse, 1991].

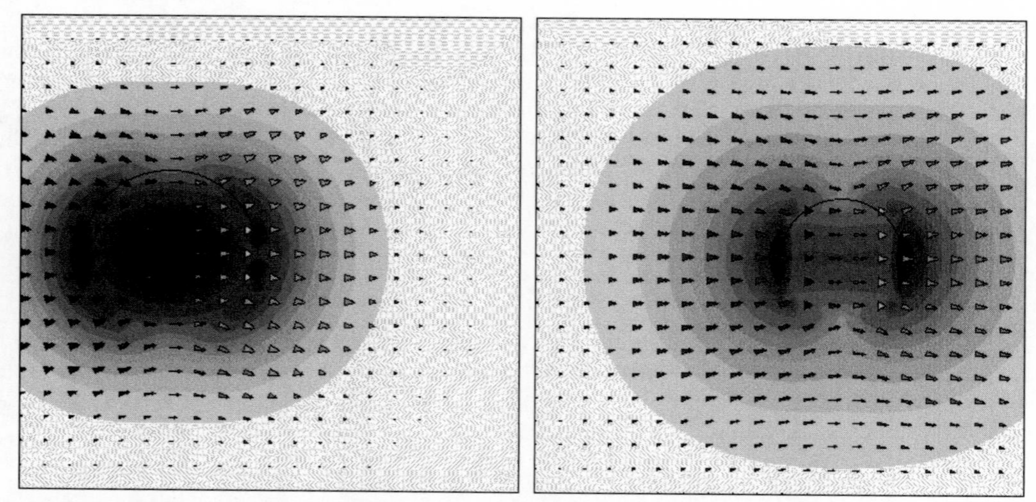

Figure 123: Electric field of HE_{11} modes of the two optical fibers shown in Figure 122, polarized in the x direction. Dark areas indicate high field strengths. The size of the baselines of the arrows indicates the strength of the longitudinal components.

Waveguide discontinuities

A waveguide discontinuity is obtained as soon as the cylindrical symmetry of a waveguide is violated. Therefore, many different problems can be considered as waveguide discontinuities [Rozzi, 1997], [Tamir, 1988]. Figure 124 illustrates a typical waveguide discontinuity with an input port and N output ports. For a complete analysis of the properties of the discontinuity, each of the output ports should once play the role of the input port. In each port one may have several propagating modes. The number M_n of propagating modes in the port number n depends on the frequency, but it is always finite. In the input port, each of the M_1 propagating modes may be considered as the excitation. Therefore, one has

$$K = \sum_{n=1}^{N} M_n \tag{9.2}$$

different excitations and for each of the excitations, one has K outgoing guided waves. The information on the amplitudes of all outgoing waves for each of the possible incoming waves (with amplitude 1) may be stored in a K by K scattering matrix. Note that one also has infinitely many evanescent modes in each port, but the field of these modes decays exponentially. Therefore, these modes are of little practical interest.

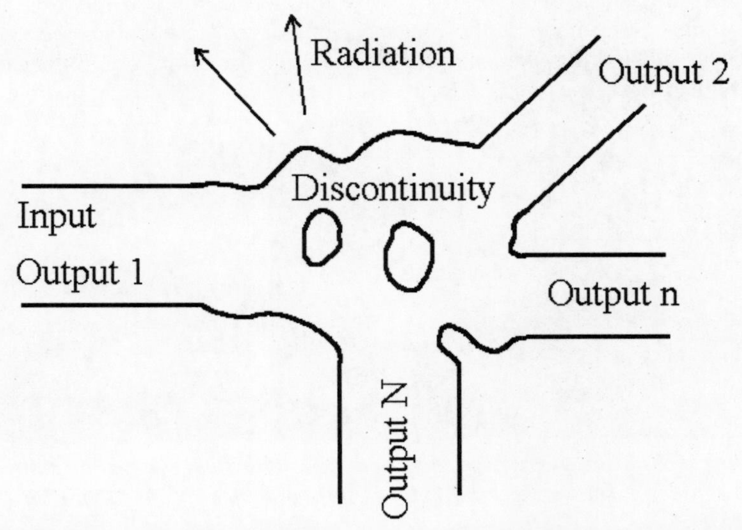

Figure 124: Schematic representation of a waveguide discontinuity with an input port and N output ports. The input port is also an output port. In each output port one may have a finite number of outgoing guided waves and infinitely many outgoing evanescent waves. The radiated waves are missing when the discontinuity has a PEC boundary.

As for the waveguides themselves, one can distinguish two different types of discontinuities: 1) discontinuities with an outer PEC boundary and 2) discontinuous with a boundary that has transparent parts which allow radiation. The first type is much simpler because the radiating modes are characterized by a continuous spectrum that can cause considerable problems.

To model waveguide discontinuities one can introduce fictitious boundaries for separating the discontinuity area and the waveguide ports. In each of the waveguides, one computes all guided waves and the lowest order evanescent waves. These waves are than connected to the discontinuity model, which is essentially handled exactly as for standard scattering problems. The details of this procedure depend on the numerical method that is applied. For a description of the MMP procedure, see the tutorial of the MaX-1 code [Hafner, 1998/S].

Figure 125 illustrates two discontinuities in a rectangular waveguide, one with two and one with three ports. These discontinuities are of the first type. Moreover, the structure is assumed to be loss-free. Therefore, the total incoming power must be equal to the total outgoing power because of the energy conservation law. As one can see, this condition is very well met.

An example of a discontinuity of the second type is shown in Figure 126. This structure is used to couple energy out of a planar waveguide. As one can see, the radiated field has a radiation pattern with several beams. The meandering pattern in the output port to the right hand side indicates that a higher mode is excited in this port.

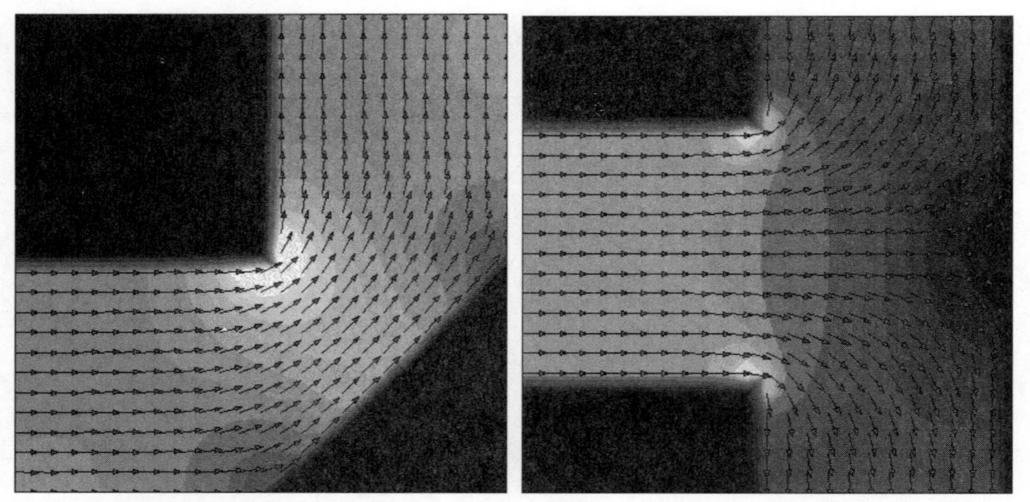

Figure 125: Time average of the Poynting vector field for two discontinuities in a rectangular waveguide of 1m width, 1.4m height (perpendicular to the plane) for an incident E_{01} wave at 140MHz.

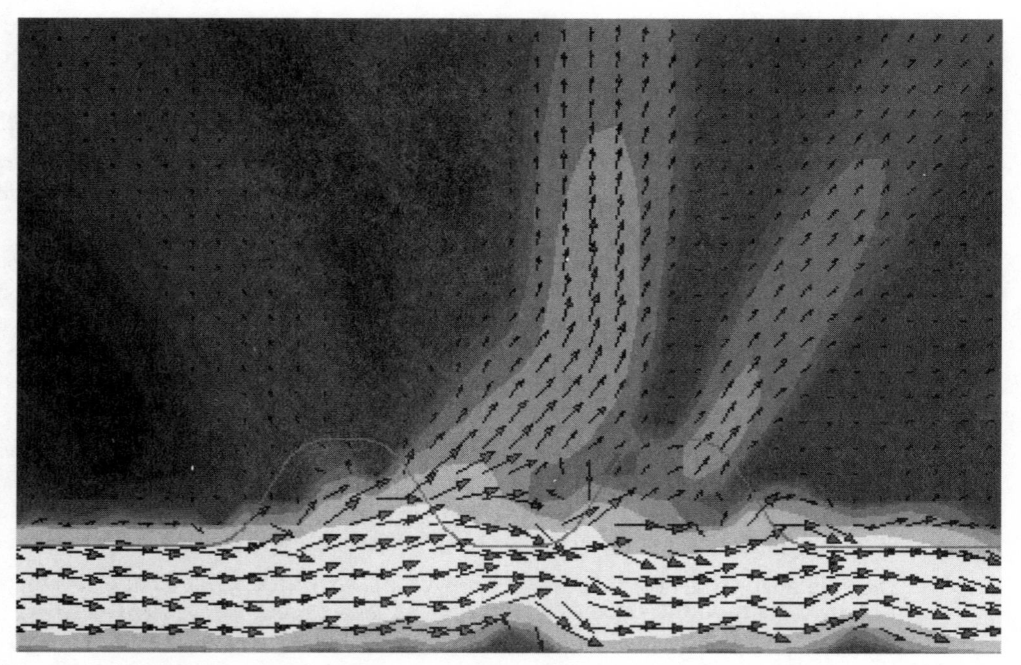

Figure 126: Time average of the Poynting vector field for a discontinuity consisting of two ribs in a planar waveguide structure. The discontinuity has two ports and two propagating modes exist in each port. The lower order mode is incident from the left hand side.

Concluding remarks

Many promising techniques that allow one to improve existing codes or to develop new codes for computational electromagnetics have been presented in this book. The MaX-1 code is a first implementation that benefits several of these techniques. Various examples in this chapter illustrate the power and capability of MaX-1. It should be mentioned that the biggest portion of MaX-1 has been implemented on a personal computer by the author within one year, i.e., the total time for the development of MaX-1 was considerably less than one man-year. Therefore, one may expect that much more powerful codes can be implemented in the near future even by relatively small groups. It is hoped that this book is an inspiring source for all who try to develop more intelligent software for electromagnetics or other areas of physics.

Some of the ideas that were presented are too time- and memory-consuming for personal computers or even for big computers. This especially holds for intelligent optimization techniques, for example, automatically optimized FD schemes. Therefore, not all of the proposed ideas have been tested in practice. Today, such techniques are neither described in traditional textbooks on computational electromagnetics nor used in commercial codes. It is expected that the rapidly growing power of computers will favor these techniques in the next millennium. Therefore, students who are starting a scientific career should get in touch with such unconventional but promising concepts.

Computational electromagnetics is a highly complex task for code designers. The usefulness of a code depends on many details of the implementation rather than on a 'big idea'. Students often prefer 'simple methods', i.e., simple main concepts that require neither much experience nor much physical knowledge. Although this often leads to relatively short codes with a simple structure, the analysis of these codes quickly leads to tricky problems that cannot be solved easily. Therefore, the improvement of such codes often turns out to be much more difficult than expected. When one observes the evolution of codes for computational electromagnetics, one obtains the impression that the complexity of such codes is rapidly growing. Growing complexity is typical of all evolutionary systems. High complexity causes enormous problems for detailed mathematical analysis or may even prevent one completely from undertaking such an analysis. As a consequence, complex codes rarely have a strong mathematical background. In practice, one may observe that such a code works very well and outputs accurate results, but it may happen that nobody really knows why and under what conditions it works. Designing complex codes can therefore be very different from conventional design. This may lead to some kind of 'experimental mathematics', where fuzzy theorems are found from experience. Maybe, these theorems will be proven some day by a mathematician, but code designers may use them without any proof. A typical example of such a theorem is the observation that matrix methods with high condition numbers have a higher probability of accurate solution.

When the main concept of a code ignores the physical aspects of the problem to be solved, one typically runs into physical problems. Also we essentially want to solve physical problems numerically; we cannot entirely replace physical knowledge and experience by computer programs. No doubt, physical knowledge is important for code designers. Many people hope that this does not have to hold for the users. They would like to have codes that can be run by a secretary or some person who does not have any physical knowledge. Although it is possible to design such codes, one should keep in mind that computational electromagnetics is very demanding also for modern computers. Codes that allow the user to bring in his knowledge may benefit from this knowledge and this may result in a much shorter computation time.

The desired features of a computer code depend very much on the application and on the reason why the code is used. Maybe, a computer simulation of an existing device is desired by the sales department for advertising. In this case, the results must be presented with nice graphics, but the accuracy of the results plays almost no role. Although the sales aspect is very important for a company, it is not the primary goal of scientists. Scientists prefer computer codes that help them improve their knowledge. Such codes must be reliable and should provide some feature to estimate the errors of the result. In this case, graphics are useful, but not most important. When the computer simulation is applied to optimize the design of a new device, accuracy is very important, while graphic output is not necessary. In this case, the computational electromagnetics code is linked with an optimization tool. This makes high automation and robustness of the code desirable. The development of such codes is obviously much more demanding than the development of codes for the sales department. Today, even expensive commercial codes do not meet all of the requirements for optimized device design, but it can be expected that such codes will soon become available and will have a profound influence on the design process as soon as they outperform experienced human designers. In this book, several tools and techniques for the development of such codes have been presented. Therefore, it is hoped that the reader of this book is prepared for codes that will become important in the next millennium.

Bibliography

Books

M. Abramowitz, I.A. Stegun: *Handbook of Mathematical Functions*, Dover Publ., New York, 1970.

G. Arfken: *Mathematical Methods for Physicists*, Academic Press, San Diego, 1985.

T. Bäck: *Evolutionary Algorithms in Theory and Practice*, Oxford University Press, New York, 1996.

R. Barrett, *et al.*: *Templates for the Solution of Linear Systems*, SIAM, Philadelphia, PA, 1994.

K.-J. Bathe: *Finite Element Procedures in Engineering Analysis*, Prentice-Hall, Englewood Cliffs, New Jersey, 1982.

G. Beer, J. O. Watson: *Introduction to Finite and Boundary Element Methods for Engineers*, John Wiley, Chichester, 1992.

K.J. Binns, P.J. Lawrenson: *Analysis and Computation of Electric and Magnetic Field Problems*, Pergamon, Oxford, 1963.

K.J. Binns, P.J. Lawrenson, C.W. Trowbridge: *The Analytical and Numerical Solution of Electric and Magnetic Fields*, John Wiley & Sons, Chichester, 1992.

L.H. Bomholt, *MMP-3D – A Computer Code for Electromagnetic Scattering Based on the GMT*, Diss. ETH No.9225, Zürich, 1990.

R.C. Booton, Jr.: *Computational Methods for Electromagnetics and Microwaves*, John Wiley, New York, 1992.

M. Born, E. Wolf: *Principles of Optics*, Pergamon Press, Oxford, 1980.

M.V.K. Chari, P.P. Silvester (Editors): *Finite Elements in Electrical and Magnetic Field Problems*, John Wiley, Chichester, 1980.

R.E. Collin, *Field Theory of Guided Waves*, IEEE Press, New York, 1991.

R. Courant, D. Hilbert: *Methods of Mathematical Physics*, John Wiley, New York, 1989.

B.N. Datta: *Numerical Linear Algebra and Applications*, Brooks/Cole Publishing, Pacific Grove, CA, 1995.

D. Erni: *Periodische und nichtperiodische Wellenleitergitter- und Laserkavitätskonzepte*, Diss. ETH No.11654, Zürich, 1996.

D.B. Fogel: *Evolutionary Computation*, IEEE Press, New York, 1995.

J. Fröhlich: *Evolutionary Optimization for Computational Electromagnetics*, Diss. ETH No.9225, Zürich, 1997.

M. Gnos: *Berechnung statischer, quasistatischer und dynamischer Feldprobleme mit MMP*, Diss. ETH No.12158, Zürich, 1997.

D.E. Goldberg: *Genetic Algorithms in Search, Optimization, and Machine Learning*, Addison Wesley, Reading, MA, 1989.

G.H. Golub, C.F. van Loan, *Matrix Computations*, John Hopkins University Press, Baltimore, MD, 1983.

I.S. Gradshteyn, I.W. Ryzhik: *Table of Integrals, Series, and Products*, Academic Press, New York, 1965.

Ch. Hafner: *Beiträge zur Berechnung elektromagnetischer Wellen in zylindrischen Strukturen mit Hilfe des 'Point-Matching'-Verfahrens*, Diss. ETH No.12232, Zürich, 1980.

Ch. Hafner: *Numerische Berechnung elektromagnetischer Felder*, Springer, Berlin, 1987.

Ch. Hafner: *The Generalized Multipole Technique for Computational Electromagnetics*, Artech House, Boston, MA, 1990.

R.F. Harrington: *Field Computation by Moment Methods*, Macmillan, New York, 1968.

H.H. Haus: *Waves and Fields in Optoelectronics*, Prentice Hall, Englewood Cliffs, NJ, 1984.

P. Henrici: *Applied and Computational Complex Analysis*, Vol.1-3, John Wiley, New York, 1974-1986.

J.H. Holland: *Adaption in Natural and Artificial Systems*, MIT Press, New York, 1975.

M.G. Imhof: *Scattering of Elastic Waves using Non-Orthogonal Expansions*, Diss. MIT, Sept. 1996.

J.D. Jackson: *Classical Electrodynamics*, John Wiley, New York, 1975.

D.S. Jones: *The Theory of Electromagnetism*, Pergamon Press, Oxford, 1964.

N.S. Kapany, J.J. Burke: *Optical Waveguides*, Academic Press, New York, 1972.

R.W.P. King: *The Theory of Linear Antennas*, Harvard University Press, Cambridge, MA, 1956.

G. Klaus: *3-D Streufeldberechnungen mit Hilfe der MMP-Methode*, Diss. ETH No.7792, Zürich, 1985.

J.A. Kong, *Electromagnetic Wave Theory*, John Wiley, New York, 1990.

J.R. Koza: *Genetic Programming*, MIT Press, Cambridge, 1992.

N. Kuster: *Dosimetric Assessment of EM Sources Near Biological Bodies by Computer Simulations*, Diss. ETH No.9697, Zürich, 1992.

L.D. Landau, E.M. Lifshitz: *Course of Theoretical Physics*, Pergamon Press, London, 1958.

K.F. Lee: *Principles of Antenna Theory*, John Wiley, Chichester, 1984.

P. Leuchtmann: *Automatisierung der Funktionenwahl bei der MMP-Methode*, Diss. ETH No.8301, Zürich, 1987.

E.G. Loewen, E. Popov: *Diffraction Gratings and Applications*, Marcel Dekker, New York, 1997.

B.B. Mandelbrot: *Fractals – Form Chance and Dimension*, Freeman, San Francisco, 1977.

D. Marcuse: *Theory of Dielectric Optical Waveguides*, Academic Press, Boston, 1991.

R. März: *Integrated Optics*, Artech House, Boston, 1995.

J.C. Maxwell: *A Treatise on Electricity and Magnetism*, Dover, New York, 1954 (republication of the 3rd edition, Clarendon Press, 1891).

C.W. Misner, K.S. Thorne, J.A. Wheeler: *Gravitation*, Freeman, San Francisco, 1973.

A.R. Mitchell, D.F. Griffiths: *The Finite Difference Method in Partial Differential Equations*, John Wiley, Chichester, 1980.

P. Moon, D. Eberle-Spencer: *Field Theory Handbook*, Springer, Berlin, 1961.

L. Novotny: *Light Propagation and Light Confinement in Near-Field Optics*, Diss. ETH No.11420, Zürich, 1996.

W.K.H. Panowsky, M. Phillips: *Classical Electricity and Magnetism*, Addison-Wesley, Reading, MA, 1975.

A. Papoulis: *The Fourier Integral and its Applications*, McGraw-Hill, New York, 1962.

R. Petit (Editor): *Electromagnetic Theory of Gratings*, Springer, Berlin, 1980.

E. Polak: *Computational Methods in Optimization*, Academic Press, New York, 1971.

W.H. Press, B.P. Flannery, S.A. Teukolsky, W.T. Vetterling: *Numerical Recipes*, Cambridge University Press, Cambridge, 1992.

D. Quagliarella, J. Périaux, C. Poloni, G. Winter (Editors): *Genetic Algorithms and Evolution Strategies in Engineering and Computer Science*, John Wiley, Chichester, 1998.

S. Ramo, J.R. Whinnery, T. van Duzer: *Fields and Waves in Communication Electronics*, John Wiley, New York, 1965.

P. Regli: *Automatische Wahl der sphärischen Entwicklungsfunktionen für die 3D-MMP Methode*, Diss. ETH No.9946, Zürich, 1992.

T. Rozzi, M. Mongiardo: *Open Electromagnetic Waveguides*, IEE, London, 1997.

P.P. Silvester, R.L. Ferrari: *Finite Elements for Electrical Engineers*, Cambridge University Press, Cambridge, 1983.

G.S. Smith: *An Introduction to Classical Electromagnetic Radiation*, Cambridge University Press, Cambridge, UK, 1997.

A. Sommerfeld: *Lectures on Theoretical Physics*, Academic press, 1950.

E. Stiefel, A. Faessler: *Group Theoretical Methods and their Applications*, Birkhaeuser, Boston, 1992.

J.A. Stratton: *Electromagnetic Theory*, McGraw Hill, New York, 1941.

A. Taflove: *Computational Electrodynamics: The Finite-Difference Time-Domain Method*, Artech House, Norwood, MA, 1995.

A. Taflove: *Advances in Computational Electrodynamics: The Finite-Difference Time-Domain Method*, Artech House, Norwood, MA, 1998.

T. Tamir: *Integrated Optics*, Springer, Berlin, 1975.

T. Tamir (Editor): *Guided-Wave Optoelectronics*, Springer, Berlin, 1988.

T. Toffoli, N. Margolus: *Cellular Automata Machines: A New Environment for Modeling*, MIT Press, New York, 1987.

J. Turunen, F. Wyrowski (Editors): *Diffractive Optics for Industrial and Commercial Applications*, Akademie Verlag, Berlin, 1997.

C. Vasallo: *Optical Waveguide Concepts*, Elsevier, Amsterdam, 1991.

I.N. Vekua: *New Methods for Solving Elliptic Equations*, North-Holland, Amsterdam, 1967.

J.R. Whiteman (Editor): *The Mathematics of Finite Elements and Applications*, John Wiley, Chichester, 1997.

G. Winter, J. Périaux, M. Malan, P. Cuesta (Editors): *Genetic Algorithms in Engineering and Computer Science*, John Wiley, Chichester, 1995.

S. Wolfram: *Cellular Automata and Complexity*, Addison Wesley, Redwood City, CA, 1994.

P. Yeh: *Optical Waves in Layered Media*, John Wiley, New York, 1988.

A.P.M. Zwamborn: *Scattering by Objects with Electric Contrast*, Delft University Press, Delft, 1991.

Books with Software

B.W. Char *et al.*: *Maple V*, Springer, New York, 1991.

J.J. Dongarra, C.B. Moler, J.R. Bunch, G.W. Stewart: *Linpack Users Guide*, SIAM, Philadelphia, 1979.

J. Gleick: *Chaos – the Software*, Autodesk, Sausalito, CA, 1991.

Ch. Hafner: *2D MMP: Two-Dimensional Multiple Multipole Software and User's Manual*, Artech House, Boston, MA, 1990/S.

Ch. Hafner, L. Bomholt: *The 3D Electrodynamic Wave Simulator*, John Wiley, Chichester, 1993/S.

Ch. Hafner: *MaX-1 A Visual Electromagnetics Platform for PCs*, John Wiley, Chichester, 1998/S.

W.J.R. Hoefer, P.P.M.So: *The Electromagnetic Wave Simulator*, John Wiley, Chichester, 1991.

MATLAB – Partial Differential Equation Toolbox, Math Works, Natick, MA, 1996.

C.R. Paul: *Analysis of Multiconductor Transmission Lines*, John Wiley, New York, 1994.

R. Rucker: *Cellular Automata Laboratory – CA Lab*, Autodesk, Sausalito, CA, 1989.

S. Wolfram: *Mathematica*, Addison Wesley, Redwood City, CA, 1991.

Papers

G.P.M. Akkermans: "Diffraction of a Plane Electromagnetic Wave by an Elliptic Cylinder – The Multi-Origin Technique", *National Technical Information Service*, Springfield, VA, Feb. 1982.

J.-P. Berenger: "Three-dimensional perfectly matched layer for the absorption of electromagnetic waves", *J. Computational Physics*, Vol. 114, pp. 185-200, 1994.

A. Boag, R. Mittra: "Complex Multipole Beam Approach to 3D Electromagnetic Scattering Problems", *J. Opt. Soc. Am. A*, Vol. 11, pp. 1505-1512, April, 1994.

F. Bogdanov, D. Karkashadze: "Conventional Method of Auxiliary Sources in the Problems of Electromagnetic Scattering by the Bodies of Complex Materials", *Proc. 3rd Workshop on Electromagnetic and Light Scattering*, Bremen, pp. 133-140, March 16-17, 1998.

S. Choi, T.K. Sarkar, J. Choi: "Adaptive antenna array for direction-of-arrival estimation utilizing the conjugate gradient method", *Signal Processing*, Vol. 45, No. 3, pp. 313-327, Sept. 1995.

B.T. Draine: "The Discrete Dipole Approximation and its Application to Interstellar Graphite Grains", *Astrophys. J.*, Vol. 333, pp. 848-872, 1988.

Y.A. Eremin, A.G. Sveshnikow: "The Discrete Sources Method in Electromagnetic Diffraction Problems", *Izdatel'stvo MGU*, Moscow, 1992 (in Russian).

Y.A. Eremin, N.V. Orlov, A.G. Sveshnikow: "The analysis of complex diffraction problems by the discrete-source method", *Comput. Math. Phys.*, Vol. 35, No. 6, pp. 731-743, 1995.

M. Gnos, P. Leuchtmann: "Curved line multipoles for the MMP-code", *10th Annual Review of Progress in Applied Computational Electromagnetics (ACES), Conference Proceedings,* Monterey, Vol. 10, Mar. 1994.

J.E. Goell: "A Circular-Harmonic Computer Analysis of Rectangular Dielectric Waveguides", *The Bell System Technical Journal*, pp. 2133-2160, Sept. 1980.

L. Greengard, V. Rokhlin: "A fast algorithm for particle simulation", *Journal of Computational Physics*, Vol. 73, pp. 325-348, 1987.

Ch. Hafner, P. Leuchtmann, R. Ballisti: "Gruppentheoretische Ausnützung von Symmetrien, Teil 1+2", *Scientia Electrica*, Vol. 27, pp. 75-100,107-138, 1981.

Ch. Hafner and R. Ballisti: "The multiple multipole method (MMP)", *COMPEL - The International Journal for Computation in Electrical and Electronic Engineering*, Vol. 2, no. 1, 1983.

Ch. Hafner: "MMP calculations of guided waves", *IEEE Transactions on Magnetics*, Vol. 21, pp. 2310-2312, Nov. 1985/1.

Ch. Hafner and G. Klaus: "Application of the multiple multipole (MMP) method to electrodynamics", *COMPEL - The International Journal for Computation in Electrical and Electronic Engineering*, Vol. 4, no. 3, 1985/2.

Ch. Hafner: "Multiple Multipole (MMP) computations of guided waves and waveguide discontinuities", *Int. J. of Numerical Modelling*, Vol. 3, pp. 247-257, 1990/1.

Ch. Hafner: "On the relationship between the MoM and the GMT", *IEEE AP-S Magazine*, Vol. 32, Dec. 1990/2.

Ch. Hafner: "On the implementation of FE and GTD concepts in the MMP code", *Archiv für Elektronik und Uebertragungstechnik*, Vol. 45, Sept. 1991.

Ch. Hafner, "On the Design of Numerical Methods", *IEEE AP-S Magazine*, Vol. 35, pp. 13-21, Aug. 1993.

Ch. Hafner, J. Waldvogel, J. Mosig, J. Zheng, and Y. Brand: "On the Combination of MMP with MoM", *Applied Computational Electromagnetics (ACES) Journal*, Vol. 9, no. 3, pp. 18-27, 1994/1.

Ch. Hafner: "MMP-CG-PET: The Parameter Estimation Technique Applied to the MMP Code with the Method of Conjugate Gradients", *Applied Computational Electromagnetics (ACES) Journal*, Vol. 9, no. 3, pp. 176-187, 1994/2.

Ch. Hafner: "On the Computation of Transmission Line Modes", *PIERS 1994 CD-ROM*, Kluver Academic Publishers, ISBN 0-7923-3019-6, Dordrecht, 1994/3.

Ch. Hafner: "MMP Computation of Periodic Structures", *Journal of the Optical Society of America*, Vol. 12, no. 5, pp.1057-1067, May 1995.

J. Haueisen, Ch. Hafner, H. Nowak, H. Brauer: "Neuromagnetic Field Computation using the Multiple Multipole Method", *International Journal of Numerical Modelling,* Vol. 9, pp. 145-158, 1996.

M.R. Hestenes, E. Stiefel: "Methods of Conjugate Gradients for Solving Linear Systems", *Journal of Research of the National Bureau of Standards*, Vol. 49, no.6, pp. 409-436, 1952.

S. Kashyap, M. Burton, A. Louie: "Stabilizing the Time-Marching EFIE Algorithm", *IEEE AP-S International Symposium*, Vol. 2, pp.1033-1036, 1995.

V. Kupradze: "About the approximate solution of problems of mathematical physics", *Success of Mathematical Sciences*, Moscow, Vol. 22, No. 2, pp. 59-107, 1967 (in Russian).

Y. Leviatan, Am. Boag, Al. Boag: "Analysis of Electromagnetic Scattering Using a Current Model Method", *Computer Physics Communications*, Vol. 68, pp. 331-345, 1991.

A. Ludwig: "A new technique for numerical electromagnetics", *IEEE AP-S Magazine*, Vol. 31, pp. 40-41, Feb. 1989.

K.K. Mei *et al.*: "Measured Equation of Invariance – a lean and fast method for field computation", *Proc. URSI Int. Symp. on Electromagnetic Theory*, 1992.

G. Mie: "Elektrische Wellen an zwei parallelen Drähten", *Annalen der Physik*, Vol. 2, pp. 201-249, 1900.

G. Mur: "Absorbing boundary conditions for the finite-difference approximation of the time-domain electromagnetic field equations", *IEEE Trans. Electromagnetic Compatibility*, Vol. 23, pp. 377-382, 1981.

L. Novotny and Ch. Hafner: "Light Propagation in a Cylindrical Waveguide with a Complex, Metallic, Dielectric Function", *Physical Review E*, Vol. 50, no. 5, pp. 4094-4106, Nov. 1994.

Y. Okuno, H. Ikuno: "Yasuura's Method, Its Relation to the Fictitious-Source Methods, and Its Advancements in the Solution of 2-D Problems", *Proc. 3rd Workshop on Electromagnetic and Light Scattering*, Bremen, pp. 223-230, March 16-17, 1998.

N.B. Piller, O.J.F. Martin: "Extension of the generalized multipole technique to three-dimensional anisotropic scatterers", *Optics Letters*, Vol. 23, No. 8, pp. 579-581, April 15, 1998.

H. Singer, H. Steinbigler, P. Weiss: "A Charge Simulation Method for the Calculation of High-Voltage Fields", *IEEE Trans. on Power Appar. Syst.*, Vol. 93, pp. 1660-1668, 1974.

H. Steinbigler: "Combined Application of Finite Element Method and Charge Simulation Method for the Computation of Electric Fields", *Third International Symposium on High Voltage Engineering*, Mailand, 28-31 August 1979, Report 11.01.

G. Tayeb, R. Petit, M. Cadilhac: "The synthesis method applied to the problem of diffraction by gratings – the method of fictitious sources", *SPIE*, Vol. 1545, pp. 95-105, 1991.

K. Yasuura, T. Itakura: "(Title in Japanese)", *Kyushu Univ. Tech. Rep.*, Vol. 38, No. 1, pp. 72-77, 1965.

K.S. Yee: "Numerical solution of initial boundary value problems involving Maxwell's equations in isotropic media", *IEEE Trans. Antennas and Propagation*, Vol. 14, pp. 302-307, 1966.

Web sites

http://alphard.ethz.ch

MMP, Max-1, GMT, GGP

http://atol.ucsd.edu/~pflatau/scatlib/index.html#t-matrix

Light Scattering Codes, T-Matrix, DDA, etc.

http://emlib.jpl.nasa.gov

EMLIB, free electromagnetics software

http://imperator.cip-iw1.uni-bremen.de/~fg01/codes2.html

Electromagnetic Scattering codes

http://www.cee.hw.ac.uk/~maria/hfss.html

HFSS, Maxwell Eminence, Finite Elements

http://www-cs-faculty.stanford.edu/~koza

Genetic Programming

http://www.cst.de

MAFIA, Finite Integral Technique

http://www.dec.tis.net/~richesop/nec

NEC, unofficial home page, Method of Moments

http://www.emclab.umr.edu/cemlinks.html

Computational Electromagnetics Links

http://www.engr.usask.ca/~macphed/finite/fe_resources/fe_resources.html

Finite Elements resources

http://www.mathworks.com

Matlab

http://www.netlib.org

Netlib, matrix solvers and many more algorithms

http://www.remcominc.com

XFDTD, Finite Differences

http://www.wjrh.ece.uvic.ca/tlm/prog.html

TLM, Transmission Line Matrix

Index

D

E

F

N

O

P

S

T

W